Telecommunications and Network Engineering

Telecommunications and Network Engineering

Edited by **Kevin Merriman**

New Jersey

Published by Clanrye International,
55 Van Reypen Street,
Jersey City, NJ 07306, USA
www.clanryeinternational.com

Telecommunications and Network Engineering
Edited by Kevin Merriman

International Standard Book Number: 978-1-63240-548-7 (Hardback)

Printed in the United States of America.

Contents

Preface

Over the recent decade, advancements and applications have progressed exponentially. This has led to the increased interest in this field and projects are being conducted to enhance knowledge. The main objective of this book is to present some of the critical challenges and provide insights into possible solutions. This book will answer the varied questions that arise in the field and also provide an increased scope for furthering studies.

Telecommunications engineering is mainly concerned with the design and development of telecommunications equipment, electronic switching systems, terrestrial radio link systems, fibre optics cabling, copper wire telephone facilities etc. It is a multidisciplinary field that primarily brings together system engineering, computer engineering, broadcast engineering and electrical engineering. Network engineering in telecom refers to developing network, hardware and software. This book contains some of the most important topics such as communications protocols, wireless networks, location based services, modulation, mobile computing systems, switching and routing, synchronization, wired communications, etc. This book is a compilation of chapters that discuss the most vital concepts and emerging trends in the fields of telecommunications and network engineering. This book is a vital tool for all researching or studying these disciplines as it gives incredible insights into emerging trends and concepts.

I hope that this book, with its visionary approach, will be a valuable addition and will promote interest among readers. Each of the authors has provided their extraordinary competence in their specific fields by providing different perspectives as they come from diverse nations and regions. I thank them for their contributions.

Editor

1

Link Layer Correction Techniques and Impact on TCP's Performance in IEEE 802.11 Wireless Networks

Purvang Dalal[1], Mohanchur Sarkar[2], Kankar Dasgupta[3], Nikhil Kothari[1]

[1]Department of Electronics and Communication, Dharmsinh Desai University, Nadiad, India
[2]Indian Space Research Organization, Ahmadabad, India
[3]Indian Institute of Space Science and Technology, Thiruvanthapuram, India
Email: pur_dalal.ec@ddu.ac.in, nil_kothari@ddu.ac.in, msarkar@sac.isro.gov.in, ksd@iist.ac.in

Abstract

TCP performance degrades when end-to-end connections extend over wireless links which are characterized by high *Bit Error Rate* and intermittent connectivity. Such degradation is mainly accounted for TCP's unnecessary congestion control actions while attempting TCP loss recovery. Several independent link loss recovery approaches are proposed by researchers to reduce number of losses visible at TCP. In this paper we first presented a survey of loss mitigation techniques at wireless link layer. Secondly performance evaluation for TCP through Type 0 *Automatic Retransmission Request* mechanism in erroneous *Wireless LAN* is presented. In particular, simulations are performed taking into account the wireless errors introduced over IEEE 802.11 link using a well-established 2-*State Markov* model. TCP performance is evaluated under different settings for maximum link retransmissions allowed for each frame. Simulation results show that, link retransmission improves TCP performance by reducing losses perceived at TCP sender. However, such improvement is often associated with adverse effect on other TCP parameters that may cost a lot in return under extreme network conditions. In this paper an attempt is made to observe impact of link retransmissions on the performance of multiple TCP flows competing with each other. The analysis presented in this paper signifies the scope for maximizing TCP's throughput at the least possible cost.

Keywords

Wireless LAN, Cross-Layer Scheme, Medium Access Control, Round Trip Time, Bandwidth Estimation

1. Introduction

In recent scenario, the IEEE 802.11 [1] or 3G/4G wireless networks represent a significant milestone in the provisioning of "*anywhere - anytime*" *Internet* connectivity to the wandering users [2] [3]. Together, *Transmission Control Protocol* (TCP) [4] has remained as the dominant transport layer protocol for majority of *Internet* applications [2] [5]. Unlike wired links, wireless links have high, bursty and random errors due to atmospheric conditions, terrestrial obstructions, fast and multi-path fading, active interference, mobility and resource constraint [3] [6]. Consequently, significant amount of efforts [7] [8] have been devoted to the provisioning of reliable TCP delivery for a wide variety of applications over different wireless infrastructures.

Impact of the losses on TCP performance has been characterized in [9] using the following Equation (1). From the view of TCP, the throughput is inversely proportional to the packet losses perceived by TCP sender.

$$B(p) \approx \left(\frac{cwnd}{RTT}, \frac{1}{RTT\sqrt{\frac{2bp}{3}} + T_0 \min\left(1, 3\sqrt{\frac{3bp}{8}}\, p\left(1+32p^2\right)\right)} \right) \tag{1}$$

Nevertheless, when the losses are seen at TCP sender due to reasons other than network congestion, end-to-end throughput is compromised due to its loss recovery attempts with detrimental transmission rate. Therefore it is advocated to endeavor initial loss recovery at link layer (link recovery) and to attempt end-to-end TCP recovery [10] [11], only if inescapable. Attempts to make wireless links resemble wired ones for high level protocols are reflected in several approaches operating at a link layer.

In this paper, some of the prominent approaches made at wireless link layer for improving TCP's performance are discussed first. In the modern *Wireless LAN* (WLAN), due to the technology progress, the physical channel condition is comparatively good enough to provide low *Packet Error Rate* (PER) and consequently higher data-rate [2] [12]. On the other hand, to guarantee the correctness of control information, the control data rate increases very slowly (e.g. in IEEE 802.11 g data-rate is raised up to 54 Mbps and control data-rate is increased up to 6 Mbps only). In this situation, the overhead of control information including the link layer acknowledgement (ACK) increases significantly. The above grounds for very low efficiency of *Automatic Retransmission reQuest* (ARQ) protocol. As a result, the effect of link recovery in such networks should be thoroughly analyzed. In earlier work [13], the performance of a series of ARQ mechanisms are analyzed in the link layer without considering the TCP layer effect. In this work, a set of simulations are performed and results are analyzed to show impact of link recovery on TCP's performance. Analysis presented in this paper not only addresses the merits and demerits of link recovery attempts but also give a new direction towards enhancing TCP's performance further.

Our results show that with light errors on the IEEE 802.11 link, better TCP performance is achieved using link layer attempts. Since, link recovery attempt increases the *Round Trip Time* (RTT) estimate at TCP, it may adversely effects on the TCP performance by deteriorating the transmission rate at TCP sender (It is controlled using TCP's internal parameters *congestion window* (*cwnd*) and RTT). In fact, even with the use of link recovery, sub-optimum TCP's performance is seen exclusively in the extreme network environments; wherein total loss recovery is not attained at link layer. The rest of the sections are as follows. Section 2 discusses the pure link layer proposals designed to mitigate wireless errors. Section 3 provides a brief summary of link recovery mechanism over IEEE 802.11 links and Section 4 discusses the simulation results to show its influence on the end-to-end TCP's performance in wireless scenario. Section 5 concludes this paper.

2. Pure Link Layer Enhancements

Addressing link errors near the site of their occurrence appears intuitively attractive for several reasons.

1) Link layer approaches are likely to respond more quickly to changes in the error environment and therefore local recovery over a point to point link is more efficient than end-to-end TCP recovery [10].

2) Since, local recovery mechanism commonly operates on exactly the link that require its rendering, the deployment of new and existing wireless link protocol is more feasible than applying novel transport layer solution.

3) The link recovery approaches do not violate the modularity of the protocol stack [13] and in this context

they are different from *Cross-layer* [14] approaches.

Various approaches have been proposed at link layer to optimize the performance of TCP in wireless networks. These approaches are broadly classified into two groups on the basis of the awareness of the transport layer protocol; (a) TCP *unaware* approaches and (b) TCP *aware* approaches.

2.1. TCP *Unaware* Approaches

IEEE 802.11 and Cellular networks employ TCP *unaware* link and physical layer mechanisms with an objective to reduce the PER to a level that would not cause significant performance degradation at TCP. In the above networks, physical layer achieves high coding gain with the use of convolution and turbo coding [15]. Additionally interleaving provides time diversity as a further safeguard against burst errors. The most remarkable TCP *unaware* link layer implementations in the above wireless technologies uses ARQ mechanism and packet scheduling to reduce effect of losses on TCP's performance. For example, *Selective Repeat* ARQ with scheduling protocols such as RLP and RLC are used in 3G1X and UMTS, respectively [15]. On the other side, IEEE 802.11 uses *stop and wait* ARQ [13] alone to recover from transmission losses locally. ARQ is a closed-loop mechanism that requires feedback and retransmission and is invoked when packets containing bit errors are discarded. Link ARQ consumes additional network resources only when a packet is retransmitted. The mechanism generally operates more efficiently for low bit rates. On a downside, the above mechanisms may cause delay variability and out-of-order delivery of packets at TCP. This can result in duplicate retransmissions from the two layers and hence inefficient utilization of the network capacity. An undesirable side effect of ARQ is that it may interfere with independent TCP mechanisms. *Forward Error Correction* (FEC) is a popular error mitigation mechanism for detection and recovery of transmission losses [16] without retransmissions, which is critical for lossy links exhibiting long delays. Unlike ARQ, FEC doesn't interfere with TCP mechanisms. FEC suffers from dead-weight overheads in favorable conditions, resulting in a waste of limited bandwidth. Furthermore FEC requires additional resources (processor, memory) and power consumption.

Based on the recommendations made by IETF and 3GPP, most of the link layer techniques for wireless lossy links, involve making the link layer reliable using hybrid FEC/ARQ [17]. Formerly a good summary of hybrid FEC/ARQ techniques for the link layer is provided in [13]. In a Type 0 scheme, recovery for losses is offered only through ARQ. This technique is applicable to links having losses subject to infrequent interference. In Type I technique, initial transmission is protected using FEC. If this fails, ARQ is used to repeat the transmission. The main characteristic of Type II schemes is that ARQ is used first followed by FEC. Typically the initial transmission has error detection built in but no error correction. Though hybrid ARQ/FEC is better than either FEC or ARQ alone, its performance also significantly degrades for higher loss rates despite unreasonably high amounts of ARQ retries, fragmentation of IP packets, FEC overhead and buffering [15]. *Asymmetric Reliable Mobile Access In Link Layer* (AIRMAIL) and *Transport Unaware Link Improvement Protocol* (TULIP) are the known link layer implementation based on TCP *unaware* link layer techniques as discussed in [7].

Barakat *et al.* [11] reported that ARQ mechanisms may vary the characteristics of the network, affecting the functionality of the upper layer protocols. Moreover lack of knowledge of the protocol operating at the transport level may results into end-to-end performance degradation in presence of independent link layer mechanisms. For instance, an approach without awareness of the transport protocol may cause local link layer retransmission of a packet, as well as duplicate acknowledgement (*dupack*), since retransmissions can be performed on both layers. This led to significant efforts in development of TCP *aware* link layer approaches [18].

2.2. TCP *Aware* Approaches

The link layer approaches in which enhancements tailored for wireless environments are made known to TCP are broadly referred as TCP *aware* link approaches. A good overview of these approaches can be found in [7] [17] [18]. The representative mechanisms employed in various approaches include *Snooping*, *Delayed Acknowledgement* and *Performance Enhancement Proxy* (*PEP*).

2.2.1. Snooping Mechanism

Most of the approaches which operate at this level rely on some intermediate point within the end-to-end connection for the introduction of performance improvement. In *Snoop* protocol [19], agent located at the *Base Station* (BS) monitors every packet that passes through the TCP connection in both directions. The *Snoop* agent suppresses the *dupacks* for lost TCP packets and retransmits locally, thereby preventing unnecessary invocation of congestion control mechanism by sender. Other enhancements proposed in similar category are *TCP SACK-aware Snoop* Protocol and SNACK-NS (*New Snoop*) [7]. This protocol may cause additional delay for TCP packets, which results into *futile* TCP retransmissions at TCP sender over the slower wireless link. However, the problem can be solved by minimizing the number of retransmissions at the link level [7].

WTCP [20], although operating in a similar way, introduces more accurate RTT estimation and thus preventing reduction in TCP throughput. WTCP conceals the difference in RTT from the sender, thus avoiding needless timeout caused by reason of local retransmissions. Both *Snoop* and WTCP must, however, have access to the header of TCP packets in order to function, which reduces their usefulness value if traffic is encrypted. The use of *Snoop* agent is also exploited in wireless TCP refinements with split connection schemes [7]. *Snoop* and WTCP increase complexity at BS, especially when transport layer per flow support is required. TCP PEPs [21] are TCP *aware* link layer mechanisms for handling bandwidth asymmetry, TCP *aware* FEC provisioning or adaptive link frame sizing [22]. Split connection approaches [7] and PEPs, though effective in many cases, break TCP end-to-end semantics and its modularity and hence not considered as pure link layer approaches.

2.2.2. Delayed Acknowledgements

Another class of work includes *Delayed Duplicate Acknowledgment* (DDA) [23], which preferred particularly when IP encryption is used. With a view to reduce interference between the TCP and link retransmissions, TCP receiver delays the third and subsequent *dupacks* for some interval "*d*". This provides enough time for link layer to recover the lost packet using local retransmission, and to prevent TCP sender from fast retransmit. Different from the *Snoop* protocol, in DDA *dupacks* are not dropped immediately but rather delayed a certain length of time. *Delayed Acknowledgements over Wireless Link* (DAWL) [24] tries to simplify the system by introducing modifications only in the ARQ scheme at the link layer and does not consider local retransmission at the BS. This design is advantageous in case of a BS crash [7]. However, in presence of congestion, the performance of these schemes is degraded, as the essential fast retransmissions are unnecessarily delayed.

Since different link layer technologies have diverse capabilities for error recovery, there is no *de-facto* standard for link layer protocols. All link layer approaches try to reduce the effect of erroneous wireless links on the performance of TCP, but do not completely shield the TCP sender from all types of wireless errors particularly in the event of lengthy disconnection or high *Bit Error Rate* (BER). The main advantage of link layer approaches is the maintenance of end-to-end semantics, without modification of higher protocol layers. This makes it possible to leave untouched the existing implementations of the protocol stack in the various operating systems.

3. IEEE 802.11 Link Recovery Mechanism

The basic channel access method in 802.11 networks is the *Distributed Coordination Function* (DCF) in which 802.11 *Medium Access Control* (MAC) layer uses the *Carrier Sense Multiple Access/Collision Avoidance* (CSMA/CA) mechanism before transmitting any frame. The IEEE 802.11 MAC layer employs Type 0 ARQ concept for loss recovery with the contention based transmission policy. The ARQ is the only error control method specified in the standard and no FEC coding is used as stated earlier.

Over an IEEE 802.11 link, whenever a wireless node notices unsuccessful transmission for a frame, it attempts local retransmission of a frame in accordance with *RetryLimit* (*RL*) [1]. To explain, as shown in **Figure 1**, a wireless node after sending a frame waits for a positive ACK from wireless receiver. The sender attempts retransmission of the frame whenever ACK is not received before expiration of ACK timeout. The node is allowed to attempt retransmissions for a particular frame maximum up to *RL*. Then after node initiates for new transmission and the previous frame is considered to be lost. Thereby the upper layer is made responsible for recovery of the frames lost after maximum retransmissions.

Such link retransmissions additionally delay the transmission of subsequent TCP packet residing at *Interface Queue* (IFQ). This *additional delay* (T_{ARQ}) is approximated using Equation (2).

$$T_{ARQ} = n \cdot \left(t' + T_{CONT}\right) \tag{2}$$

T_{ARQ} increases with (a) increase in the number of unsuccessful retransmissions (n) and (b) increase in link delay (t'). Here T_{CONT} signifies the contention delay over a shared medium [1]. TCP's RTT estimation includes wired link delay and processing delay at nodes (the combined is referred as RTT_0), delay due to network con-

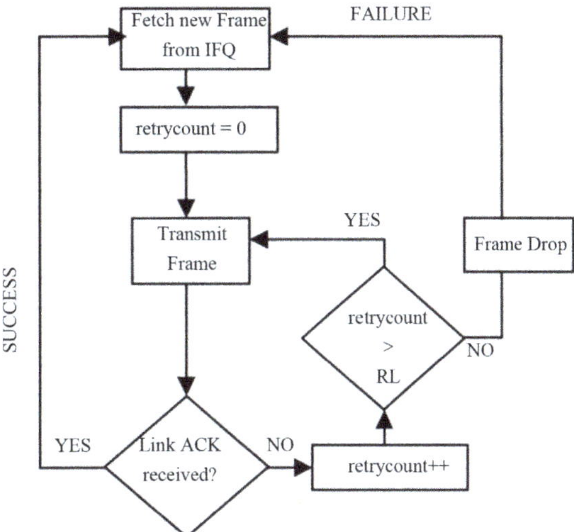

Figure 1. IEEE 802.11 link loss recovery mechanism.

gestion (d_c) and T_{ARQ} as shown in Equation (3).

$$\mathrm{RTT} = \mathrm{RTT}_0 + d_c + T_{ARQ} \tag{3}$$

In short, link recovery mechanisms shield TCP from the wireless losses and hence prevent undue reduction of *cwnd*, but at the same time increase RTT estimation at TCP sender. The earlier effect prevents sacrificed transmission rate and the later accounts for diminished growth in transmission rate. In the next section, an analysis of results obtained through simulations is presented to illustrate impact of link recovery on TCP's performance.

4. Simulation Results and Analysis

4.1. Simulation Scenario

A network topology illustrated in **Figure 2** was used for analyzing *Medium Access Control* (MAC) and TCP behavior using a TCP connection between sender (S1) and receiver (D). Wireless losses over an IEEE 802.11b link between S1 and BS were introduced using 2-*State Markov* model. The queue length at intermediate nodes is set in accordance to *Bandwidth Delay Product* (BDP) in the network.

A TCP/FTP flow was introduced for 100 sec. To investigate the impact of link recovery on the efficiency of link and transport layer protocols, the simulations were performed using different value for *RL* and *Frame Error Rate* (FER) in the range mentioned along with results. TCP's RTT estimation and layer efficiency are the measurable parameters used for performance evaluation. Layer efficiency is defined as the ratio of total number of successful transmissions to the total transmissions made by that layer during simulation interval. Nevertheless the most critical parameter under observation is (*cwnd*/RTT). For better understanding, analysis of results is presented with different traffic conditions; (a) a single TCP flow and (b) multiple TCP flows competing over a bottleneck link. In the mentioned topology, TCP New Jersey [25] (TCP NJ) has demonstrated better throughput compared to that achieved using other well-known TCP variants (*i.e.* NewReno, Vegas and Westwood [26]). Therefore the analysis is presented for TCPNJ as TCP variant. However similar trend in the results is observed with other TCP variants as well.

4.2. Analysis with Single TCP Flow

Figures 3-8 present analysis based on the results obtained from the experiments.

As shown in **Figure 3(a)**, with link recovery (RL = 7), increase in FER resulted into increase in link retransmissions in order to combat transmission losses locally. Pl. note that even with 0% FER (introduced using error model), losses were witnessed due to channel contention (due to transmission of TCP data and acknowledgement packets on the same channel in reverse direction). It led to noticeable MAC retransmissions for recovering from such losses. During simulations with link recovery (RL = 7), % of wireless TCP drops were significantly

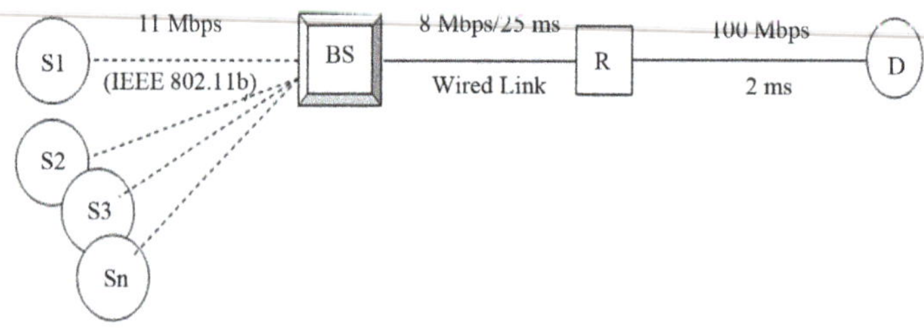

Figure 2. Network Topology.

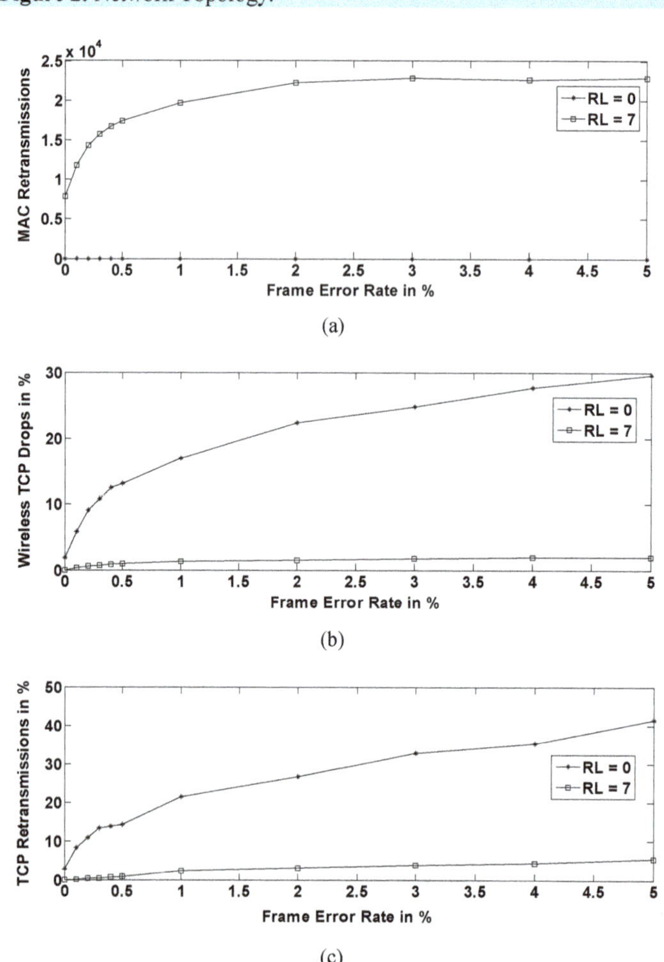

Figure 3. (a) Impact of link ARQ on MAC retransmissions, (b) Impact of link ARQ on Wireless TCP Drops in %, (c) Impact of link ARQ on TCP Retransmissions (in %).

reduced together with increase in FER compared to those witnessed during simulations without link recovery (RL = 0) (refer **Figure 3(b)**). Since, the error model introduces losses in % of total transmissions; in order to demystify the impact of link loss recovery correctly the analysis is presented in terms of % of TCP drops. Reduction in wireless TCP drops has caused significant reduction in end-to-end TCP retransmissions (analysis is presented in % of total TCP transmission due to similar reasons stated earlier), as shown in **Figure 3(c)**.

The simulations were performed using single TCP flow and therefore the packet losses were attributed only to the transmission errors on wireless link. During simulations TCP flow encountered timeouts due to either loss of retransmissions or due to loss of TCP acknowledgement in presence of errors on the wireless links, those are very costlier to TCP. **Figure 4** represents comparison for number of TCP timeouts with RL = 0 and RL = 7. As

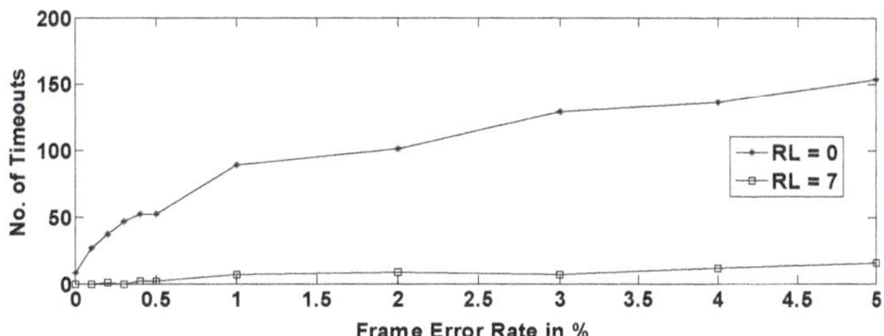

Figure 4. Impact of link ARQ on TCP timeouts.

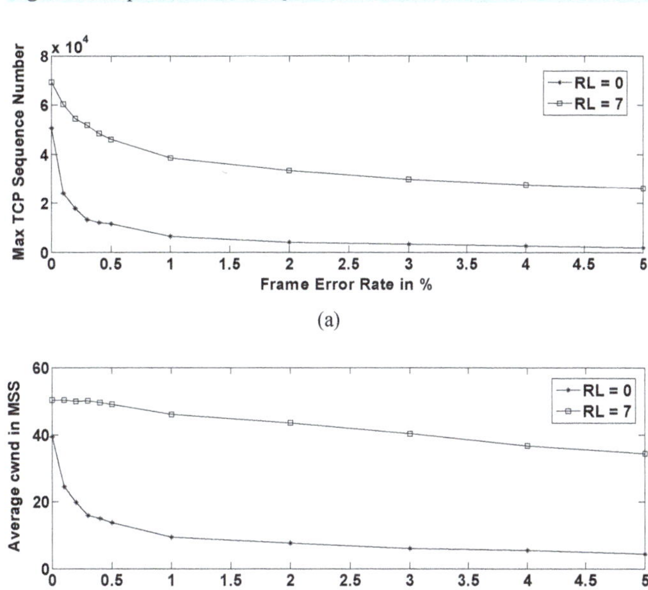

Figure 5. (a) Impact of link ARQ on TCP's maximum sequence number, (b) Impact of link ARQ on *congestion window*.

seen from the Figure, number of timeouts were considerably condensed when link retransmissions were enabled (*i.e.* RL = 7).

In absence of any support for loss discrimination, TCP reacts to timeout with drastic reduction in *cwnd*, which is inappropriate for wireless losses. **Figure 5(a)** shows that without link recovery (RL = 0), TCP's maximum sequence number dropped rapidly with increase in FER; since loss of TCP segments has reduced the *cwnd* and slowed down TCP's transmissions per RTT. However with link recovery, significantly higher value of TCP's maximum sequence number is obtained, relative to that obtained with *RL* = 0.

This improvement is additionally attributed to the massive reduction in costlier TCP timeouts, in presence of link loss recovery attempts (RL = 7). It is apparent that link ARQs can improve the performance of TCP by shielding it from wireless losses as much as possible. Comparison for average *cwnd* with and without link loss recovery is shown in **Figure 5(b)**. When RL = 7, TCP's average *cwnd* and maximum sequence number decreased linearly with increase in FER.

Figure 6(a) and **Figure 6(b)** represent transmission efficiency for MAC and TCP protocols with RL = 0 and RL = 7. As anticipated, link recovery (RL = 7) improves efficiency for both the layers. It must be noted that without link recovery (RL = 0), TCP's efficiency is reduced to a great extent with higher FER. Hence, link recovery improves end-to-end performance in the network by improving TCP's transmission efficiency on large scale. In **Figure 6(b)**, about 30% of improvement for TCP's efficiency at 5% FER is recorded. Improved efficiency of MAC and TCP protocols result into significant improvement in TCP's throughput as shown in **Figure 6(c)**.

In **Figure 7(a)**, comparison for TCP's RTT estimated with RL = 0 and RL = 7 is presented for different FER.

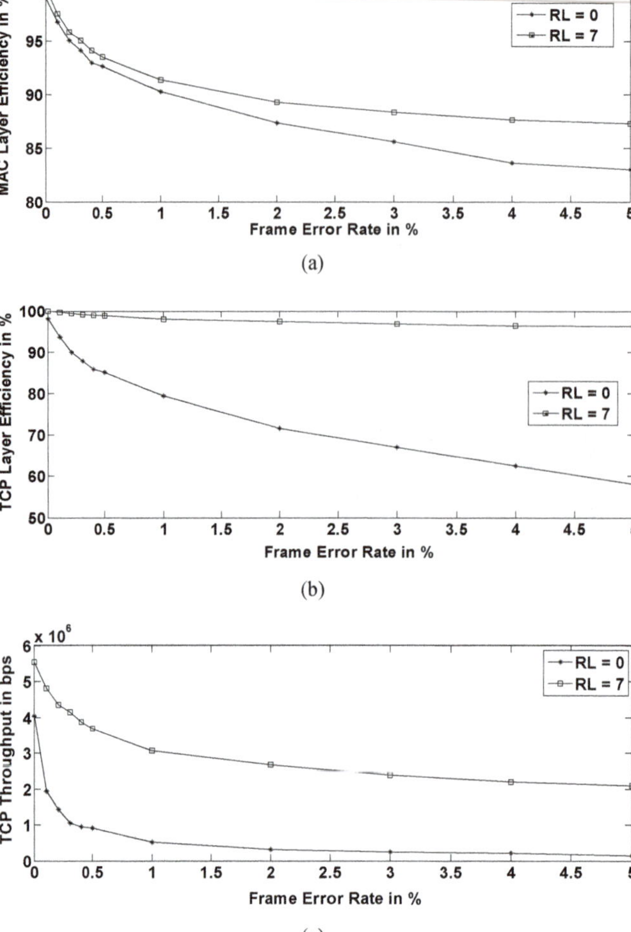

Figure 6. (a) Impact of link ARQ on MAC layer efficiency, (b) Impact of link ARQ on TCP layer efficiency, (c) Impact of link ARQ on TCP's throughput.

Without link recovery (RL = 0), the average RTT (say RTT_0) decreases with increase in FER. As shown in **Figure 5(b)** earlier, $cwnd_0$ has reduced sharply with increase in FER up to 2%. This reduction in $cwnd_0$ grounded for less number of TCP segments at an IFQ and hence as expected RTT_0 estimated at sender was also lowered sharply. It must be noted that the RTT_0 is found very close to the theoretical RTT in the network. A bit higher value is attributed to the contention delay at wireless interface and processing delay at nodes. In contrary, with link recovery (RL = 7), average RTT (say RTT_7) is increased along with increase in FER; as the link layer on average took a longer time for error recovery before delivering the TCP segments or acknowledgements in either directions. **Figure 7(a)** shows that the average RTT is increased with increase in FER until it reaches about 2%. With increase in FER above 2%, there is insignificant change in MAC retransmissions (refer **Figure 3(a)**) and hence similar was the impact on RTT_7 estimation. With FER above 2%, marginal reduction in RTT_7 was observed due to reduction in *cwnd.*

Based on the foremost observation it seems that the TCP with link layer recovery (RL = 7) utilizes network with higher value for average *cwnd* ($cwnd_7$) compared to that accomplished without link recovery ($cwnd_0$). Consequently, it leads to enhanced TCP's performance. In fact TCP with link recovery has achieved higher value of average *cwnd* at the cost of additional rise in average RTT estimation.

Since TCP's throughput is proportionate to its effective sending rate (*i.e. cwnd*/RTT), it is apparent that the net improvement in end-to-end TCP throughput is realized only when ($cwnd_7/RTT_7$) > ($cwnd_0/RTT_0$). In **Figure 7(b)**, average value of $cwnd_7/RTT_7$ and $cwnd_0/RTT_0$ is plotted for different value of FER. It is revealed that the

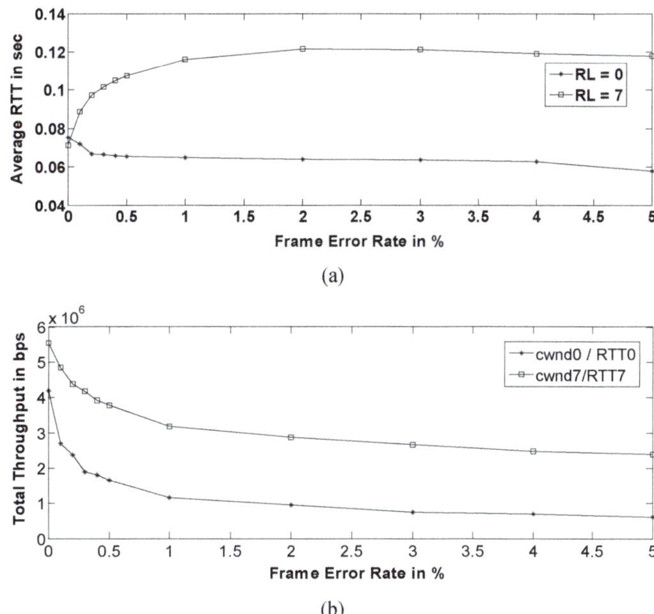

Figure 7. (a) Impact of link ARQ on TCP's RTT Estimate, (b) Impact of link ARQ on TCP's RTT Estimate.

improvement in TCP's throughput is following the net improvement for average *cwnd*/RTT. Pl. note that even with 0% FER (introduced using error model), transmission losses were witnessed due to channel contention (due to transmission of TCP data and ack packets on the same channel in reverse direction). Link ARQs recovered from majority of contention losses and hence higher TCP's throughput is seen compared to that observed without link recovery.

4.3. Analysis with Multiple TCP Flows

After analyzing impact of link ARQ on TCP's performance in non-congested erroneous wireless network, our simulations were extended for further investigations in a congested wireless network. For that number of simultaneous TCP senders in the networks were increased in the range of 2,4,8 and 16. The statistics obtained for FER of 0.001% are as shown in **Figure 8**.

As seen from **Figure 8(a)**, with increase in number of competing flows, effective value of *cwnd* (for all flows over a bottleneck link) reduces with or without loss recovery as anticipated. However, for given number of competing flows, effective value of average *cwnd* is found much higher when link recovery was enabled. With increase in the number of TCP flows, congestion in the network increases.

Moreover link recovery in presence of transmission losses gave additional rise in RTT (as shown in **Figure 8(b)**). This aggravated for congestion in the network and consequently large number of congestion drops are witnessed, particularly with increase in the competing flows (for a given error rate). This has remarkably pulled down growth in *cwnd* due to rise in RTT, which can be seen from the **Figure 8(c)**. As mentioned in the Figure, with link recovery (RL = 7), improvement in *cwnd*/RTT is seen when there was a single TCP flow. However with increase in number of TCP flows improvement for *cwnd*/RTT using link recovery is found insignificant. In fact when number of competing TCP flows increased upto 4 or above, link recovery failed in protecting *cwnd*/RTT and consequently, as presented in **Figure 8(d)**, TCP performance is found deteriorated.

This gives insight to the issue related to the TCP's performance over an IEEE 802.11 network, even in presence of link recovery mechanism. The problem really occurs when link layer mitigation substantially diminishes growth in *cwnd* by increasing RTT to a much higher value with increase in FER. The similar problem may occur when link layer mitigation fails due to substantial underlying error conditions. In fact use of link layer approaches may degrade performance especially in presence of highly variable error rates. In the stated situation, reduction in *cwnd* and rise in RTT is unavoidable at sender on account of failed ARQs. This adversely effects on the throughput conceived by TCP.

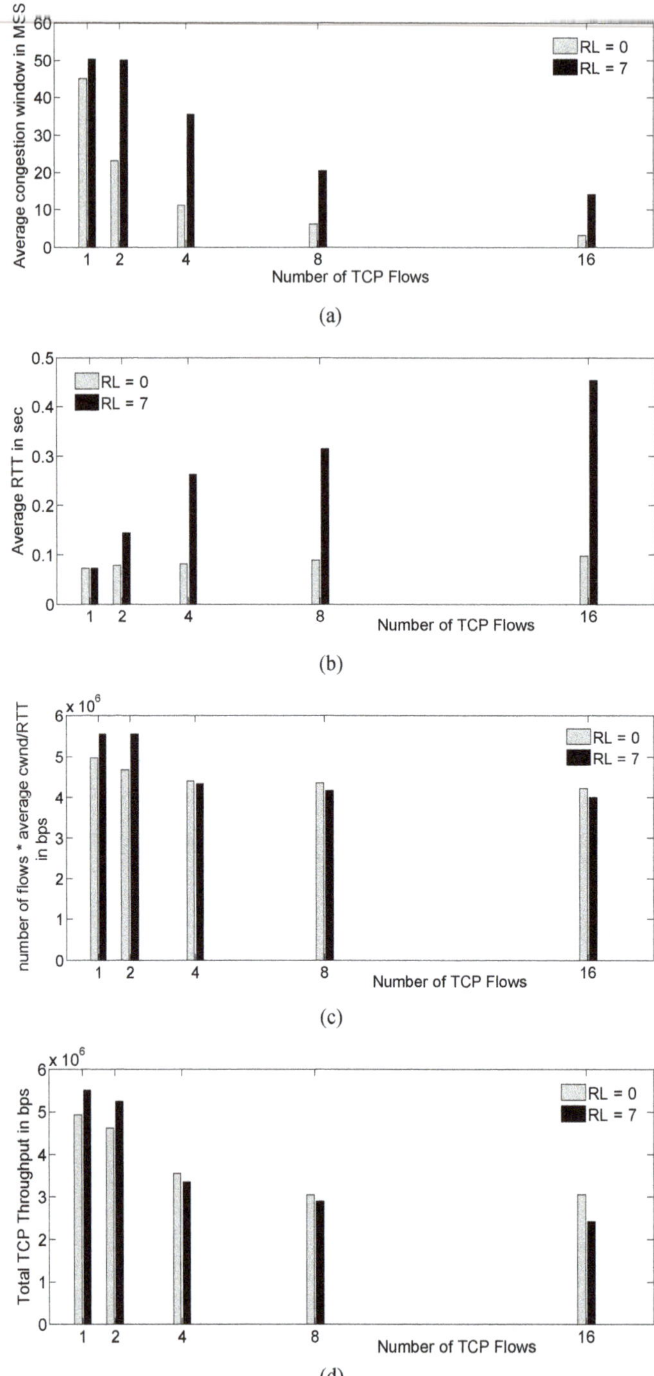

Figure 8. (a) Impact of link ARQ on average *congestion window*, (b) Impact of link ARQ on average RTT, (c) Impact of link ARQ on effective *cwnd*/RTT, (d) Impact of link ARQ on total TCP throughput.

5. Conclusions

We have performed a detailed analysis of using link ARQs to improve the end-to-end performance of TCP in wireless networks. The activities are concluded with the following outcomes.

1) Link recovery attempts using ARQ in IEEE 802.11 networks effects on two important TCP parameters; *i.e.* *cwnd* and RTT. TCP performance is decided jointly by both of the above parameters; *i.e.* always attributed to the

improvement in the *cwnd*/RTT value.

2) Link ARQs in absence of network congestion and light wireless errors yield superior value of *cwnd*/RTT in comparison to that observed without link ARQs. This in turn results in better TCP performance.

3) Recovering from all types of wireless losses using link mitigation techniques is not always possible; particularly with high and bursty error environments. This may lead to performance degradation due to unavoidable cutback in *cwnd* and increase in RTT.

4) Link ARQs may degrade performance in congested networks even with very small amount of wireless errors. The performance degradation is attributed to any of the following reasons:

- When link level error mitigation fails, reduction in *cwnd* at TCP sender is unavoidable. Since loss recovery attempts increase RTT and unable to protect *cwnd*, TCP''s end-to-end performance is compromised.
- If multiple TCP flows are sharing common wireless channel for transmission, loss recovery attempts made for one TCP flow may affect RTT of all competing flows and therefore overall network performance is found compromised.

This shows possibility for further improvement in *cwnd*/RTT (and hence TCP throughput) using a corrective mechanism that makes RTT unaffected from link recovery attempts. Nevertheless, any correction that reduces RTT inappropriately may cause the sender to timeout earlier when the retransmission is being performed on the wireless link. Therefore, a well thought upon approach for RTT correction may be considered for maximizing TCP's throughput in existence of link layer ARQs. This highlights the scope for further performance enhancement of TCP proposals those advocate for retaining of *cwnd* based on loss discrimination.

References

[1] (2003) IEEE Standard for Information Technology-Telecommunications and Information Exchange between Systems-Local and Metropolitan Area Networks-Specific Requirements-Part 11: Wireless LAN Medium Access Control (MAC) and Physical Layer (PHY) Specifications. *ANSI/IEEE Std* 802.11, 1999 Edition (R2003), I-513. http://dx.doi.org/10.1109/IEEESTD.2003.95617

[2] Inamura, H., Ludwig, R. and Khafizov, F. (2003) TCP over Second (2.5G) and Third (3G) Generation Wireless Networks. *IETF RFC* 3481.

[3] Lohier, S., Doudane, Y.G. and Pujolle, G. (2007) Cross-Layer Loss Differentiation Algorithms to Improve TCP Performance in WLANs. *Springer US Journal of Telecommunication Systems*, **36**, 61-72.

[4] Floyd, S., Henderson, T. and Gurtov, A. (2004) RFC 3782: The NewReno Modification to TCP's Fast Recovery Algorithm. *IETF RFC* 3782.

[5] Kliazovich, D., Gerla, M. and Granelli, F. (2007) Performance Improvement in Wireless Networks using Cross Layer ARQ. *The International Journal of Computer and Telecommunications Networking*, **51**, 4396-4411.

[6] Leung, K.-C. and Li, V.O.K. (2006) Transmission Control Protocol (TCP) in Wireless Networks: Issues, Approaches and Challenges. *IEEE Communications Surveys & Tutorials*, **8**, 64-79. http://dx.doi.org/10.1109/COMST.2006.283822

[7] Sardar, B. and Saha, D. (2006) A Survey of TCP Enhancements for Last-Hop Wireless Networks. *IEEE Communications Surveys & Tutorials*, **8**, 20-34. http://dx.doi.org/10.1109/COMST.2006.253273

[8] Tian, Y., Xu, K. and Ansari, N. (2005) TCP in Wireless Environments: Problems and Solutions. *IEEE Communications Magazine*, **43**, s27-s32.

[9] Padhye, J., Firoiu, V., Towsley, D. and Kurose, J. (1998) Modeling TCP Throughput: A Simple Model and Its Empirical Validation. *Proceedings of SIGCOMM*, 303-314.

[10] Ludwig, R. and Gurtov, A (2005) The Eifel Response Algorithm for TCP. *IETF RFC* 4015.

[11] Barakat, C. and Altman, E. (2002) Bandwidth Tradeoff Between TCP and Link-level FEC. *Computer Networks*, **39**, 133-150. http://dx.doi.org/10.1016/S1389-1286(01)00305-X

[12] Singh, J.P., Li, Y., Bambos, N., Bahai, A., Xu, B. and Zimmermann, G. (2007) TCP Performance Dynamics and Link-Layer Adaptation Based Optimization Methods for Wireless Networks. *IEEE Transactions on Wireless Communications*, **6**, 1864-1879.

[13] Barakat, C. and Fawal, A.A. (2004) Analysis of Link-Level Hybrid FEC/ARQ-SR for Wireless Links and Long-Lived TCP Traffic. *Performance Evaluation*, **57**, 453-476. http://dx.doi.org/10.1016/j.peva.2004.03.002

[14] Cheng, R.S. and Lin, H.T. (2008) A Cross-Layer Design for TCP End-to-End Performance Improvement in Multi-Hop Wireless Networks. *Computer Communications*, **31**, 3145-3152. http://dx.doi.org/10.1016/j.comcom.2008.04.017

[15] Chan, M.C. and Ramjee, R. (2004) TCP/IP Performance over 3G Wireless Links with Rate and Delay Variation.

INFOCOM.

[16] Luby, M., Vicisano, L., Gemmell, J., Rizzo, L., Handley, M. and Crowcroft, J. (2002) RFC 3452: Forward Error Correction (FEC) Building Block. *IETF RFC* 3452.

[17] Kim, D., Choi, Y., Jin, S., Han, K. and Choi, S. (2008) A MAC/PHY Cross-Layer Design for Efficient ARQ Protocols. *IEEE Communications Letters*, **12**, 909-911. http://dx.doi.org/10.1109/LCOMM.2008.081259

[18] Scott, J. and Mapp, G. (2003) Link Layer-Based TCP Optimization for Disconnecting Networks. *ACM SIGCOMM Computer Communications Review*, **33**, 31-42.

[19] Sun, F., Li, V.O.K. and Liew, S.C. (2004) Design of SNACK Mechanism for Wireless TCP with New Snoop. *IEEE Wireless Communications and Networking Conference*, **5**, 1046-1051.

[20] Sinha, P., Nandagopal, T., Venkitaraman, N., Sivakumar, R. and Bharghavan, V. (2002) WTCP: A Reliable Transport Protocol for Wireless Wide-Area Networks. *Wireless Networks*, **8**, 301-316. http://dx.doi.org/10.1023/A:1013702428498

[21] Border, J., Kojo, M., Griner, J., Montenegro, G. and Shelby, Z. (2001) RFC 3135: Performance Enhancing Proxies Intended to mitigate Link Related Degradation. *IETF RFC* 3135.

[22] Tickoo, O., Subramanian, V., Kalyanaraman, S. and Ramakrishnan, K.K. (2005) LT-TCP: End-to-End Framework to Improve TCP Performance over Networks with Lossy Channels. *IEEE International Workshop on Quality of Service*.

[23] Vaidya, N. and Mehta, M. (2002) Delayed Duplicate Acknowledgments: A TCP-Unaware Approach to Improve Performance of TCP over Wireless. *Wireless Communications and Mobile Computing*, **2**, 59-70. http://dx.doi.org/10.1002/wcm.33

[24] Kliazovich, D. and Granelli, F. (2005) DAWL: A Delayed-ACK Scheme for MAC-Level Performance Enhancement of Wireless LANs. *ACM/Kluwer Journal on Mobile Networks and Applications* (*MONET*), **10**, 607-615.

[25] Kim, J., Koo, J. and Choo, H. (2007) TCP NJ+: Packet Loss Differentiated Transmission Mechanism Robust to High BER Environments. *Ad Hoc and Sensor Networks, Springer Wireless Networks, Next Generation Internet Lecture Notes in Computer Science*, **4479**, 380-390.

[26] Marcondes, C., Sanadidi, M.Y., Gerla, M. and Shimonishi, H. (2008) TCP Adaptive Westwood-Combining TCP Westwood and Adaptive Reno: A Safe Congestion Control Proposal. *ICC '08, IEEE International Conference on Communications*, Beijing, 19-23 May 2008, 5569-5575. http://dx.doi.org/10.1109/ICC.2008.1044

2

Markov Model Based Jamming and Anti-Jamming Performance Analysis for Cognitive Radio Networks

Wednel Cadeau, Xiaohua Li, Chengyu Xiong

Department of Electrical and Computer Engineering, State University of New York at Binghamton, Binghamton, USA
Email: wcadeau1@binghamton.edu, xli@binghamton.edu, cxiong1@binghamton.edu

Abstract

In this paper, we conduct a cross-layer analysis of both the jamming capability of the cognitive-radio-based jammers and the anti-jamming capability of the cognitive radio networks (CRN). We use a Markov chain to model the CRN operations in spectrum sensing, channel access and channel switching under jamming. With various jamming models, the jamming probabilities and the throughputs of the CRN are obtained in closed-form expressions. Furthermore, the models and expressions are simplified to determine the minimum and the maximum CRN throughput expressions under jamming, and to optimize important anti-jamming parameters. The results are helpful for the optimal anti-jamming CRN design. The model and the analysis results are verified by simulations.

Keywords

Cognitive Radio, Dynamic Spectrum Access, Jamming, Markov Model, Throughput

1. Introduction

Cognitive radio networks (CRNs) have attracted great attention recently because they can potentially resolve the critical spectrum shortage problem [1]. Under the umbrella of dynamic spectrum access, CRN accesses the spectrum secondarily, *i.e.*, as long as it can guarantee no interference to any primary user (PU) who is using this spectrum at this time in this location [2]. This means that the cognitive radios need to periodically sense the spectrum to detect the PU's activity. They have to vacate the channel immediately whenever the PU activity is detected.

While cognitive radios can realize more flexible spectrum access and higher spectrum efficiency, malicious users can also exploit them to launch more effective attacks, in particular jamming attacks. As a matter of fact, CRN is extremely susceptible to jamming attacks because of its unique requirements in the physical- and MAC-layers, such as the requirement of channel vacating when detecting any PU signals. On the other hand, it is believed that the capability of hopping among many channels gives CRN a unique advantage in improving their anti-jamming performance.

The anti-jamming performance of CRN is a new and interesting research topic that is critical for the secure and reliable CRN design [2]-[4].Conventionally, anti-jamming study is conducted in the Physical-layer via some anti-jamming modulations, such as spread spectrum, or in the layers above MAC via channel switching. However, even if the CRN has an anti-jam Physical-layer transmission scheme, it may still be sensitive to jamming attacks because of the unique property of CRN in spectrum sensing and spectrum vacating [5]-[7]. In addition, channel switching in CRN is costly considering the required timing/frequency synchronization, channel estimation, handshaking for information exchange and network setup. In particular, the information about the available channels may not be identical among the CRN nodes because of the asynchronous spectrum sensing and the inevitable sensing errors. Extensive handshaking is necessary, which can be extremely timing/bandwidth consuming. Considering the complexity of jamming and anti-jamming interactions in CRN, game theory has also been adopted in anti-jamming research [8]-[10]. Nevertheless, the cost of channel switching has not been addressed sufficiently in these studies.

In this paper, we study the anti-jamming performance of CRN against jammers that are also equipped with similar cognitive radios. We focus on evaluating quantitatively some best jamming parameters as well as some optimal anti-jamming parameters, in particular the effect of number of white space channels. We conduct a cross-layer analysis of both the jamming capability and the anti-jamming capability. To address the CRN specific properties such as channel switching more accurately, we use a Markov chain to model the CRN operations. This provides us an efficient way to analyze the role of channel switching in mitigating jamming attacks. Although Markov model has been widely used for CRN performance analysis [11], its application in anti-jamming study is relatively less. In contrast, we addressed them in [12]-[14], which form the foundation for this paper.

The organization of this paper is as follows. In Section 2, we give the models of the CRN and the cognitive-radio-based jammers. Then in Section 3, we derive the jamming and anti-jamming performance expressions. In Section 4, we analyze and optimize important parameters that are critical for anti-jamming design. Simulations are conducted in Section 5. Conclusions are then given in Section 6.

2. Models of CRN and Jammers

We consider a generic cognitive radio transmission model that includes three states: spectrum sensing, data transmission and channel switching, as illustrated in **Figure 1(a)**. The working sequence of a cognitive radio always begins with the spectrum sensing. If the spectrum sensing indicates that the channel is available for secondary access, then the cognitive radio transmits a data packet, and the model shifts into the data transmission state. If the spectrum sensing indicates the channel is not available, the cognitive radio conducts channel switching, and the model shifts into the channel switching state.

We use the Markov chain in **Figure 1(b)** to model the operation of the CRN, where p_s, p_d, p_c are the probabilities of the CRN in the spectrum sensing, data transmission and channel switching modes, respectively. The transitional probabilities p_{js}, p_{jd}, p_{jc} are the probabilities that the spectrum sensing, data transmission and channel switching procedures are jammed, respectively.

The durations of spectrum sensing slot, data transmission slot, and channel switching slot are T_s, T_d, and T_c. Usually the spectrum sensing duration T_s is much smaller than both T_d and T_c.The CRN has M white space channels to select from. The availability of each channel depends on the activity of the PU and the jammers. The large number of channels is one of the primary advantages of CRN to combat jamming.

We use signal-to-noise-and-interference ratio (SINR) to measure the signal and jamming levels. For the data transmission slot, we assume that the minimum workable SINR is Γ_d. SINR less than Γ_d means that the CRN's data packet transmission is jammed. For the spectrum sensing slot, we assume that if the SINR is larger than the detection threshold Γ_s, then the cognitive radio will make a decision that the channel is occupied by the PU, and is thus not available. Γ_d is usually much larger than Γ_s. We also assume that the minimum SINR for the channel switching procedures is Γ_c, which is usually smaller than Γ_d because the CRN may adopt

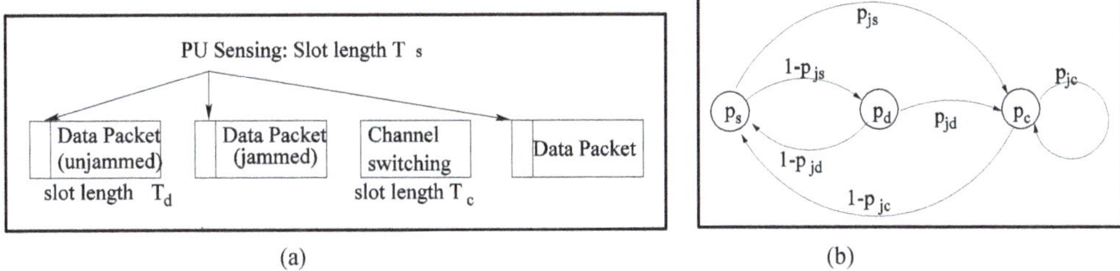

Figure 1. (a) Illustration of cognitive radio transmissions. (b) Markov model for cognitive radio transmissions under jamming.

some more reliable transmission techniques (albeit with lower data rate) such as spread spectrum modulations to increase the reliability of channel switching.

We make the following assumptions about the jammers. 1) There are J jammers. 2) Each jammer uses devices that have similar capabilities as a CRN node, including spectrum sensing and RF transceiving. 3) The jammers do not know the secret keys that the CRN is using for channel selection and communication. Therefore, the jammers do not know which channel the CRN is using. The only way left for the jammers is to randomly select some channels to jam.

Since the jammers can fastly switch among the channels, they may choose to jam multiple channels simultaneously. In this paper the jamming strategies are described by two parameters: the jamming signal duration T_j and the number of channels that are jammed simultaneously. We assume that each jammer has the same transmission power $P_j = P_s$ as the CRN, where P_s is the CRN node's transmission power. The demodulation and signal detection of the CRN receiver depend on the average SINR received during the entire slot. If the jamming duration T_j is smaller, the overall jamming signal power in this slot is lower. But the jammers can jam more channels simultaneously with smaller T_j. We assume that all the jammers adopt the same jamming parameter T_j.

3. CRN Throughput under Multiple Uncooperative Jammers

3.1. Jamming Probabilities

Consider a CRN where a pair of CRN transmitter and receiver is conducting transmission at unit data throughput. A group of J jammers want to jam the CRN transmission so as to reduce the throughput.

First, we consider a data packet transmission slot with slot length T_d and SINR requirement Γ_d. If there are k jamming signals falling in this slot, each with duration T_j, then the SINR is

$$\gamma_d(k) = \frac{P_s \alpha_s^2 T_d}{\sum_{\ell=1}^{k} P_j \alpha_\ell^2 \min\{T_d, T_j\} + N T_d} \tag{1}$$

where α_s^2 is the power gain of the Rayleigh flat fading channel of the CRN, α_ℓ^2 is the power gain of the Rayleigh flat fading channel of the ℓth jamming signal, N is the power of the additive white Gaussian noise (AWGN). We assume that α_s^2 and α_ℓ^2 are independent exponential random variables with unit mean. A successful jamming means that $\gamma_d(k) < \Gamma_d$.

The number of jamming signals k is limited to $0 \le k \le K_d$, where $K_d \triangleq J\lceil T_d/T_j \rceil$ and $\lceil x \rceil$ denotes the minimum integer that is no less than x. The probability that there are k jamming signals in this slot follows the binomial distribution

$$\mathbb{P}_d[k] = \binom{K_d}{k} p_j^k (1-p_j)^{K_d-k}, \tag{2}$$

where $p_j = 1/M$ is the probability that a jammer chooses the same channel as the CRN. For simplicity, we do not consider the white space detection errors of the CRN and jammers. White space detection errors may make the available white space channels less for CRN. However, for the jammers, a safer approach might be just to jam every one of the M white space channels.

Proposition 1. If there are k jamming signals with the same jamming duration T_j injected into a data transmission slot of duration T_d, the probability that the data packet transmission is jammed is

$$\mathbb{P}\left[\gamma_d(k)<\Gamma_d\right]=1-\mathrm{e}^{-\frac{N\Gamma_d}{P_s}}\left(1+\frac{P_j\min\{T_d,T_j\}}{P_sT_d}\Gamma_d\right)^{-k}. \tag{3}$$

Proof. Using (1), we can change $\gamma_d(k)<\Gamma_d$ into $z<\Gamma_d$ with

$$z=\frac{1}{N}P_s\alpha_s^2-\frac{P_j\min\{T_d,T_j\}}{NT_d}\Gamma_d\sum_{\ell=1}^{k}\alpha_\ell^2. \tag{4}$$

Since α_s^2 is an exponential random variable with unit mean, its probability density function is $f_{\alpha_s^2}(x)=\mathrm{e}^{-x}$, for $x\geq 0$. In addition, $Y=\sum_{\ell=1}^{k}\alpha_\ell^2$ has Erlong distribution with probability density function $f_Y(y)=y^{k-1}\mathrm{e}^{-y}/(k-1)!$, for $y\geq 0$. Due to the independence assumption, their joint distribution is

$$f_{\alpha_s^2,Y}(x,y)=f_{\alpha_s^2}(x)f_Y(y)=\begin{cases}\mathrm{e}^{-x}\dfrac{y^{k-1}\mathrm{e}^{-y}}{(k-1)!}, & x\geq 0, y\geq 0\\ 0, & \text{else}\end{cases}. \tag{5}$$

Then, the probability $\mathbb{P}\left[\gamma_d(k)<\Gamma_d\right]$ can be evaluated as

$$\begin{aligned}\mathbb{P}[z<\Gamma_d]&=\iint_{z<\Gamma_d}f_{\alpha_s^2,Y}(x,y)\mathrm{d}x\mathrm{d}y=\int_0^{\infty}\frac{y^{k-1}\mathrm{e}^{-y}}{(k-1)!}\mathrm{d}y\int_0^{\frac{N}{P_s}\Gamma_d\left(1+\frac{P_j\min\{T_d,T_j\}}{NT_d}y\right)}\mathrm{e}^{-x}\mathrm{d}x\\&=\int_0^{\infty}\frac{y^{k-1}\mathrm{e}^{-y}}{(k-1)!}\left[1-\mathrm{e}^{-\frac{N\Gamma_d}{P_s}}\mathrm{e}^{-\Gamma_d\frac{P_j\min\{T_d,T_j\}}{P_sT_d}y}\right]\mathrm{d}y=1-\frac{\mathrm{e}^{-\frac{N\Gamma_d}{P_s}}}{(k-1)!}\int_0^{\infty}y^{k-1}\mathrm{e}^{-\left(1+\frac{P_j\min\{T_d,T_j\}}{P_sT_d}\Gamma_d\right)y}\mathrm{d}y\end{aligned} \tag{6}$$

The last integration in (6) can be changed to the integration of the Erlong probability density function. According to the property of the Erlong distribution, we can derive (3). ■

Averaging over all possible k, the probability that the data transmission slot is jammed is

$$p_{jd}=\sum_{k=0}^{K_d}\mathbb{P}\left[\gamma_d(k)<\Gamma_d\right]\mathbb{P}_d[k], \tag{7}$$

which can be evaluated by using (2) and (3).

Proposition 2. For the channel switching slot with duration T_c and required SINR Γ_c, the probability of being jammed is

$$p_{jc}=\sum_{k=0}^{K_c}\mathbb{P}\left[\gamma_c(k)<\Gamma_c\right]\mathbb{P}_c[k], \tag{8}$$

where the maximum number of jamming signals in this slot is $K_c=J\lceil T_c/T_j\rceil$, the jamming probability under k jamming signals is

$$\mathbb{P}\left[\gamma_c(k)<\Gamma_c\right]=1-\mathrm{e}^{-\frac{N\Gamma_c}{P_s}}\left(1+\frac{P_j\min\{T_c,T_j\}}{P_sT_c}\Gamma_c\right)^{-k}, \tag{9}$$

and the probability of having k jamming signals is

$$\mathbb{P}_c[k]=\binom{K_c}{k}p_j^k(1-p_j)^{K_c-k}. \tag{10}$$

Proof. We can derive (8)-(10) by following the proof of the Proposition 1, and by replacing T_d and Γ_d with T_c and Γ_c, respectively. ■

The spectrum sensing slot is different from either the data packet slot or the channel switching slot because the SINR $\gamma_s(k)$ is in fact the interference (jamming) to noise ratio

$$\gamma_s(k)=\frac{\sum_{\ell=1}^{k}P_j\min\{T_s,T_j\}\alpha_\ell^2}{NT_s}. \tag{11}$$

Usually the CRN is highly sensitive in PU sensing, which means that there is an extremely small sensing threshold Γ_s. By making $\gamma_s(k)\geq\Gamma_s$, the jammers disguise the PUs to force the CRN to conduct channel switching, which defines the jamming of the sensing slots.

Proposition 3. For the channel sensing slot with duration T_s and sensing threshold Γ_s, the probability of being jammed is

$$p_{js}=\sum_{k=0}^{K_s}\mathbb{P}[\gamma_s(k)\geq\Gamma_s]\mathbb{P}_s[k], \tag{12}$$

where the maximum number of jamming signals in this slot is $K_s=J\lceil T_s/T_j\rceil$, the probability of having k jamming signals is

$$\mathbb{P}_s[k]=\binom{K_s}{k}p_j^k(1-p_j)^{K_s-k}, \tag{13}$$

and the jamming probability under k jamming signals is

$$\mathbb{P}[\gamma_s(k)\geq\Gamma_s]=1-\frac{\gamma(k,a)}{(k-1)!}=\sum_{n=0}^{k-1}\frac{1}{n!}\mathrm{e}^{-a}a^n, \tag{14}$$

where $a=\dfrac{NT_s\Gamma_s}{P_j\min\{T_s,T_j\}}$ and $\gamma(k,a)$ is the lower incomplete Gamma function.

Proof. The equation (13) can be derived similarly as $\mathbb{P}_d[k]$ in (2) by replacing T_d with T_s. To derive (14), from the SINR definition (11) and the Erlong distribution of $Y=\sum_{\ell=1}^{k}\alpha_\ell^2$, we have

$$\mathbb{P}[\gamma_s(k)\geq\Gamma_s]=\mathbb{P}\left[Y\geq\frac{NT_s\Gamma_s}{P_j\min\{T_s,T_j\}}\right]=1-\int_0^{\frac{NT_s\Gamma_s}{P_j\min\{T_s,T_j\}}}f_Y(y)\mathrm{d}y. \tag{15}$$

From the property of the Erlong distribution, the integration of (15) leads to (14). ■

3.2. Throughput of CRN under Jamming

With the jamming probabilities p_{jd}, p_{jc}, p_{js} derived in Equations (7), (8) and (12), we can calculate the steady state probabilities of the three states p_s, p_d and p_c of the Markov model in **Figure 1(b)** by solving the equation

$$\begin{bmatrix}-1 & 1-p_{jd} & 1-p_{jc}\\ 1-p_{js} & -1 & 0\\ p_{js} & p_{jd} & p_{jc}-1\end{bmatrix}\begin{bmatrix}p_s\\ p_d\\ p_c\end{bmatrix}=\begin{bmatrix}0\\0\\0\end{bmatrix}. \tag{16}$$

We need the constraint $p_s+p_d+p_c=1$ for (16) to have a unique solution.

From (16) we can find that the system stays in the data packet transmission slots with probability

$$p_d=\frac{(1-p_{js})(1-p_{jc})}{2-p_{jc}+(p_{jd}-p_{jc})(1-p_{js})}. \tag{17}$$

However, some of the data packets are lost due to jamming. Considering the jamming probability p_{jd}, the data packet transmission is successful with probability $p_d(1-p_{jd})$.

We define the normalized average throughput of the CRN transmission as

$$R = \frac{p_d\left(1-p_{jd}\right)T_d}{p_sT_s + p_dT_d + p_cT_c}. \tag{18}$$

Proposition 4. Considering the throughput definition (18), the throughput of CRN under jamming is

$$R = \frac{\left(1-p_{js}\right)\left(1-p_{jc}\right)\left(1-p_{jd}\right)T_d}{\left(1-p_{jc}\right)T_s + \left(1-p_{js}\right)\left(1-p_{jc}\right)T_d + \left(p_{jd}+p_{js}-p_{jd}p_{js}\right)T_c}. \tag{19}$$

Proof. From (16), we can describe p_d and p_c by p_s as

$$p_d = \left(1-p_{js}\right)p_s, \qquad p_c = \frac{p_{jd}+p_{js}-p_{jd}p_{js}}{1-p_{jc}}p_s \tag{20}$$

Substituting (20) into (18), we can get (19). ■

4. Analysis of Anti-Jamming Performance

4.1. A Closed-Form Throughput Expression

Although (19) can be used to evaluate numerically the anti-jamming performance, further analytical study is difficult. In this section, we first adopt some reasonable simplifications to simplify the jamming probability and the throughput expressions. Then, we analyze the anti-jamming performance by deriving the maximum and minimum throughputs under jamming.

For the jammer model, instead of considering J non-cooperative jammers that randomly inject k jamming signals with duration T_j and power P_j, we assume in this section that they cooperatively inject a single jamming signal of duration T_J (which equals to the CRN slot lengths without loss of generality). The total power is $P_J = JP_j$. We assume that the jammers can jam K channels simultaneously, and the jamming power sent to each channel is $P_K = P_J/K$. We define $P_0 = 0$. We also assume that the maximum number of channels that these jammers can jam simultaneously is K_J, which means $0 \le K \le K_J$.

For the data transmission slot, if there is jamming signal, the cognitive radio's received signal's SINR is $\gamma_d = P_s\alpha_s^2/\left(P_K\alpha_j^2 + N\right)$. If there is no jamming signal, the SINR becomes $\gamma_{d'} = P_s\alpha_s^2/N$.

Since there are M white space channels, if the jammers randomly select K channels to jam, then they have probability K/M of sending a jamming signal to the channel being used by the CRN. Therefore, the probability that the data transmission is jammed (including the case that the channel gain is too small to transmit data successfully) can be written as

$$p_{jd} = \mathbb{P}\left[\gamma_d < \Gamma_d\right]\frac{K}{M} + \mathbb{P}\left[\gamma_{d'} < \Gamma_d\right]\left(1-\frac{K}{M}\right). \tag{21}$$

Note that $\mathbb{P}\left[\gamma_d < \Gamma_d\right] = \mathbb{P}\left[P_s\alpha_s^2 - P_K\Gamma_d\alpha_j^2 < N\Gamma_d\right]$. Similar to the proof of the Proposition 1, we can find

$$\mathbb{P}\left[\gamma_d < \Gamma_d\right] = 1 - \mathrm{e}^{-\frac{N\Gamma_d}{P_s}}\frac{P_s}{P_s + P_K\Gamma_d}, \qquad \mathbb{P}\left[\gamma_{d'} < \Gamma_d\right] = 1 - \mathrm{e}^{-\frac{N\Gamma_d}{P_s}}. \tag{22}$$

Therefore, the probability that the data transmission is jammed can be derived from (21)-(22) as

$$p_{jd} = 1 - \mathrm{e}^{-\frac{N\Gamma_d}{P_s}}\left(1 + \frac{P_J\Gamma_d}{M\left(P_s + P_K\Gamma_d\right)}\right). \tag{23}$$

Next, we consider the channel sensing slots. In case of absence of PU, if there is jamming signal in this sensing slot, then we have SINR $\gamma_s = P_K\alpha_j^2/N$. Otherwise, the SINR becomes simply $1/N$. For jamming probability, we just need to consider γ_s. The probability of having jamming signal in this sensing slot is similarly K/M. Therefore, the probability that the sensing slot is jammed can be found as

$$p_{js} = \mathbb{P}\left[\gamma_s \ge \Gamma_s\right]\frac{K}{M} = \frac{P_J}{MP_K}\mathrm{e}^{-\frac{N\Gamma_s}{P_K}} \tag{24}$$

because γ_s has exponential distribution.

Finally, because channel switching is usually more jamming-resistant than channel sensing and data transmission, we let $p_{jc} = 0$. Furthermore, without loss of generality, we let $T_c = T_d$, which means that the CRN waits for a full data slot before switching to a new channel. Then the throughput (19) can be readily deduced into a close-form expression

$$R = \frac{\mathrm{e}^{-\frac{N\Gamma_d}{P_s}}}{1+\frac{T_s}{T_d}}\left(1+\frac{\Gamma_d P_J}{MP_s + MP_K\Gamma_d}\right)\left(1-\mathrm{e}^{-\frac{N\Gamma_s}{P_K}}\frac{P_J}{MP_K}\right). \tag{25}$$

4.2. Anti-Jamming Throughput Analysis

One of the major parameters for the jammers to adjust jamming attacks is the jamming signal strength P_K, or equivalently, the number of channels to jam simultaneously $K = P_J / P_K$. In contrast, one of the major parameters for the CRN to mitigate jamming is the number of white space channels M.

If considering just P_K and M, the optimal anti-jamming performance of CRN can be found from the max-min optimization $\max_{M>0} \min_{0\le P_K \le P_J} R$.

First, let us analyze the jammer's best strategy to minimize the CRN throughput. Define $y = P_K / P_J = 1/K$, and rewrite the throughput (25) into

$$R(y) = \frac{\mathrm{e}^{-\frac{N\Gamma_d}{P_s}}}{1+\frac{T_s}{T_d}}\left(1+\frac{\frac{\Gamma_d}{M}}{\frac{P_s}{P_J}+y\Gamma_d}\right)\left(1-\mathrm{e}^{-\frac{N\Gamma_s}{P_J y}}\frac{1}{My}\right). \tag{26}$$

Note that the range of y is $1/K_J \le y \le 1$. If y is extremely small, the item $\mathrm{e}^{-N\Gamma_s/P_J y}/My \approx 0$ can be omitted from $R(y)$. In this case, the derivative

$$\frac{\partial R(y)}{\partial y} \approx -\frac{\mathrm{e}^{-\frac{N\Gamma_d}{P_s}}}{1+\frac{T_s}{T_d}}\frac{\Gamma_d^2}{M\left(\frac{P_s}{P_J}+\Gamma_d y\right)^2} < 0, \tag{27}$$

which means that $R(y)$ is a monotone decreasing function of y when y is extremely small.

When y is not so small, because $N\Gamma_s/P_J$ is usually a very small number, we let $\mathrm{e}^{-N\Gamma_s/P_J y} \approx 0$. In this case, by taking the derivative of (26) with respect to y, we can easily find that

$$\frac{\partial R(y)}{\partial y} \approx \frac{\mathrm{e}^{-\frac{N\Gamma_d}{P_s}}}{1+\frac{T_s}{T_d}}\frac{1}{M}\left(\frac{1}{\frac{P_s}{P_J\Gamma_d}+y}+\frac{1}{y}\right) > 0, \tag{28}$$

which means that the throughput $R(y)$ becomes a monotone increasing function for relatively large y. Therefore, the minimum $R(y)$ should happen with some extremely small y values, whereas the maximum throughput happens either when $y = 1$ or $y = 1/K_J$. The former means that the jammers just jam one channel at a time, while the latter means the weakest jamming signal is used. The maximum throughput is thus $R_{\max} = \max\{R(1), R(1/K_J)\}$, which can be calculated from (26).

The optimal jamming parameter y for the jammers to minimize the CRN throughput is shown below.

Proposition 5. For the jammers, the (approximately) optimal jamming parameter is

$$y_o = \max\left\{\frac{1}{K_J}, \frac{N\Gamma_s}{P_J}\right\} \tag{29}$$

which minimizes the CRN throughput into $R_{\min} = R(y_o)$.

Proof. From (26), we can take the derivative $\partial R(y)/\partial y$ and let it be zero to find the optimal y. After some straightforward deductions, we can get

$$\left(1-\frac{1}{My}e^{-\frac{N\Gamma_s}{P_J y}}\right)\frac{\Gamma_d}{M\left(P_s/P_J+\Gamma_d y\right)^2}-\left(1+\frac{\Gamma_d/M}{P_s/P_J+\Gamma_d y}\right)e^{-\frac{N\Gamma_s}{P_J y}}\frac{1}{My^2}\left(1-\frac{N\Gamma_s}{P_J y}\right)=0. \tag{30}$$

Unfortunately, (30) is too complex to find closed-form solutions to y. Therefore, as an approximation, we consider the major items only. Because the minimum $R(y)$ happens when y is extremely small, we can consider only those items in (30) involving $O(y^{-2})$ and $O(y^{-3})$. Then (30) can be approximately simplified to

$$\left(1+\frac{\Gamma_d/M}{P_s/P_J+\Gamma_d y}\right)e^{-\frac{N\Gamma_s}{P_J y}}\frac{1}{My^2}\left(1-\frac{N\Gamma_s}{P_J y}\right)=0 \tag{31}$$

which gives solution $y=N\Gamma_s/P_J$. Considering the practical range of y and the monotone property of $R(y)$, we can derive (29). The throughput can be obtained by applying y_o into (26). ■

If M is large and $y_o=N\Gamma_s/P_J$, then the minimum throughput becomes

$$R_{\min}=\frac{e^{-\frac{N\Gamma_d}{P_s}}}{1+\frac{T_s}{T_d}}\left(1+\frac{P_J/M}{P_J\Gamma_d+N\Gamma_s}\right)\left(1-\frac{e^{-1}P_J}{MN\Gamma_s}\right)\le\frac{e^{-\frac{N\Gamma_d}{P_s}}}{1+\frac{T_s}{T_d}}\left(1-\frac{P_J^2}{M^2N^2\Gamma_s^2}\right). \tag{32}$$

On the other hand, if M is small so the jammer can jam all the channels with $y=1/M$, then the throughput becomes

$$R_{\min}=\frac{e^{-\frac{N\Gamma_d}{P_s}}}{1+\frac{T_s}{T_d}}\left(1+\frac{\Gamma_d}{M\frac{P_s}{P_J}+\Gamma_d}\right)\left(1-e^{-\frac{MN\Gamma_s}{P_J}}\right). \tag{33}$$

From (32)-(33), it can be seen that M should be extremely large (e.g., several hundreds) for moderately high throughput. The CRN can increase M to mitigate jamming. Unfortunately, from (32) we can see that the throughput increases according to $O(1-1/M^2)$ only, which means larger M only brings smaller throughput increase, or the throughput increase tends to saturate at large M. Alternatively, the CRN may reduce the length of spectrum sensing slot T_s or increase the spectrum sensing threshold Γ_s to increase the throughput. But this may increase interference to PUs. Therefore, anti-jamming CRN design is a challenging issue.

5. Simulations

In this section, we use simulations to verify the analysis results derived in Sections 3 and 4. Specifically, the normalized average throughput R and the probability of transmitting unjammed data packets $p_d(1-p_{jd})$ were evaluated. The following parameters were used: $M=100$, $J=10$, $T_d=5$, $T_c=10$, $T_s=0.25$, $\Gamma_d=15$ dB, $\Gamma_c=10$ dB, $\Gamma_s=-15$ dB, $P_s=P_j=-80$ dBm, and $N=-100$ dBm.

First, we verify the results in Section 3. For the jammers, we tested the jamming signals with duration T_j from 0 (jamming-free) up to the duration of T_c since T_c was the longest slot length. For the theoretical results, we used (19) to calculate R and used (7) and (17) to calculate $p_d(1-p_{jd})$.

The simulation results are shown in **Figure 2**. From **Figure 2(a)**, we can see that the theoretical results fit well with the simulated results, which demonstrated the validity of the modelling and analysis. Compared to the jamming-free throughput $(T_j/T_c=0)$ which was near unity, the throughput drastically reduced to just below 0.3 when facing 10 jammers that used small T_j. Even with 100 channels to hop from, the CRN throughput still suffered from detrimental effect of jamming.

In **Figure 2(b)**, we evaluated the anti-jamming performance of CRN when the CRN could hop among a large number of white space channels. It can be seen that while increasing the channel number M could drastically increase the anti-jamming capability of CRN, such a benefit tended to saturate after tens of channels had been used. Even with 1000 white space channels, the average throughput were still just around 0.6. In contrast, the jammers could reduce this benefit by just using a few more jammers.

Next, we use simulations to verify that the analysis results of Section 4. For the jammer, we evaluated the jamming parameters y from 0 (jamming-free) up to 0.1. The simulation results are shown in **Figure 3**. From the

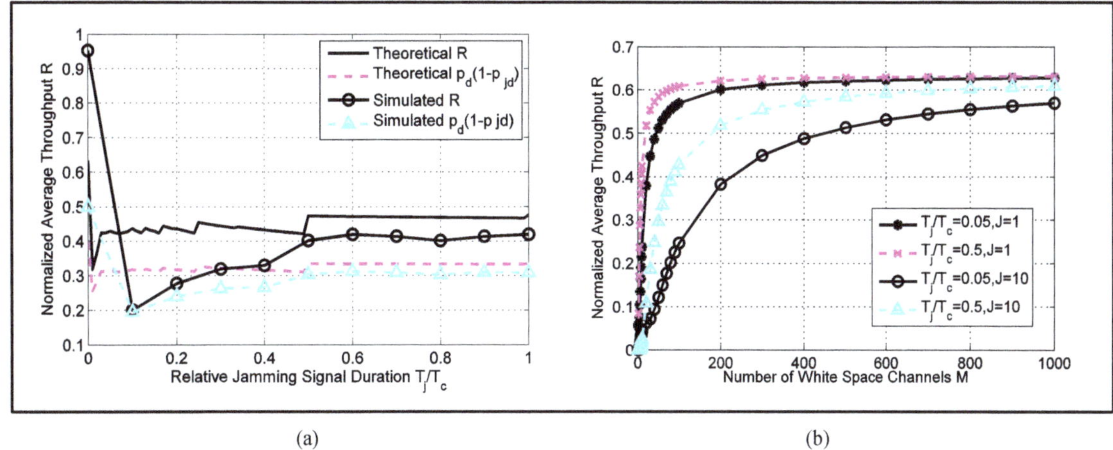

(a) (b)

Figure 2. (a) Average throughput and probability of transmitting unjammed data packets under various jamming parameters. (b) Average throughput as function of number of white space channels, under various jamming parameters.

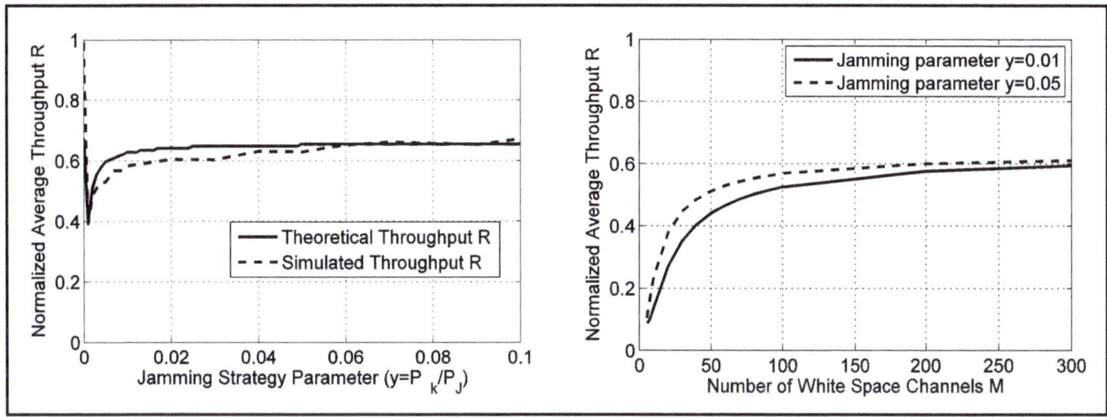

Figure 3. Comparison of simulation results to the analysis results of the average throughput.

results, we can see that the analysis results fit well with the simulated results, which demonstrated the validity of the analysis. It clearly showed that the throughput reduced with y when y was extremely small, but increased with y when y became larger.

6. Conclusion

In this paper, with a Markov model of the cognitive radio transmissions, both the jamming performance of the cognitive-radio-based jammers and the anti-jamming performance of the CRN are analyzed. Expressions of the CRN average throughput and jamming probabilities are derived. Some optimal jamming parameters and anti-jamming parameters are analyzed, in particular the number of white space channels, which are verified by simulations. The results indicate that the CRN is extremely susceptible to jamming attacks, and it suffers from a saturation effect when combating jamming attacks by increasing the number of white space channels.

References

[1] Akyildiz, I.F., Lee, W.-Y., Vuran, M.C. and Mohanty, S. (2006) NeXt Generation/Dynamic Spectrum Access/Cognitive Radio Wireless Networks: A Survey. *Computer Networks*, **50**, 2127-2159. http://dx.doi.org/10.1016/j.comnet.2006.05.001

[2] McHenry, M., Livsics, E., Nguyen, T. and Majumdar, N. (2007) XG Dynamic Spectrum Access Field Test Results. *IEEE Xplore*: *Communications Magazine*, **45**, 51-57. http://dx.doi.org/10.1109/MCOM.2007.374432

[3] Cordeiro, C., Challapali, K., Birru, D. and Shankar, S. (2006) IEEE 802.22: An Introduction to the First Wireless

Standard Based on Cognitive Radios. *Journal of Communication*, **1**, 38-47.

[4] Clancy, T.C. and Goergen, N. (2008) Security in Cognitive Radio Networks: Threats and Mitigation. *International Conference on Cognitive Radio Oriented Wireless Networks and Communications*, Singapore.

[5] Chen, R., Park, J.-M. and Reed, J.H. (2008) Defense Against Primary User Emulation Attacks in Cognitive Radio Networks. *IEEE Journal on Selected Areas in Communications*, *Special Issue on Cognitive Radio Theory and Applications*, **26**, 25-37. http://dx.doi.org/10.1109/JSAC.2008.080104

[6] Wang, Q., Ren, K. and Ning, P. (2011) Anti-Jamming Communication in Cognitive Radio Networks with Unknown Channel Statistics. 2011 19*th IEEE International Conference on Network Protocols* (*ICNP*), 393-402.

[7] Pietro, R.D. and Oligeri, G. (2013) Jamming Mitigation in Cognitive Radio Networks. *IEEE Network*, **27**, 10-15. http://dx.doi.org/10.1109/MNET.2013.6523802

[8] Li, H. and Han, Z. (2009) Dogfight in Spectrum: Jamming and Anti-Jamming Inmultichannel Cognitive Radio Systems. *Proc. of IEEE GLOBECOM*, 1-6.

[9] Wang, B., Wu, Y., Liu, K.J.R. and Clancy, T.C. (2011) An Anti-Jamming Stochastic Game for Cognitive Radio Networks. *IEEE Journal on Selected Areas in Communications*, **29**, 877-889. http://dx.doi.org/10.1109/JSAC.2011.110418

[10] Chen, C., Song, M., Xin, C. and Backens, J. (2013) A Game-Theoretical Anti-Jamming Scheme for Cognitive Radio Networks. *IEEE Network*, **27**, 22-27. http://dx.doi.org/10.1109/MNET.2013.6523804

[11] Tumuluru, V.K., Wang, P., Niyato, D. and Song, W. (2012) Performance Analysis of Cognitive Radio Spectrum Access with Prioritized Traffic. *IEEE Transactions on Vehicular Technology*, **61**, 1895-1906. http://dx.doi.org/10.1109/TVT.2012.2186471

[12] Li, X. and Cadeau, W (2011) Anti-Jamming Performance of Cognitive Radio Networks. *Proceedings of the* 45*th Annual Conf. on Information Sciences and Systems* (CISS), Johns Hopkins Univ., Baltimore.

[13] Cadeau, W. and Li, X. (2012) Anti-Jamming Performance of Cognitive Radio Networks under Multiple Uncoordinated Jammers in Fading Environment. *Proceedings of the* 46*th Annual Conf. on Information Sciences and Systems* (CISS), Princeton Univ., Princeton. http://dx.doi.org/10.1109/CISS.2012.6310843

[14] Cadeau, W. and Li, X. (2013) Jamming Probabilities and Throughput of Cognitive Radio Communications against a Wideband Jammer. *The* 47*th Annual Conference on Information Sciences and Systems* (CISS), Johns Hopkins University.

3

Community Detection in Dynamic Social Networks

Nathan Aston, Wei Hu*

Department of Computer Science, Houghton College, Houghton, USA
Email: *wei.hu@houghton.edu

Abstract

There are many community detection algorithms for discovering communities in networks, but very few deal with networks that change structure. The SCAN (Structural Clustering Algorithm for Networks) algorithm is one of these algorithms that detect communities in static networks. To make SCAN more effective for the dynamic social networks that are continually changing their structure, we propose the algorithm DSCAN (Dynamic SCAN) which improves SCAN to allow it to update a local structure in less time than it would to run SCAN on the entire network. We also improve SCAN by removing the need for parameter tuning. DSCAN, tested on real world dynamic networks, performs faster and comparably to SCAN from one timestamp to another, relative to the size of the change. We also devised an approach to genetic algorithms for detecting communities in dynamic social networks, which performs well in speed and modularity.

Keywords

Community Detection, Dynamic Social Networks, Density, Genetic Algorithms

1. Introduction

Social networks, such as Facebook and Twitter, have been rapidly growing in recent years. Such a network can be represented as a graph, where a node represents a user and an edge represents their affiliation with others. **Figure 1** illustrates this idea. These affiliations can represent friendships or likes, as in Facebook, or followings, as in Twitter. Nodes with similar affiliations tend to group into densely knit communities to form network structures. Moreover, research has discovered three characteristics of social network structure. First, the small world phenomenon can be described as any two nodes that are related to each other through only a small number of other nodes. Second, the power law is the distribution of node degree following the pattern of a power func-

*Corresponding author.

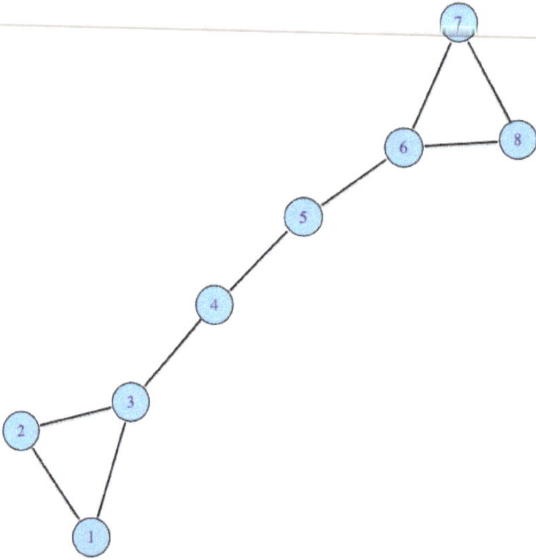

Figure 1. Illustrates a graph with nodes connected to other nodes by edges.

tion. The third is the observed community structures within a network [1]. The fast growth of social networking sites in the past decade has caused a need to analyze its community structures.

A community in a network is a group of nodes that are densely connected inside and sparsely outside. Community detection can reveal information about the users in each community, such as common interests, relationships, or views. Similarities between individuals within a community can be used in advertisements, where one user's likes can be applicable for the other users within the same community; behavioral trends between individuals of a common community; or even similarities between genes in a genetic sequence.

There is no fixed order or form to network structures, as they arise randomly in different shapes and sizes [1], leading to difficulties in detecting communities accurately. There are many community detection algorithms in use today, ranging from label propagation [1] to density analysis [2]. Many of these algorithms are designed to discover communities in static networks and do not scale well. Networks today include millions of nodes and billions of edges and are continually changing their structure. Community detection in dynamic networks involves the process of incorporating the community model of a previous timestamp, or snapshot of a network structure, into the detection of the next to improve the efficiency of detecting the new community structure.

Additionally, another aspect about community structure is that many community detection algorithms discover the best possible community membership for each node in the network, but some nodes are too distant from all other nodes and should be considered outliers. Also, there are those nodes that bridge two or more communities, which may or may not be members of the communities they bridge. These bridging nodes are called hubs.

2. Related Work

A general density-based clustering algorithm DBSCAN, Density Based Spatial Clustering of Applications with Noise, was proposed by Ester *et al.* [3]. DBSCAN forms communities from individual nodes labeled core points. These core points must satisfy a user-defined number of neighbors within a given radius. A node with a neighborhood that is too small is labeled a noise point, unless it falls within the neighborhood of a core point, which then it is labeled a border point. Core points connect to other nearby core points to form the center of a community. Ester *et al.* [3] devise the means to group these types of densely connected communities.

The SCAN algorithm (Structural Clustering Algorithm for Networks) [2], derived from DBSCAN, is capable of discovering communities, hubs, and outliers in a network. A community is grown from a group of centralized nodes which all satisfy a given neighborhood size. To define the neighborhood of a node, a user-defined threshold, ε, is introduced. Instead of looking at a node's immediate neighbors, SCAN uses the ε neighborhood of a node and groups it with those who share a common set of neighbors [2]. A structural similarity measure is used to calculate the similarity between two nodes. Its time complexity is O(n) due to the one time pass through of the set of nodes in the network.

An incremental updating process was described in the DENGRAPH-IO algorithm [4]. This algorithm uses its own variation of DBSCAN, called DENGRAPH, to update the current community structure of a network from a previously detected structure and its changes over time. The changes in nodes and edges are eventually reflected in the edges alone. Incrementally, each edge change, whether added or removed, is incorporated into the previous structure of the network, and an update to the community structure is performed in relation to the edge change. This algorithm defines six possible outcomes that an edge change in the network would cause to a community structure. An added feature to DENGRAPH, compared to SCAN, is that it can discover overlapping communities, by allowing each node to inherit multiple community labels instead of one. Also, to define a density-based neighborhood of a node, DENGRAPH uses the distance between two nodes, while SCAN uses neighborhood similarity.

Similar to community detection by node density is the idea of edge density. This density measures the ratio of the number of inter community edges to the number of intra community edges [1]. Darts *et al.* introduce the study of edge density in community detection and provided insight into its implementations with other community detection approaches [5]. They introduced the density measure of actual number of edges divided by the possible number of edges, where a community absorbs nodes that allow a specified edge density threshold to remain satisfied.

A common benchmark algorithm used in many comparisons is the Label Propagation Algorithm (LPA) [6]. This algorithm starts out with each node having a unique label, or community membership. It progresses through multiple iterations, which consist of each node acquiring the label from each of its neighbors and taking on the label of the majority; in the case of a tie, a label is randomly selected. Iterations are performed until a convergence is discovered or a designated number of iterations is reached.

LPA is a simple and effective algorithm that can only find disjoint communities, with no overlap. So a variation of the LPA algorithm was proposed that helped to incorporate community overlap. SLPA (Speaker—Listener Label Propagation Algorithm) [7] incorporates LPA with the allowance of each node to hold more than one label to discover overlapping communities.

Genetic algorithm (GA) is a technique that imitates the process of gene inheritance, mutation, selection, and crossover to optimize the solution to a problem. For the use of GA with community detection, the genes represent all individual nodes and their corresponding community. To find a gene that satisfies a good community structure of a network, the gene must maximize a given fitness function, such as modularity. In finding this maximized gene, a population is used consisting of multiple genes, each representing a different community structure. A user-defined number of iterations is performed over the population which mutates each gene into a cross between two or more genes that produce the highest values due to the fitness function. At the end of the iterations, the gene with the highest value from the fitness function is used as the final community structure. More on the use of genetic algorithms in networks can be read at [8].

3. Density Based Community Detection

Ester *et al.* [3], in their formulation of the DBSCAN algorithm, proposed definitions on how to relate nodes to one another in a density-based structure. Density-based community detection is defined where each node in a community share a greater number of connections with each other than with those outside. Let $G = \{V, E\}$ be the network graph, where V is the set of nodes and E is the set of edges. Also let $N(v)$, where $v \in V$, be the set of neighbors of node v, including v, and let $E(v)$ be the set of all edges incident to v. There are several definitions that must be covered to describe the structure of communities in a density-based network. **Figure 2** illustrates these points.

3.1. Epsilon-Neighborhood of Nodes

The use of a node's neighbor is insufficient to define a densely connected community, because a densely connected community must have a high similarity with neighbors in its own community. So to handle these more closely related neighbors, the idea of an ε-neighborhood was formulated.

In DBSCAN, Ester *et al.* [3] uses a distance function to relate the closeness between two nodes. Their ε-neighborhood incorporated all neighbors that meet a user-defined threshold ε. Given any two nodes u and v:

$$\varepsilon\text{-}Nei(v) = \{u \in N(v) \mid dist(v,u) \le \varepsilon\}$$

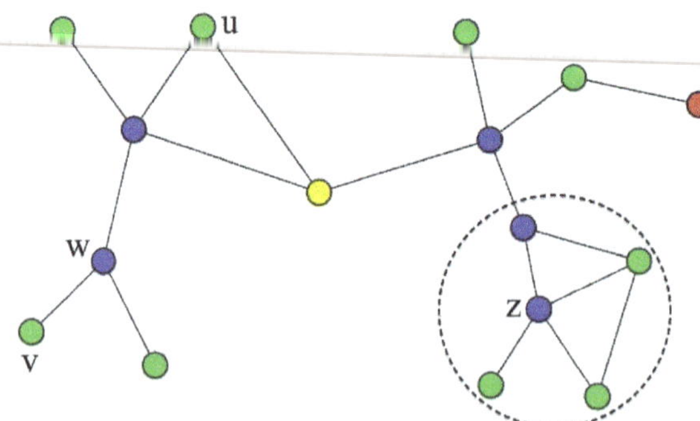

Figure 2. Shows the definitions of density structure. Node z is shown with its ε-neighborhood, v is directly reachable from w, u is structurally reachable from w, and u and v are structurally connected. The single yellow node is a hub node between the two communities and the red node is an outlier.

SCAN takes a different approach when relating nodes closeness. They define a similarity function that finds a ratio between the number of neighbors they both share in common to the number of neighbors that they each have:

$$sim(v,u) = \frac{|N(v) \cap N(u)|}{\sqrt{|N(v)||N(u)|}}$$

The more neighbors they both have in common, the greater their similarity will be; in the range of (0, 1). So Xu *et al.* uses a slightly modified version of the ε-neighborhood definition:

$$\varepsilon\text{-}Nei(v) = \{u \in N(v) | sim(v,u) \geq \varepsilon\}.$$

3.2. Core Nodes

A node that is considered a core node is one that contains an ε-neighborhood that has a size that satisfies a user-defined threshold of μ nodes: $|\varepsilon\text{-}Nei(v)| \geq \mu$. According to Xu *et al.*, a μ of 2 is recommended for analysis.

3.3. Direct Structure Reachability

A node u is directly reachable from node v, if v is a core node and u is in the epsilon-neighborhood of v:

$$u \in \varepsilon\text{-}Nei(v) \wedge |\varepsilon\text{-}Nei(v)| \geq \mu.$$

3.4. Structure Reachability

A node u is structurally reachable from node v if there is a chain of nodes that are all directly reachable from the previous until u is reached. This is only capable if the chain of nodes are core nodes, seen by the definition of directly reachable: $\{v_1,\cdots,v_n\} \in V$; $v_1 = v \wedge v_i = u$; $\forall i \in \{2,\cdots,n\}$: *Directly Reachable* (v_{i-1}, v_i).

3.5. Structural Connectivity

Two nodes (v, u) are structurally connected if there exists a node x from which v and u are structurally reachable. The two nodes, in this case, can be border nodes, and the node that connects them is a core: $\exists x \in V$; *Structure Reachable*(x, v) ∧ *Structure Reachable*(x, u).

3.6. Hubs and Outliers

By SCAN, a node v is a hub node if it does not belong to any community, but it does have more than two neighbors belonging to different communities: $\exists(x, y) \in N(v)$; $x \neq y \wedge Community(x) \neq Community(y)$.

A node v is an outlier if it does not belong to any community, and it does not contain more than one neighboring node that belongs to different communities: $\sim\ \exists(x,y)\in N(v)\,;\, x\neq y \wedge Community(x)\neq Community(y)$.

4. Community Structure Reformation Due to Network Changes

In this section, we will introduce the process of updating a dynamic network and the six scenarios that may arise.

4.1. Updating Similarity Value Between Nodes

A change in the structure of a network, such as an edge being added or removed, would affect the similarities between the two nodes that are incident to this edge. Instead of checking the similarity of all edges in the network, only those edges that are related to these two nodes need to be updated. Edge similarities to recalculate due to an edge change of nodes v and u are $\{e\in E \mid e\in E(v) \text{ or } e\in E(u)\}$.

4.2. Effects of a Network Change

Falkowski [4] introduces six possible outcomes that a change in a network would produce. Each of these six scenarios is split into two groups by how that changes the community structure. There is the idea of a positive change, where a change will create a new community, a community will receive a new member, or two communities merge. Then there is a negative change, which could remove a community, remove a member from a community, or split a community into two [4]. These two groups with three scenarios within are the fundamental ideas for updating a dynamic network.

Within a positive change, a creation of a community, an addition to a community or a merge of two communities may occur. A creation of a new community would result when nodes become cores and have no ε-neighbors that were cores previously; these new core nodes would establish a new community with its ε-neighborhood. An addition to a community will occur when a node, which previously had not belonged to a community, obtains a similarity with a core that is greater than ε, thus it now belongs to the community of this core. Then a merging of two communities will happen when a core of one community obtains a similarity with a core of another community that is greater than ε, and then the communities of these cores will become one.

In a negative change, a split of a community, removal from a community, or removal of a community may occur. A split will occur when a node, that was once a core, now is no longer a core, and it creates a gap between the other cores of the community. This gap will create a division between two or more chains of cores causing each chain of cores to form separate communities. A removal from a community occurs when a node no longer has a strong connection, or high enough similarity, to a core of its community and is no longer labeled as a member of that community. A removal of a community happens when all cores of a community are reduced from their core status and the community is removed.

To determine whether a change will produce a positive or negative change to a network is not an easy task. Any change in a network will involve an addition or removal of an edge. Then by these two principles, [4] relates the two groups of scenarios with edge changes as: a new edge produces a positive scenario, and the removal of an edge produces a negative scenario. This idea may work for the distance-based approach for Falkowski, but according to our trials on a similarity based approach; this is not always the case. The next section illustrates our algorithm, which handles the edge changes of a network.

5. DSCAN: Improvements to SCAN

Our study is based on the algorithm SCAN [2] and applies improvements to its density based detection. Our modifications to SCAN allow it to form communities without the user-defined threshold of epsilon and can update a dynamic network of timestamps. We propose the algorithm DSCAN (Dynamic SCAN), which can handle these dynamic networks.

SCAN's threshold of ε, in the range of (0, 1), defines the minimum similarity between two adjacent nodes that must exist for the two nodes to be ε neighbors. This ε-neighborhood is what defines the community structure. Having to rely on user specifications for ε can decrease performance if an incorrect ε is used, and performing multiple runs with different epsilons is costly. In our research with testing various networks, we have discovered that possible good epsilons for any network fall in the range of (0.4, 0.8). An ε too low will produce few large communities, while a larger ε will result many small communities. For our study, to calculate the ε-neighborhood of a node, we perform a check ε the range (0.4, 0.8) and take an average of those results. This approach produces comparable results and alleviates the need to run multiple times to check all ε values. Xu *et al.* found a good value of μ to be 2, which is the value that we have also used.

Algorithm 1 illustrates the dynamic updating of a network with DSCAN. Given a sequence of timestamps of a network, perform SCAN on the first timestamp. Then for all consecutive timestamps, obtain the difference in edges between the two timestamps. The network is updated from the nodes of each edge change of the networks, illustrated in Method 1. This update on the network handles a node that either becomes a core or was a core and now no longer is. When a change in the network is detected and needs to be updated, an existing cluster id or a new cluster id is propagated through all structurally connected nodes to form a new community. Method 2 explains the propagation of an id through this new community. One final change to the network is when a node no longer belongs to the cluster. Method 3 shows how to handle these nodes and make them either a hub or an outlier.

```
Algorithm 1: DSCAN
Input: Graph timestamps (0 … T), ε, μ
run SCAN for G_0
for t in 1 thru T:
    ΔE = edge changes between G_t-1 and G_t
    G_t = G_t-1
    for e in ΔE:
        if e is an addition:
            add e to G_t
        else:
            remove e from G_t
        updateNetwork (G_t, e, ε, μ)
```

```
Method 1: updateNetwork
Input: Graph G, edge changed e, ε, μ
u, v = nodes of e
for x in (u, v):
    if x == u:
        x' = v
    else:
        x' = u
    if x is a core:
        propagateId(x, generateNewId(), ε, μ)
        if e was a removal:
            if x' is not density connected from x and id(x') == id(x):
                if x' is a core:
                    propagateId(x', generateNewId(), ε, μ)
                else:
                    makeHubOrOutlier(x')
    else:
        if x was a core:
            for nei in neighborhood of x:
                if id(nei) == id(x) and nei has not been changed:
                    propagateId(x', generateNewId(), ε, μ)
            if x was not changed:
                makeHubOrOutlier(x)
                for nei in neighborhood of x:
                    if nei is not a core and does not neighbor a core:
                        makeHubOrOutlier(nei)
```

```
Method 2: propagateId
Input: Vertex v, id, ε, μ
if v is a core:
    push v onto the queue
while |queue| > 0:
u = pop from top of queue
    if u was not checked already:
        id(u) = id
        if u is a core:
            push all ε-neighbors of u to the queue
```

```
Method 3: makeHubOrOutlier
Input: Vertex v
ids = ids of neighbors of v
if ||ids|| > 1:
    make v a hub
else:
    make v an outlier
```

6. Genetic Algorithms Applied to Dynamic Networks

In addition to SCAN, we also researched the topic of genetic algorithms and their use with dynamic networks. Genetic algorithm (GA) is a technique that imitates the process of gene inheritance, mutation, selection, and crossover to optimize the solution to a problem. For the use of GA with community detection, the genes represent all individual nodes and their corresponding community. To find a gene that satisfies a good community structure of a network, the gene must maximize a given fitness function, such as modularity. In finding this maximized gene, a population is used consisting of multiple genes, each representing a different community structure. A user-defined number of iterations is performed over the population which mutates each gene into a cross between two or more genes that produce the highest values due to the fitness function. At the end of the iterations, the gene with the highest fitness function value it used as the final community structure. More on the use of genetic algorithms in networks can be read at [8].

Genetic algorithms start with a random structure for each gene of the population, and from there slowly selects the best genes to mutate into fitter ones. For the use of genetic algorithms in dynamic networks, if only a small change to the network has occurred, then the previous community structure will produce a good value for the fitness function. The idea leads to needing fewer iterations over the population to find a good structure than is needed when starting with the beginning random population.

To make use of this idea of fewer iterations, we introduce a user-defined number of iterations, not for the total number of population mutations, but the number of consecutive iterations where no better gene is found. This parameter allows for a better termination of genetic algorithms, when there is a clear time in the iterations that no more genes could be discovered rather than terminating in the middle of a sequence of better genes being discovered.

We propose the algorithm GAD (Genetic Algorithm Dynamic) for the use of genetic algorithms with dynamic networks. GAD uses the previously found "best" gene of the previous timestamp as a gene in the population of the next. This results in the next timestamp starting with a gene in the population with a relatively good fitness. If this gene happens to still be the best gene for the current timestamp, then the genetic algorithm will not mutate the population an unnecessary amount of times.

7. Dynamic Network Datasets

In our measuring the performance of our algorithms, we experiment on three real world networks of varying size.

arXiv HEP-TH Physics Network. This high energy physics citation network contains up to 27,770 nodes and 352,807 edges. Each node corresponds to a particular paper in a database, and each edge represents a citation from one paper to another. The network's data includes a period from January 1993 to April 2003 and the timestamp for each citation on when it was added to the database [9]. This information made it possible to create timestamps for the network's change over time. We converted this information into 844 timestamps by what consecutive days the graph had changed.

MIT Social Evolution Call-SMS Network. This is a considerably smaller graph composed of up to 80 nodes and 273 edges. This network illustrates the mobile phone usage between undergraduate students in a dormitory. Mobile phone usage involves calls or SMS texts between students in the same dormitory. This growing network depicts the relationships and connections that establish from October 2008 until May 2009 [10]. Our timestamps for this network are by days that show new communication between two individuals.

MIT Friends and Family Network. This dataset was created to study the decision process of individuals in response to social interaction. We focused on the SMS network of this dataset, depicting the communication between individuals, similar to the Call-SMS network. The network consists of up to 80 nodes and 150 edges. We created timestamps in the same way as the Call-SMS network, by days that show a change in communication.

8. Results

We have proposed two algorithms to handle community detection in dynamic social networks, DSCAN and GAD. Both algorithms detect communities faster than their original static community detection counterparts with little to no decrease in overall modularity, which we used as our measure for evaluating the quality of the community detection of our algorithms. GAD shows a considerable increase in both speed and modularity over GA.

8.1. DSCAN on Dynamic Social Networks

Our algorithm DSCAN performs community detection on dynamic networks faster than SCAN, with little to no performance decrease. Our implementation of averaging over the range of 0.4 to 0.8 for a good epsilon value produces results very similar to the best epsilon value chosen from the same range on SCAN.

DSCAN is capable of updating a network in a more efficient manner than SCAN can on the whole network. This is done by recalculating the new community structure only in the areas of the network that showed a change in structure. In **Figure 3**, the runtime of the arXiv network is displayed between SCAN detecting the community structure on each timestamp versus DSCAN updating the network for each timestamp. It is clear that DSCAN can perform the update faster than SCAN does the whole network on most timestamps. The timestamps where DSCAN displays a longer runtime is related to the number of changes that occurred in that timestamp, as shown in **Figure 4**. A large number of changes causes DSCAN to update the whole network and not be restricted to a certain area. This demonstrates that DSCAN is best performed on smaller changes to increase runtime. On average, using a machine with a 2.7 GHz quad-core AMD processor with 32GB of memory, SCAN calculated all timestamps in 74 minutes and DSCAN ran in only 55 minutes. The other two networks were too small to see a significant increase in speed between the two algorithms.

Modularity is the most used measure in evaluating the quality of communities found by community detection algorithms. Modularity [11] is simply defined as:

$$Q = \frac{1}{2M} \sum_{u,v} \left(A_{uv} - \frac{k_u k_v}{2M} \right) s_{uv}$$

where A is the adjacency matrix of the network, k_i is the degree of node i, and s_{ij} is 1 if nodes i and j have the same community membership, and 0 otherwise. Modularity is the standard for calculating the goodness of the community structure of a network, butmodularity performs best when all nodes are grouped into their ideal-communities. With SCAN, and DSCAN, not all nodes are members of a community. Hubs and outliers do not belong to any community, which decreases the modularity. With this idea, it is difficult to compare SCAN and DSCAN with other algorithms that do not incorporate the idea of hubs and outliers. **Figure 5** shows the modularity of the communities for each timestamp of the arXiv network with SCAN, DSCAN, and the non-hub and outlier algorithm Infomap. From **Figure 6**, the number of hubs and outliers increases over time, decreasing the modularity shown in **Figure 5**. **Figure 7** also gives some more information on the impact that the hubs and out-

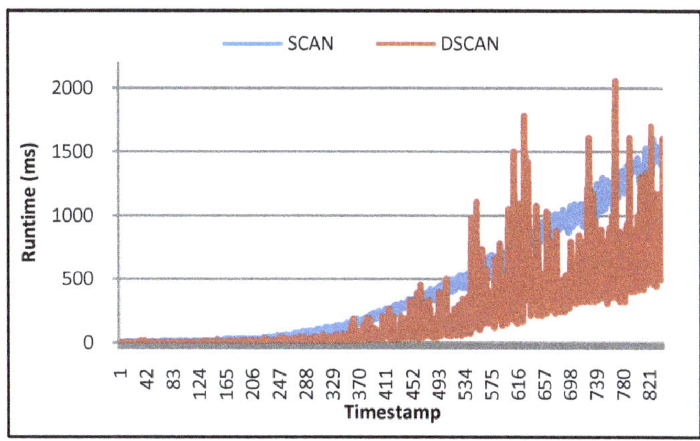

Figure 3. Runtime (ms) of SCAN and SCAND on arXiv Network for each timestamp.

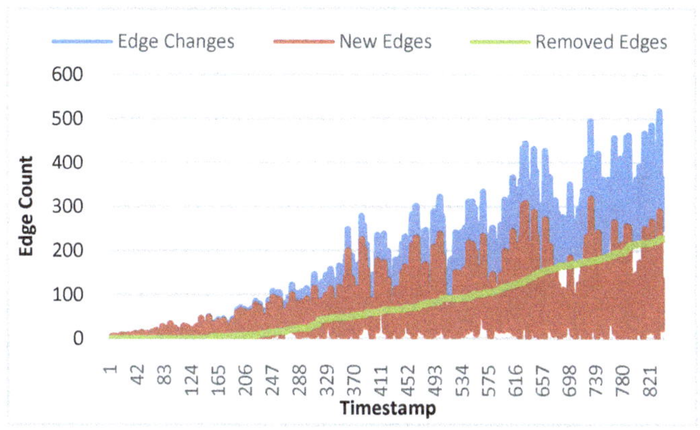

Figure 4. Number of New Edges (red), Removed Edges (green), and Edge changes in total (blue) of arVix Network for each timestamp.

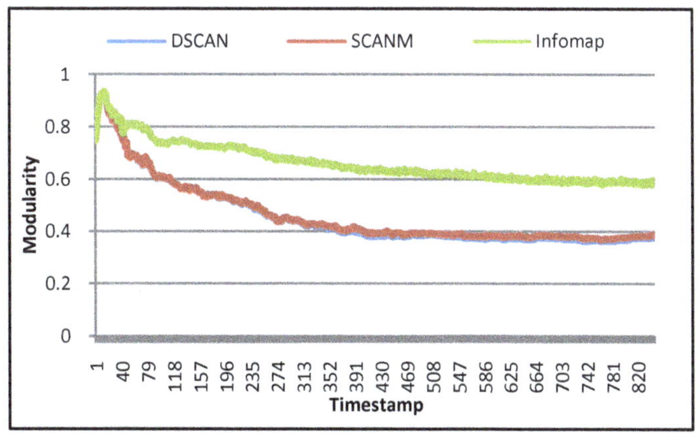

Figure 5. Modularity of arXiv Network for each timestamp with DSCAN, SCAN with multiple epsilon averaging (SCANM), and Infomap.

liers have on the arXiv network's modularity. Since modularity is based upon the edges of a network, the more edges with nodes not in the same community, the lower modularity becomes. **Figure 6** shows the average degree of hubs and outliers for each timestamp of arXiv. With this information, the number of edges that affect the value of modularity is the average degrees from **Figure 7** times the number of hubs and outliers from **Figure 6**.

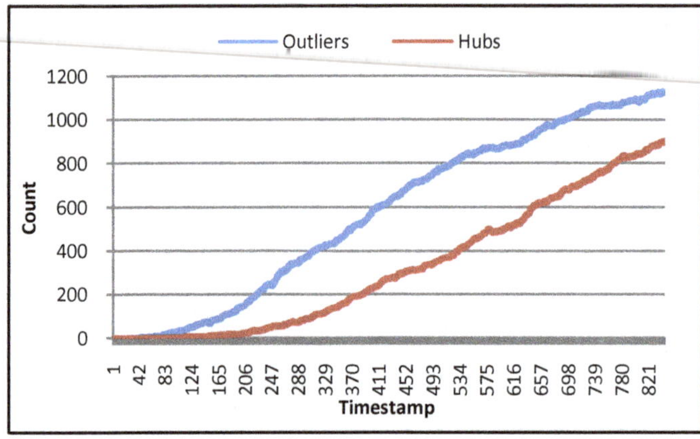

Figure 6. Number of outliers and hubs of arXiv Network for each timestamp.

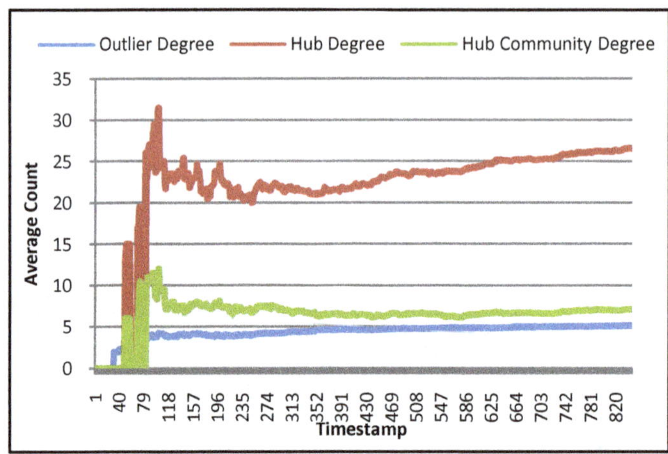

Figure 7. Average Degree of outliers, hubs, and hub communities of arXiv Network for each timestamp.

This number is approaching 3000 edges near the final timestamps of arXiv.Comparing SCAN and DSCAN by themselves with modularity will still be an effective measure.

Our implementation of SCAN using an average of epsilon values performs almost exactly as the best epsilon value would from normal SCAN. **Figure 8** and **Figure 9** of the modularity on the SMS and Call-SMS networks respectively, show that SCAN with multiple epsilons performs nearly the same as SCAN with using the epsilon with the highest modularity. There are very insignificant fluctuations between the timestamps on both networks, which is acceptable compared to having to run SCAN multiple times to check which epsilon value is the best to use. **Figure 5**, **Figure 8**, and **Figure 9** also show that DSCAN performs similarly to SCAN as well. The modularity of each timestamp is almost the same between SCAN, SCAN with multiple epsilon, and DSCAN.

We have shown that DSCAN performs community detection on dynamic networks almost similarly to what SCAN would produce by doing the entire network. This dynamic updating a community structure is an important concept when dealing with networks that grow to a large size.

8.2. GAD on Dynamic Social Networks

GAD produces a more stable runtime of genetic algorithms through each timestamp, while reducing the number of iterations needed to find a good final community membership. To run a comparison between GAD and normal genetic algorithms (GA), we ran GAD as normal over all timestamps and ran GA on each timestamp individually. **Figure 10** and **Figure 11** show the modularity of communities in the final timestamp compared between GA and GAD on the Call-SMS and SMS networks. As is seen from the figures, GAD produced stable results in

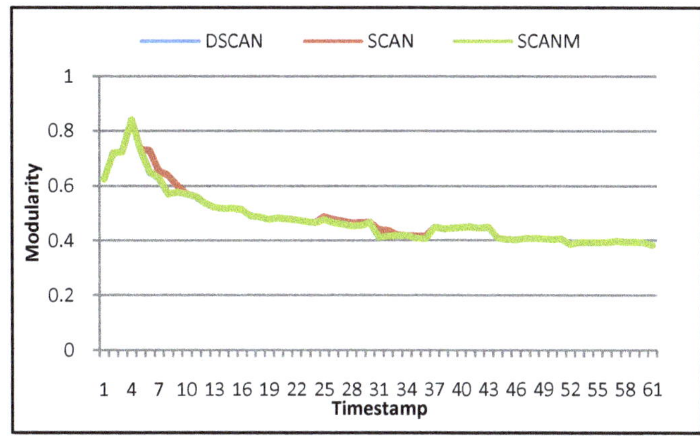

Figure 8. Modularity of SMS for each timestamp with DSCAN, SCAN using the best epsilon, and SCAN with multiple epsilon averaging (SCANM).

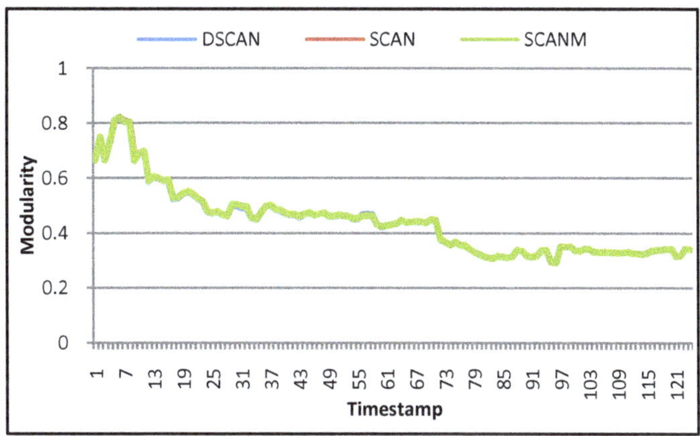

Figure 9. Modularity of Call-SMS for each timestamp with DSCAN, SCAN using the best epsilon, and SCAN with multiple epsilon averaging (SCANM).

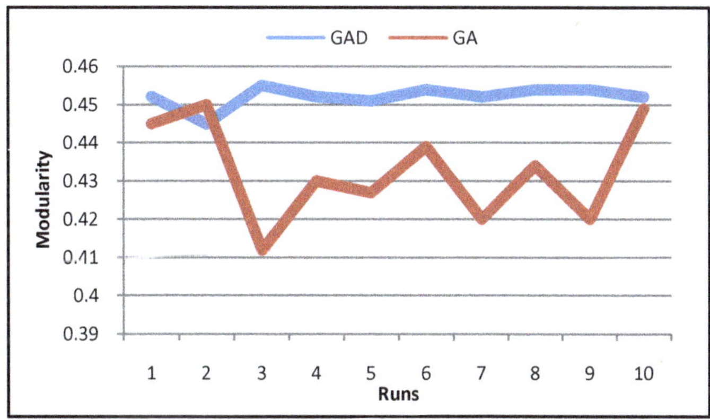

Figure 10. Modularity of SMS Network of ten runs of GA and GAD.

terms of modularity for every run. GA does not always produce a good community structure each time it runs, unless the number of iterations was increased more, increasing the runtime.

Figure 12 and **Figure 13** show the runtime of GAD over all timestamps and the sum of GA runtimes of each timestamp on the Call-SMS and SMS networks. GAD decreases the iteration count of the Call-SMS network by

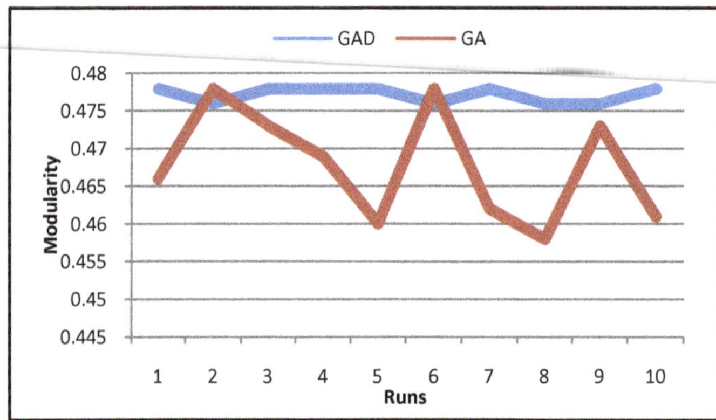

Figure 11. Modularity of Call-SMS Network of ten runs of GA and GAD.

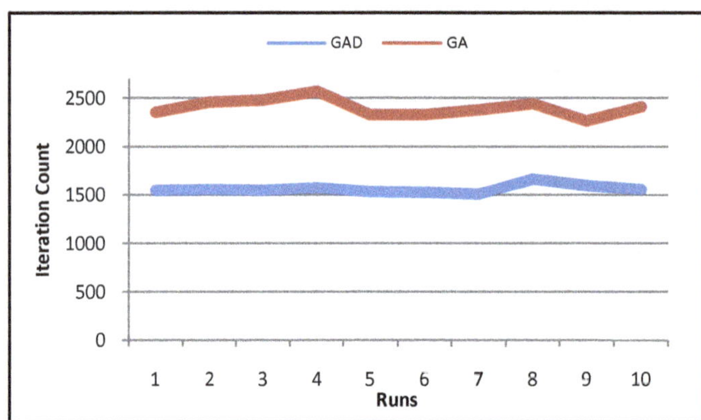

Figure 12. Number of Iterations of SMS Network of ten runs of GA and GAD.

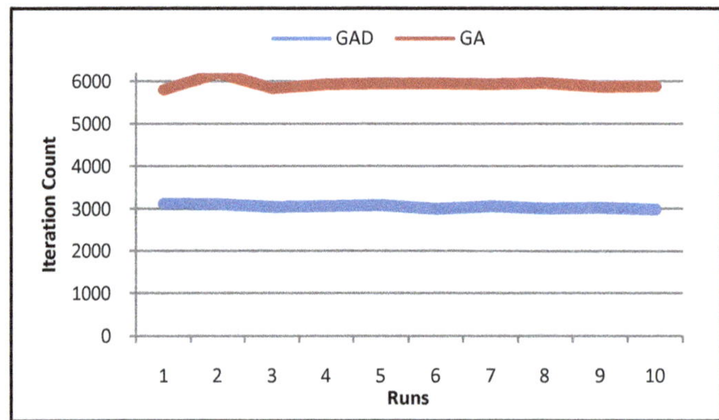

Figure 13. Number of Iterations of Call-SMS Network of ten runs of GA and GAD.

about 50% and the SMS network by about 35%. With the performance stability and the iteration decrease, it is clear that the techniques of GAD improve genetic algorithms tremendously for dynamic networks.

9. Conclusions

We have shown that DSCAN and GAD are faster in runtime and comparable, if not better, in modularity to their static community detection counterparts. This speed increase and low loss in modularity is ideal when large dynamic social networks are involved, which are constantly changing their structure. But as we have shown with our experiments on the arXiv network, the number of changes from one timestamp to another is a deciding factor in the runtime of a DSCAN. With this, it is feasible to run DSCAN after only a few changes have occurred in the network, keeping the community structure up to date more often.

Techniques like those shown in DSCAN and GAD,may possibly be used in other community detection algorithms for converting them to dynamic community detection algorithms. With the rapid increase in the size of social networks, it is costly and even infeasible to use static community detection techniques to analyze these networks. Community detection needs to start swaying to the side of dynamic network community detection to keep up with modern networks. This is an important area of future work in the field of community detection.

Acknowledgements

We thank Houghton College for its financial support. We also want to thank Dr. Xu for his SCAN algorithm code.

References

[1] Fortunato, S. (2010) Community Detection in Graphs. *Physics Reports*, **486**, 75-174. http://dx.doi.org/10.1016/j.physrep.2009.11.002

[2] Xu, X., Yuruk, N., Feng, Z. and Schweiger, T. (2007) SCAN: A Structural Clustering Algorithm for Networks. KDD'07. ACM, 824-833. http://dx.doi.org/10.1145/1281192.1281280

[3] Ester, M., Kriegel, H.-P., Sander, J. and Xu, X. (1996) A Density-Based Algorithm for Discovering Communities in Large Spatial Databases with Noise. *Proceedings of 2nd International Conference on Knowledge Discovery and Data Mining* (KDD-96).

[4] Falkowski, T. (2009) Community Analysis in Dynamic Social Networks. PhD. Otto-von-Guericke-University, Magdeburg.

[5] Ronhovde, R.K., Peter, R. and Nussinov, Z. (2013) An Edge Density Definition of Overlapping and Weighted Graph Communities. arXiv preprint arXiv:1301.3120.

[6] Raghavan, U.N., Albert, R. and Kumara, S. (2007) Near Linear Time Algorithm to Detect Community Structures in Large-Scale Networks. *Physical Review E*, **76**, 036106. http://dx.doi.org/10.1103/PhysRevE.76.036106

[7] Xie, J., Szymanski, B.K. and Liu, X. (2011) SLPA: Uncovering Overlapping Communities in Social Networks via a Speaker-Listener Interaction Dynamic Process. *Proc. ICDM Workshop*, 344-349.

[8] Pizzuti, C. (2008) GA-Net: A Genetic Algorithm for Community Detection in Social Networks. PPSN, Volume 5199 of Lecture Notes in Computer Science, pages 1081-1090. Springer.

[9] Leskovec, J., Kleinberg, J. and Faloutsos, C. (2005) Graphs over Time: Densification Laws, Shrinking Diameters and Possible Explanations. ACM SIGKDD International Conference on Knowledge Discovery and Data Mining (KDD).

[10] Dong, W., Lepri, B. and Pentland, A.S. (2011) Modeling the Co-Evolution of Behaviors and Social Relationships Using Mobile Phone Data, Media, 134-143,

[11] Newman, M.E.J. (2006) Modularity and Community Structure in Networks. *Proceedings of the National Academy of Sciences of the United States of America*, **103**, 8577-8858. http://dx.doi.org/10.1073/pnas.0601602103

Internet as a Growing and Dynamic Network: An Economic View

Pasquale Lucio Scandizzo[1], Alessandra Imperiali[2]

[1]University of Rome Tor Vergata, Center for Economic and International Studies, Rome, Italy
[2]University of Rome Tor Vergata, Rome, Italy
Email: scandizzo@uniroma2.it, Alessandra.Imperiali@uniroma2.it

Abstract

The past few decades have witnessed renewed interest and research efforts on the part of the scientific community. After spending decades to disassemble nature, focusing the attention on its components, scientists have shifted their attention on complex networks. These basic structures constitute a wide range of systems in nature and society, but their design is irregular, evolves dynamically over time and their components can fit in a large multiplicity of alternative ways. Nevertheless, the most recent studies of networks have made remarkable progresses by investigating some critical issues of structure and dynamics, thereby improving the understanding of the topology and the growth processes of complex networks. From an economic point of view, networks are especially interesting because they can be considered as a problem of allocation of a critical resource, information, under multiple constraints. They can also be viewed as forms of poliarchies that reproduce, for many aspects, the market paradigm, with surprising properties of self-organization and resilience, which go much beyond the characteristics that are generally attributed to general equilibrium structures. In this paper we first address the major results achieved in the study of complex network and then focus our attention on two specific, highly dynamic and complex networks: Internet and the World Wide Web.

Keywords

Network Analysis, Internet, Social Networks

1. Introduction

Since the 60's, the desire to understand the properties of the networks has prompted scientists from different fields to investigate the mechanisms that determine the topology of a variety of systems, ranging from biology to the structure of social relations. In the past two decades the availability of large databases on the topology of various real networks and the increased availability of computing power have offered scientists the chance to investigate networks of millions of nodes. Motivated by these circumstances, many new and important concepts about the topology of the interactions between different components have been proposed.

In the words of E. O. Wilson [1]:

> "The greatest challenge today not just in cell biology and ecology but in all of science, is the accurate and complete description of complex systems. Scientists have broken down many kinds of systems. They think they know most of the elements and forces. The next task is to reassemble them, at least in mathematical models, that capture the key properties of the entire ensemble."

In our analysis, we first introduce the basic framework for the treatment of complex networks and then investigate the mechanism determining the structural properties and topologies of real networks.

2. Graph Theory

Graph Theory is the natural framework for the treatment of complex networks, since a graph can be considered the natural representation of the topology of a complex network [2]. Leonhard Euler, one of greatest mathematicians of all time, pioneered graph analysis and made important discoveries both from the point of view of its theory and its applications. Euler spent most of his life in the city of Konigsberg, Prussia, which was set on the Pragel River and included two large islands, connected to each other and the mainland by seven bridges. In 1776 Euler solved the famous Konigsberg Bridge problem, consisting in finding a path that traversed each of the seven bridge of the Prussian city of Konigsberg exactly once and returned to the starting point. He offered a rigorous mathematical proof that the problem had no solution [3]. This result is considered the first theorem of graph theory, specifically of "planar graph theory". Since then, graph theory has undergone many more interesting developments and today represents the basis for the thinking about real networks.

In order to better understand graph theory, we must define some of its basic concepts. A graph is composed by a pair of sets G = (V,E), where V is a set of n vertices (also called points or nodes) and E is a set of K edges (links or lines) which connects two elements of the V. The graphs are constituted by a set of dots, which represents the nodes, and where two dots may be joined together by a line representing a link. In graph theory how these dots and lines are drawn is irrelevant, the only thing that matters is which pairs of nodes form a link and which do not. While it is clear that a graph may represent a network as a set of nodes and links, concepts such as arcs, paths and a number of other characteristics have been developed and superimposed, so to say, to the definition of the original definition of a graph. In this regard, a graph can be considered also a representation of a "structure", a more general concept, which more naturally lends itself to be described by additional, more abstract properties.

Graphs are also representations of market structures, in that they may depict trade flows, or other systems of individuals and linkages that have economic relevance. Contractual relations, for example, can be represented as connections among contracting partners, and can be analyzed through graphs that represent not only bilateral obligations, but also the interdependencies that are created, as a consequence of bilateral deals, in a system of production and exchange.

3. Hierarchical Clustering

Lin (1999) [4] noticed that hierarchical position and network location facilitate the access to embedded resources such as wealth, status and power of social ties. As an alternative to community networks, hierarchical clustering appears to be also a powerful basis of organization and power in social networks according to two distinct strategies: 1) agglomerative clustering and 2) divisive clustering. The agglomerative model provides for a series of fusions of the single nodes into groups. It can be defined as the result of a "bottom up" approach, where each node starts in its own cluster and then pairs of clusters are merged as one moves up the hierarchy. Divisive methods, on the other hand, separate the single objects and correspond to a "top down approach", where all nodes start in one cluster and then splits are performed as one by one they move down the hierarchy. The hierarchical structure of clusters can be graphically represented by dendograms, or hierarchical trees, which are often used to display the clusters which are produced at each step of agglomeration.

The importance of clusters for economic theory arises from Coase's approach to the theory of the firm. According to Coase (1988) [5], in fact, the existence of the firm is the attempt to reorganize contracts of exchange in alternative to the market, by economizing on transaction costs. The value of the firm thus derives from a peculiar configuration of "rights". This depends on its "dedicated hierarchical nature", that is, its specialized cluster structure, which is the consequence of both agglomerative and divisive clustering, and the assignment of different "rights" to its various stakeholders [6], with ownership and control embedded into residual "rights" of shareholders. Because most economic activities can be interpreted as "enterprises", *i.e.* as business ventures of a sort, the Coasian approach and the further developments of the neo-institutional school imply that the cluster paradigm (the enterprise as a cluster of contracts) may be applied to a wide variety of situations and, in particular, to the configuration of different actors presented by Internet and WWW. The concept of the enterprise as a cluster of contracts and of the parties involved as stakeholders has forced economists to face the issue of the plurality and heterogeneity of economic agents, especially in the new forms of "user based" enterprises.

4. Short Path Lengths

While the random graph model is the basis for the study of the formation of long-range connectivity in random systems (which is studied by percolation theory), it does not consider the random growth of complex structures, consisting of different points and connections of different types to obtain more complex real growth processes. In its original formulation [7], furthermore, it only considers static networks in which the number of vertices is fixed, thus neglecting the fact that in reality many networks evolve with the continuous addition of new elements to the system. For any given network, however, the random graph model has the merit of identifying an underlying random structure which can be very useful to establish some basic properties that do not depend on complexity and/or on growth.

The so called "small world property" is the most important network property in this respect. It was discovered by Milgram (1967) [8], who proposed an experiment, where randomly selected people in Nebraska would be asked to send letters to a distant individual in Boston, identified only by name, occupation and rough location, so that the letter could only be sent to someone presumably closer to him. Milgram was tracking the letters and found a surprising result. The average number of links needed to find the targeted person was found to be only six. This result is called "the six degrees of separation" and is the consequence of the fact that two individuals who don't know each other may nevertheless be linked by a common acquaintance.

The small world property is important because it reveals properties of the performance of a given action that depend on the underlying structure rather than on any special procedure of optimization. It also shows that local information may be conducive to global success and that social networks, regardless of their size and complexity, exhibit two key characteristics: 1) a plurality of short path lengths and, 2) a structure that enables individuals to find short path lengths even in absence of a global knowledge of the network. These aspects are very important not only for social systems, but also for the World Wide Web, the traffic way-finding in a city, and the transport of information packets on the Internet and the diffusion of signaling molecules in biological cells [2]. The small world property also suggests that whereas complexity may make more difficult to comprehend the properties of a network, it carries with itself an increase in connectivity that makes easier to cover seeming longer distances with relatively short paths. This property is the basis, for example, of the so called strength of weak ties. This argument, put forward by Granovetter (1973) [9], asserts that "Our acquaintances (weak ties) are less likely to be socially involved with one another than are our close friends (strong ties). Thus the set of people made up of any individual I and his or her acquaintances comprises a low-density network (one in which many of the possible relational lines are absent) whereas the set consisting of the same individual and his or her close friends will be densely knit (many of the possible lines are present)". Complex networks can thus be conceived as sets of simpler closely connected networks of strong ties (such as cliques) loosely connected by weak ties. The small world property would be the result of this local-global structure, whereby weak ties would be the means to bridge the gaps between two or more densely connected "strongly tied" networks.

5. Small-World Networks

The small world property (SWP) was the object of different analysis on the structure of real networks, especially in biological and technological networks. Watts and Strogatz (1998) [10] extensively analyzed SWP in their "Collective dynamics of small-world networks", where they found the existence of a relationship between the "small-world" networks and a high value of the clustering coefficient. Analytically, the clustering coefficient is a parameter introduced by the authors in 1999, to characterize the structure of complex networks. It represents a measure of the degree to which nodes in a graph tend to cluster together. Two versions of this measure exist: a local and a global clustering coefficient. The local clustering coefficient gives an indication of the clustering around a single node inside the network, while the global clustering coefficient is used to define an overall indication of the clustering inside the network.

The small-world property is also important for economics, where technological development, trade growth and the so called globalization phenomenon can be interpreted, as a consequence of "global clustering", as reducing the distance between people and make the world smaller. Goyal, van der Leij and Moraga-Gonzalez (2005) [11] studied the evolution of social distance among economists who publish in journals in the period from 1997 to 2000, to show that, despite the fact that the number of economists has more than doubled in this period, the distance between any two of them has declined.

6. Post-Structuralism and Hypertexts

Complex networks find their most recent and egregious incarnation in Internet and the World Wide Web, two constructs whose complexity, because of their continuous, never ending growth, appear boundless. It is an extreme level of present and expected convexity that, paradoxically, stimulates the search for paradigm that cut across the intricacy and multiplicity of links, to discover drastic simplifications. The small world property is one of these, but a more pervasive idea is that of the personal classification embedded in the so called hypertext. This and other ideas appear to incarnate many of the intuitions of the so called post-structuralism, a strain of thought concentrating on a set of themes on the transmission of meaning through language, the role of networks of signification, and the perpetuation of power. These themes parallel and to some extent predict hypertexts and some other features of Internet and the World Wide Web and, in particular, share with the originators of hypertext and the Web the notions about the structure and workings of text, and of the network as the coordinating principle behind the transmission of meaning through texts. Post-structuralist theory challenges the assumption that organizing structures can be imposed on information in a neutral and objective fashion. This is a similar mistrust to discrete set approaches to information organization and retrieval influencing the innovators of hypertext and the Web.

Post structuralism economics may also be interpreted as an extension of Coase's theory (1988) [7], which sees the power of the firm arising from its capacity to tie its stakeholders in a multiplicity of explicit and implicit contractual knots. As in a post-structuralist Foucault, 1980 [12] in the case of language and power, the neo-institutional school that emerged after Coase's work interprets economic power as the cause, rather than the consequence of the economic structure. Thus, for example, as Coase firmly establishes, in the presence of transaction costs, the distribution of property rights, by empowering one particular set of stakeholders rather than another one, determines the ensuing structure of the market and, as such, the particular clustering of contracts that characterizes the firms and their relations. This approach has also caused a "new theory of corporations" to emerge [13] in the school of law and legal analysis, whereby corporations are considered networks that lock in equity investors' initial capital contributions by making it far more difficult for those investors to subsequently withdraw assets from the firm.

These characteristics of economic enterprises and their networks appear also important to understand the post-structuralist nature of much of the World Wide Web. Foucault, [12] as a key theorist of the post-structuralist movement, describes the properties of an interdependent system by exploring the role of power within a special category, which he calls "discourse". Discourse for Foucault is a framework through which knowledge is transmitted and exploited, and, what is more, a framework regulated by power relations. Those power relationships are evident both between individuals, and more importantly between groups. Through language, customs, classifications and other more subtle means that impose a structure on knowledge, discourse therefore manifests its power by delimiting what it is acceptable and even possible to say about given subjects at given times. Contrary to Bacon who believed that knowledge was essentially empowering, Foucault argued that power defines what can be considered knowledge.

As a hypertext defined by essentially free associative relations, which can be traced by constructing a highly personalized and unpredictable chain of links among texts of different levels, the WWW appears an apparently successful attempt to overcome the dictatorship of an exogenously established discourse, which determines the extent and the nature of the knowledge that can be gained. The WWW is free from the arbitrary and tendentious nature of the classifications used to index and navigate the system of traditional texts and lends itself to be exploied, without having to use the scaffolding of analogical categories that are the base of all dictionaries, encyclopedias and library classification systems.

The economic side of this analysis resides in the nature of the World Wide Web as a system of information management, which arose from Enquire, a personal information retrieval system developed by Tim Berners-Lee, [14], who recognized its potential as a global information system [14] from the outset. Berners-Lee attempted to overcome the formal hierarchical structures imposed on information management solutions, because of their essential sub-optimality in retrieving and organizing information. The basic idea of the new system, which has important economic implications, is very remindful of the emergence of Coase's enterprise, and consists in the intuition that self-organizing clusters of knowledge would come more efficiently from textual networks connected by semantic and associative relationships. Contrary to the formal structures dominating traditional text indexing and retrieval algorithms, hypertext and the Web could thus progressively emerge from an underlying loser structure of random networks, by creating dynamic clusters of associative relationships emerging from the texts of an information collection. This would in turn give rise to self-organizing associative networks of information, which would dynamically optimize information search and retrieval.

7. Scale-Free Networks

The year 1999 can be considered a turning point in the analysis of complex networks because scientists found that networks don't show static scale-free graphs but expand continuously by the addition of new vertices. The network models discussed by Erdos and Renyi and by Watts and Strogatz assume that the number of vertices inside the network remains fixed. In this way, static scale-free graphs are models in which growth or aging processes do not play a dominant role in determining the structural properties of the network. [2] In reality many real networks are ruled by the dynamical evolution of the whole system. In this respect, Barabasi and Albert [15] observed that most real networks are open systems which grow by the continuous addition of new nodes.

The Barabasi and Albert (BA) [16] model was inspired by the topology structure of the World Wide Web that constitutes a network in continuous evolution and where the number of sites increases dynamically. By exploring several large databases describing the topology of large networks, AB found that, for most large networks, the degree distribution deviates from the Poisson law and that, in most of cases, it follows a power-law for large K. Since power-laws are independent of the unit of measure, these networks are called "scale-free" [17] [18]. This topological characteristic is determined by two mechanisms that interact inside the network: growth and preferential attachment. In contrast with the static models, the scale-free model describes a dynamic system which grows by the continuous addition of new vertices, as for example does the World Wide Web, which grows by the continuous addition of new Web pages. So, growth means that the number of nodes increases over time. The algorithm of this mechanism can be represented as an algorithm which starts with a small number of nodes m_0 and at each time step, adds new nodes with $m \leq m_0$ edges, with each new node being linked to the m nodes that are already present in the system. The algorithm is also non-random in the connectivity for a node inside the network, and dependent on the node's degree. This means that, when choosing the vertices to which the new node connects, the probability that a new node will be connected to node i depends on the degree K_i (the number of nodes already connected) of this node. New vertices attach preferentially to already well connect ones. An example of preferential attachment is represented by the hyperlinks of the Web page that will have a higher probability to include links to the more popular documents than to less-known ones. New pages link preferentially to hubs, very well-known sites such as Google, rather than to less-known pages. In this way, older vertices increase their connectivity, leading to a rich-gets-richer phenomenon that can easily be identified inside real networks.

Growth and preferential attachment represent two important mechanisms of the networks, and both lead to the discovery of the networks with a power-law degree distribution [2]:

$$P(K) = AK^{-\gamma} \quad (1)$$

The exponent takes different values with respect to different networks, within a relatively narrow range (2.1 to 4): for example for the World Wide Web the value is approximately 3. K stands for the average degree of a node *i*, that is the number of edges incident with the node, while *P* stands for the probability that a node chosen at random has degree *K*.

BA investigated two different variants of the model: one with growth and no preferential attachment and one with preferential attachment without growth. In both cases no scale free structure emerged. Thus, both properties are needed to empower the network to self-organize according to a stationary power law distribution. The scale-free nature of networks, which has been widely accepted by most scientists, forces us to acknowledge that networks constantly change over time. The evidence comes from better maps and data sets but also from the agreement between the empirical data and the analytical models that predict the network structure. In his book "Linked" (2002), Barabasi [3] states that "*power-laws are at the heart of some of the most stunning conceptual advances in the second half of the twentieth century, emerging in fields like chaos, fractals and phase transitions. Spotting them in networks signaled unsuspected links to other natural phenomena and placed networks at the forefront of our understanding of complex systems in general. The fact that the networks behind the Web, Hollywood, scientists, the cell, and many other complex systems all obey to a power law allowed us to paraphrase Pareto and claim for the first time that perhaps there were laws behind complex networks.*"

It was the well know Italian economist Wilfredo Pareto who, at the end of the nineteenth century, noticed that a few quantities in nature follow a power law. As a careful observer of economic inequalities, Pareto noticed that 80 per cent of the money is earned by 20 per cent of the population and also that 80 per cent of his peas were produced by only 20 per cent of the peapods. Pareto's rule is a power-law degree distribution and appears to approximately hold for many networks, including the World Wide Web, where around 80% of the links on the Web point to only 15% of the Webpages. As Barabasi [19] (2000) puts it, "*power laws mathematically formulate the fact that in most real networks the majority of the nodes coexist with a few big hubs, nodes with an anomalously number of links. The few links are not sufficient to connect the entire network, but this function is secured by the rare hubs.*"

8. Conclusions

The analysis of complex and dynamic networks is at the heart of several new fields of scientific inquiry and the basis of an interpretation of reality that cuts across several disciplines. As a method to understand Internet and its economic significance, modern network theory appears especially relevant, even though most of its disciplinary and interdisciplinary connections are yet to be discovered. The Social Accounting Matrix and the input output systems are an early application of network theory, even though their development in economics has been autonomous and mostly centered on the quantitative implications of impact analysis. On a different front, as information management systems, both Internet and the Web are environments of enterprise creation that recall Coase's original theories and the subsequent outgrowth of institutional economics. The basic idea here is that markets can be viewed as a loose network of connections with the ensuing emergence of denser sub-networks as hierarchical clusters of contracts and other types of relationships. In this respect, both the Internet and the Web enlarge the horizons of Coase's original theory much beyond the classical concept of the firm to the idea of an enterprise that can be recursively and completely defined in terms of its internal and external relations, and whose organization and production is largely dependent on the contribution of a plurality of users/stakeholders.

A second important element of Internet as a social and economic system, which is related to its clustered nature, is the fact that it is a system of small worlds, or, to cite a phrase that has become popular also in other contexts, a system of strong and weak ties that make possible communications of different types and intensities within and across communities. While the determinant factor of strong and weak ties for internet is built in its physical configuration, the small world characteristics of the web depend on its nature of a dynamic clustering system and the scale free property of the distribution of its links. This property is most intriguing, because it seems to denote a form of accumulation of "network capital", whose distribution is based on a more than proportional connection reward to the nodes that already have a higher number of connections. The Web seems thus characterized by increasing benefits of accumulating information in a few privileged hubs, without correspondent increases in congestion costs. As in Coase's model, this property may be itself the consequence of the tendency of self-organizing clusters to reduce transaction costs.

As an information management tool, the Web presents itself as a network of relations clustering around the principle of cognitive gain from free association. The hypertext results from the possibility to navigate among different texts without the limitations imposed by external classifications. As such, it is a source of allocative efficiencies that deserves further analysis. In principle, not only it allows exploiting more fully the information contained in the texts examined, but it also frees the reader from the dictatorship of the framework superimposed by any existing authority, which may effectively assert its power by limiting the extent and the form of knowledge that can be acquired.

References

[1] Strogatz, S.H. (2001) Exploring Complex Networks. *Nature*, **410**, 268-276. http://dx.doi.org/10.1038/35065725

[2] Boccaletti, S., Latora, V., Moreno, Y., Chavez, M. and Hwang, D.-U. (2006) Complex Networks: Structure and Dynamics. *Physics Reports*, **424**, 175-308.

[3] Barabasi, A.L. (2002) Linked: the New Science of Networks. Perseus Books Group, New York City.

[4] Lin, N. (1999) Building a Network Theory of Social Capital. *Connections*, **22**, 28-51.

[5] Coase, R. (1988) The Firm, the Market, and the Law. University of Chicago Press, Chicago.

[6] Ackerman, B.A. and Alstott, A. (1999) The Stakeholder Society. Yale University Press, New Haven.

[7] Erdos, P. and Renyi, A. (1959) On Random Graphs. *Publicationes Mathematicae* (*Debrecen*), **6**, 290.

[8] Milgram, S. (1967) The Small World Problem. *Psychology Today*, **160**.

[9] Granovetter, M.S. (1973) The Strength of Weak Ties. *Sociol.*, **78**, 1360.

[10] Watts, D.J. and Strogatz, S.H. (1998) Collective Dynamics of Small-World Networks. *Nature*, **393**, 440-442.

[11] Goyal, S., Van der Leij, M. and Moraga-Gonzàlez, J.L. (2005) Economics: An Emerging Small-World. *Journal of Political Economy*, **114**, 403-412.

[12] Foucault, M. (1980) War in the Filigree of Peace Course Summary. *Oxford Literary Review*, **4**, 15-19.

[13] Stout, L. (2004) On the Nature of Corporations. *Deakin Law Review*, **9**, 775-789.

[14] Berners Lee, T.J. and Fiaschetti, M. (1999) Weaving the WEB, Harper, San Francisco.

[15] Albert, R. and Barabasi, A.L. (2002) Linked: How Everything Is Connected to Everything Else and What It Means for Business, Science and Everyday Life. Plumes, New York City.

[16] Albert, R. and Barabasi, A.L. (2002) Statistical Mechanics of Complex Networks. *Review of Modern Physics*, **74**, 47-97. http://dx.doi.org/10.1103/RevModPhys.74.47

[17] Barabasi, A.L. and Albert, R. (1999) Emergence of Scale in Random Networks. *Science*, **286**, 509. http://dx.doi.org/10.1126/science.286.5439.509

[18] Barabasi, A.L., Albert, R. and Jeong, H. (1999) Scale Free Characyeristics of Random Networks: The Topology of the World Wide Web. *Physica* A, **272**, 173.

[19] Albert, R. and Barabasi, A.L. (2000) Topology of Evolving Networks: Local Events and Universalities. *Physical Review Letters*, **85**, 5234. http://dx.doi.org/10.1103/PhysRevLett.85.5234

5

Job Scheduling for Cloud Computing Using Neural Networks

Mahmoud Maqableh[1], Huda Karajeh[2], Ra'ed (Moh'd Taisir) Masa'deh[1]

[1]Management Information Systems, Faculty of Business, The University of Jordan, Amman, Jordan
[2]Computer Information Systems, King Abdullah II School for Information Technology, The University of Jordan, Amman, Jordan
Email: maqableh@ju.edu.jo, h.karajeh@ju.edu.jo, r.masadeh@ju.edu.jo

Abstract

Cloud computing aims to maximize the benefit of distributed resources and aggregate them to achieve higher throughput to solve large scale computation problems. In this technology, the customers rent the resources and only pay per use. Job scheduling is one of the biggest issues in cloud computing. Scheduling of users' requests means how to allocate resources to these requests to finish the tasks in minimum time. The main task of job scheduling system is to find the best resources for user's jobs, taking into consideration some statistics and dynamic parameters restrictions of users' jobs. In this research, we introduce cloud computing, genetic algorithm and artificial neural networks, and then review the literature of cloud job scheduling. Many researchers in the literature tried to solve the cloud job scheduling using different techniques. Most of them use artificial intelligence techniques such as genetic algorithm and ant colony to solve the problem of job scheduling and to find the optimal distribution of resources. Unfortunately, there are still some problems in this research area. Therefore, we propose implementing artificial neural networks to optimize the job scheduling results in cloud as it can find new set of classifications not only search within the available set.

Keywords

Cloud Computing, Job Scheduling, Artificial Intelligence, Artificial Neural Networks

1. Introduction

Cloud computing is an emerging paradigm that accesses network and shares computing resources with convenient and minimal management efforts, see **Figure 1**. It is one of the smart technologies that will reshape the world and shifts Information Technology infrastructure to third party to be available to the customers as commodities [1] [2]. The computing environment of cloud computing can be outsourced to another party to use the computing power or resources via Internet. Emerging of this technology moves the computing power and data from personal computer and portable devices into large data centers. End-users access and use all the services without knowing the physical location and the configuration of the system at the providers' sides [3] [4].

This paper is organized as follows. First, the researchers will introduce and discuss cloud computing deployment, characteristics, models and advantages. Then, the researchers will briefly discuss and introduce genetic algorithm and artificial neural networks. After that, the cloud computing job scheduling techniques are explained in details. Subsequently, the researchers will provide a literature review of job scheduling in cloud computing. Finally, conclusion and future works are discussed.

2. Cloud Computing Deployment, Characteristics, Models and Advantages

Cloud computing has wide acceptance due to its characteristics such as fulfilled customization, portability, availability on demand and isolation [6]. Moreover, it attracts the users due to the reduction of the cost of the provided services and at the same time improving the outcome [7]. Companies that use cloud computing do not need to invest in new infrastructure and training your employee. Using cloud computing, Small and Medium Businesses (SMB) can access to the best applications and resources at very low cost [8]. In the Information Technology industry, cloud computing is growing very fast at the same time many concerns are growing about the environment safety [8]-[10].

There are four types of cloud computing deployment: public cloud, private cloud hybrid cloud and community cloud. In the public cloud, the users access the cloud via interfaces using the web browsers. Thus, the user needs to pay only for the time duration of service usage. This will reduce the operation costs. On the other hand, public clouds are less secure compared to other clouds models, as all the software and data on this model are more vulnerable to various attacks [11]. In the private cloud, all the operations of this model are within an organization's data centers. This model is similar to the Intranet. The main advantage is that it is easy to manage the security and the maintenance and upgrades are more controlled. Compared to the public cloud where all the services and the applications are located outside the organization, in private model these services and applications are available at the organization level [4].

The hybrid model is a combination of both public cloud and private cloud. In this model, a private cloud is linked to one or more external cloud services. It enables the organization to meet its need in the private cloud, if some occasional needs occur; it asks the public cloud for intensive computing resources. Finally, the community cloud occurs when many organization jointly construct and share the cloud infrastructure, the requirements and polices [4] [11].

According to M. Malathi, 2011, cloud computing system has many attractive characteristics. Cloud computing users can access the resources via the Internet regardless of the users' location or the machine type at the minimum cost [4]. Its implementation and configuration are required the minimum skills. Moreover, using cloud

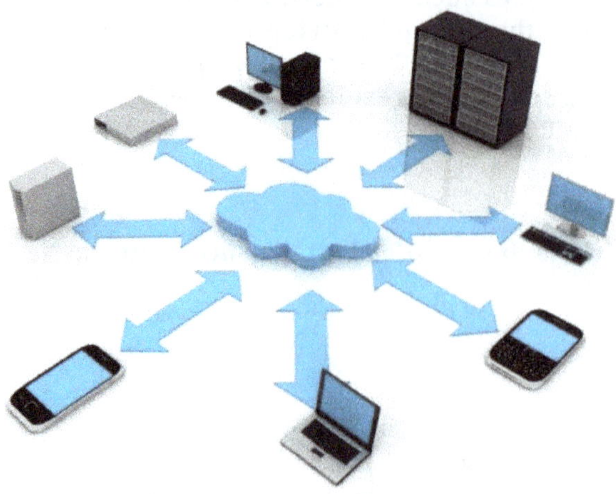

Figure 1. Cloud computing paradigm [5].

computing is reliable due to multiple sites service delivery. In addition, cloud computing resources utilization is efficient due to its sharing and scheduling between several customers. Cloud computing users do not need to concern about resources and system maintenance, which is being performed at the provided side. Finally, the security issues in cloud computing can be solved easier than the issues in the traditional systems that is being solved by specialized people and resources at the provider side using several traditional security methods such as encryption methods and Hash functions [10] [12]-[16].

Furthermore, cloud computing system consists of two main parts that are connected to each other via the Internet: front-end and back-end. The front-end is the part that the user see and has on the machine with the required applications to connect to the cloud computing. The back-end part is the cloud system with all resources and services such as software, servers and data storages. Three different cloud computing models have been proposed to gain its benefits: Software as a Service (SaaS), Infrastructure as a Service (IaaS), and Platform as a Service (PaaS), see **Figure 2** [6]. Software as a Service (SaaS) model offers finished application to the end-users via the Internet. Thus, end-users do not need to install the programs and applications on their machines that are controlled and managed by centralized authority. Platform as a Service (PaaS) provides an operating system, programming languages and software development through the cloud infrastructure. Infrastructure as a Service (IaaS) provides the required infrastructure as a service such as processing, data centers and network resource.

Cloud computing aims to maximum the benefit of distributed resources, and aggregate them to be able solve large scale computation problems [4]. It provides the computing services for users as public utility, which is available to organizations and individual [9]. In this technology the customers do not have the physical infrastructure, but they use the resources as a service and only pay when they need to use a resource [4]. Service providers provide the services to the subscribers on contractual basis. They charge the subscribers according to the provided services. Users can pay for the provided service with the payment system "pay as you go" [18]. Thus, cost reduction is considered one of the main advantages of using cloud computing. Moreover, service providers guarantee the quality of the provided services such as data processing, data storage and data access.

Another advantage is the ease of management as the maintenance of the infrastructure (software or hardware) is simplified. Furthermore, the applications that needs huge amount of storage are easier to use in the cloud

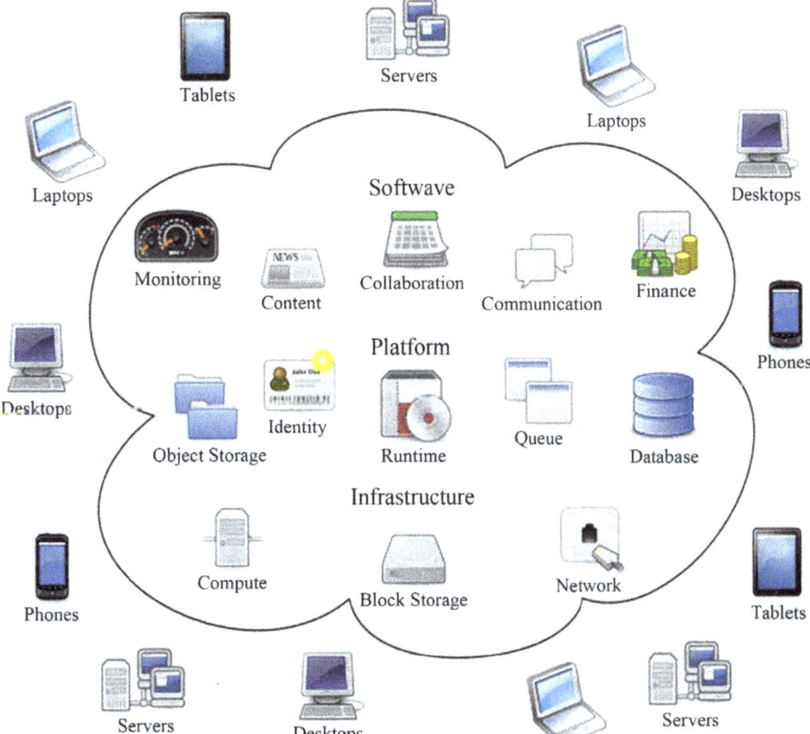

Figure 2. Cloud computing services architecture [17].

computing environment. At the user level, the user just only needs a web browser with Internet connection to use the cloud computing system [4]. Furthermore, the uninterrupted service of the reliability of the provided service is another advantage. Finally, in case of disaster, an offsite backup is always helpful. In the cloud computing system frequently backed up the data in case of disaster occurred [3].

3. Genetic Algorithm and Artificial Neural Networks

Genetic Algorithms (GAs) are search algorithms that mimic the processes of natural selection and natural genetics that use to find estimated solutions to difficult problems, see **Figure 3** [19] [20]. The principles of genetic algorithm were introduced in 1962 by Holland [21]. Genetic algorithm population is competing with each other to evolving the beat candidate for the problem solution, which will be selected based on the fitness function [20]. The main genetic algorithms steps are initial population, fitness function, selection, crossover and mutation [22] [23]. The initial population is composed of all the individuals that are used in the genetic algorithm to find out the optimal solution. Every single solution in the population is called as an individual. And every individual is called as a chromosome to make it suitable for the genetic algorithm. The individuals are selected from the initial population and some operations are applied on those to generate the next generation. The selection operation of mating chromosomes is based on some specific criteria.

A fitness function is used to measure the quality of the selected individuals from the population according to a specific optimization objective. The fitness function can be different, in some cases the fitness function can be based on maximization some factors while in other cases it can be based on minimization other factors. The mutation means that the values of some gene that is located in the chromosome code were replaced by the other gene values in order to generate a new individual in the population. In every generation, the population individuals are evaluated scouring to some defined quality measures. The chromosomes are selected based on the fitness value. Two parents' genes are allowed to be exchanged to generate new children generation, which will replace their parents. The current population becomes the new population and the old generation is removed. Then, the current population is examined to find the solution suitability. This operation will be iterated a number of times or until the desired result is obtained. GAs has been adopted in many different disciplines to optimize problem the solution such as scheduling algorithms, game playing, cognitive modeling, and salesman problems. Genetic algorithms have been used in some aspects of neural networks design as it can find the optimal solution. GAs can be view as a way of job scheduling that is based on the biological concept of population generation [22].

Artificial Neural Network (ANN) is an information processing paradigm that simulates the human brain neural, see **Figure 4** [25]. It designed to mimic the way that the human brain execute a specific task or function [20]. The adaptive nature of this network is consider one of the most important feature, where "learning by example" is used to solve the problems [19]. Thus, this model is used to solve complex or ambiguous systems problems, pattern classification and recognition. These problems would be very difficult to be extracted using many other computer techniques. ANN can give very great result when it used with complex systems that has not fully understandable relationships or chaotic properties [26]. ANN model has three main issues: network topology, transfer function, and training algorithm. ANN consists of processing units, weighted connections, activation rule, and learning rules. Neural network consist of three or more layers and each layer has number of processing unit that called neurons [27]. It has input layer, output layer and hidden layers. ANN link the input layers with the output layers using hidden layers with nonlinear transformation function and weighted connections. Artificial neural networks can have different number of layers and different number of nodes. The nature of the problem and the degree of complexity are controlled the number of hidden layers and their neurons. The nonlinear transformation functions give an advantage over the predictable functions.

ANN are trained using different learning rates, parameters and different propagation methods such as feed-forward and back propagation [25]. It learns by changing the connections between the input and output layers. The ANN output accuracy is depending on the parameters that used to train the system. Networks performance is affected by number of layers, number of nodes and training algorithms. ANNs are trained by iterating the recombination, mutation and fitness selection until development of chromosomes with accurate ANN. After neural network has been trained in certain information collection, it can be used to predict new situation and to model various non-linear applications [20]. The suitable output is generated at the output layer at the end of the learning or training process. Better results can be achieved by using neural network architecture with proper selection of input variable and training set [26].

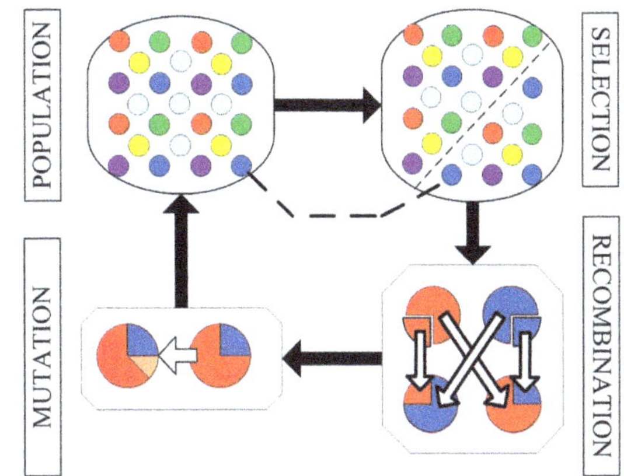

Figure 3. Simple genetic algorithm operation [24].

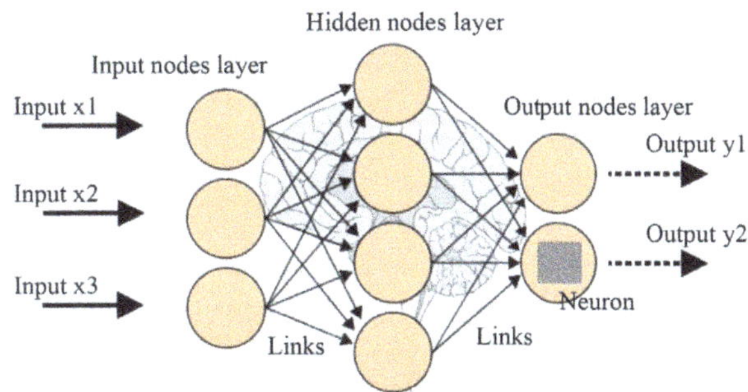

Figure 4. Basic structure of an artificial neural network (ANN) [28].

4. Cloud Computing Job Scheduling

As mentioned earlier, cloud computing is a new technology, and it becomes so popular because of its great characteristics. By using this technology, everything such as software, hardware and platform are provided as a service. The users of these services pay for every use of them. The cloud provider in cloud computing provides services based on the clients' requests [22]. One of the biggest issues in cloud computing is job scheduling. It is a hot research area in cloud and grid computing. It plays the same role in cloud and grid computing. Job scheduling of users' requests means how to allocate resources to these requests. Therefore, the required tasks can be finished in minimum time according to time defined in user request. The main task of job scheduling system is to find the best resources in a cloud for the cloud computing user's jobs, taking into consideration some statistics and dynamic parameters restrictions of users' jobs. Most researches that used in grid computing can be used in cloud computing environment [29].

However, scheduling in cloud computing can be divided into two main views: from the cloud computing users and from the cloud computing provider. From the user's view, the scheduling algorithm should minimize both the execution time and user's budget. On the other hand, from the cloud provider view, the scheduling algorithm should improve the resource utilization and reduce the cost of maintenance and energy consumption [30]. Job scheduling is a combinational problem. It cannot be considered as linear programming and it is impossible to find a global optimal solution by using a simple algorithm or rule. It is well known as NP-complete problem. In order to solve this problem, some kind of branch and bound and other approximation method are proposed, but the result is unpredictable and needs a lot of time that is not practical in cloud environment. Moreover, the goal of cloud computing maybe too complex and depends on the business orientation of cloud environment, by which it is impossible to be solved in linear time by using traditional scheduling algorithm [31]. Indeed, many researches in the literature tried to solve the issue of job scheduling in cloud computing. All of them share the same goal in mapping the user jobs onto a computing resource to achieve the maximum benefit, and satisfying the various quality of service (QoS) of user's jobs is the main goal of cloud provider. In the following section, the researchers will discuss the literature review of job scheduling in cloud computing.

5. Literature Review of Job Scheduling in Cloud Computing

Recently, many researchers studied job scheduling in cloud computing [22] [29]-[41]. In [22], the authors discussed three scheduling algorithms: Min-Min, Max-Min and genetic algorithm. Further, they propose a new scheduling algorithm in which Min-Min and Max-Min can be combined in genetic algorithm. The Min-Min algorithm starts with a set of all unassigned tasks. Firstly, the minimum completion time for all tasks is calculated. Then, among these calculated minimum times, the minimum value is selected. After that, the task is scheduled on the corresponding machine. Then, the execution time for all other tasks is added to the execution time of the assigned task, and the assigned task is removed from the list. Then, again and again the same operation is repeated until all tasks are assigned on the resources. The Max-Min algorithm is approximated the same as Min-Min algorithm except of the following: after computing the minimum execution times, the maximum value is selected that is the maximum time between all the tasks on any resource. After that and according to maximum time, the task is scheduled on the corresponding machine. Then, the execution time of the assigned task is added to the execution time all other tasks on that machine, and the assigned task is popped out from the list. Then the same operation is repeated until all tasks are assigned on the resources.

[22] used the proportion selection operator to determine the probability of various individuals genetic to be chosen to the next generation in population. The proportional selection operator means the probability which is selected and genetic to next generation groups is proportional to the size of the individual's fitness. They also used a single-point crossover operator. Single-point crossover means only one position was chosen in the individual code, at that point part of the pair of individual chromosomes is exchanged. The mutation means that the values of some gene that is located in the chromosome code were replaced by the other gene values in order to generate a new individual in the population. The authors in [22] proposed a new technique that is based on genetic algorithm which generates the initial population by using Min-Min and Max-Min can provide better initial population than if they choose the initial population randomly. The experimental results show that the improved genetic algorithm maximizes the utilization of the resources effectively than the original genetic algorithm. They used the makespan as a fitness function for checking the fitness of the scheduling results. The idea can be further extended in which they can use the execution cost of the resource as fitness criteria. This method can be modified in existing cloud computing systems for decreasing makespan and better resource utilization.

In [34] the authors proposed a cloud task scheduling policy based on Load Balancing Ant Colony Optimization (LBACO) algorithm. The main contribution of this algorithm is to balance between the entire system load while trying to minimizing the makespan of a given tasks set. The authors used the Cloud Sim toolkit package in order to simulate the new scheduling algorithm. The experimental results show that the proposed LBACO algorithm outperformed FCFS (First Come First Serve) and the basic ACO (Ant Colony Optimization).

In [31] the authors presented a genetic algorithm approach to cost based multi QoS job scheduling. The authors also proposed a model for cloud computing environment and some popular genetic cross over operators, like PMX, OX, CX and mutation operators, swap and insertion mutation are used to produce a better schedule. The algorithm guarantees the optimal solution in finite time. The experimental results show that this approach for job scheduling guarantees the QoS requirement of customer job, and also make best profit of cloud providers.

The authors in [35] presented private cloud characteristics that are used for e-Learning purposes along with a genetic algorithm that being used to optimize the scheduling of the e-Learning workloads according to a set of factors that are imposed by the underlying virtualization technology such as memory over-commitment and IOPS rate distribution. The experimental results show that the genetic algorithm is an efficient technique for enabling co-existence of Planned Scheduling Requests and One-Off Scheduling Requests, by enabling a high and uniform utilization of the cloud resources. Also, the solutions generated by the genetic algorithm generate the optimal co-scheduling of workloads based on the workload profile.

An Improved Differential Evolution Algorithm (IDEA) is proposed by Tsai *et al.* [41] to optimize task scheduling and resource allocation on cloud computing organization. The proposed algorithm improves the Differential Evolution Algorithm (DEA) by using Taguchi method to generate improved offspring. Two models are developed to minimize the total cost and the time in task scheduling. The processing and receiving cost are included in the cost model, while time model takes into account the receiving, processing, and waiting time. The effectiveness of the proposed algorithm is tested using two scenarios of the cloud environment which are; five-task five-resource scenario and ten-task ten-resources scenario. In both scenarios, the proposed algorithms

(IDEA) outperform the other scheduling algorithms in the literature (DEA/NSGA). Moreover, Gantt chart is used to show the efficiency of the proposed algorithm in task scheduling in term of having smaller cost and time. In addition, this approach can help the decision makers to choose the correct decision in case of object conflicting.

A new fully distributed scheduling framework for uncoordinated federated cloud environment is proposed by Palmerieri *et al.* [36]. This scheduling schema based on independent and self-organized agents which do not depend on any kind of centralized control that coverage towards Nash equilibrium solution with taking into account the possible contradiction between the client and the service provider interests in the cloud environment. An implicit coordination is forced by applying a marginal cost on agent behaviour. The effectiveness of the proposed schema is tested on the simulator for the cloud environment that emulates the service provider, agents, and the used protocols. The experimental results show that the proposed approach provides a good solution in terms of scalability and quality. In addition, it gains a high performance in according to the completion time. Due to the efficient partitioning strategy of the complex task into smaller one, this approach had great benefits in a very large cloud organization that have a lot of nodes with large number of tasks to be served.

Scheduling algorithms for highly available applications on cloud computing is proposed by Marc Frîncu [37]. This algorithm ensures the applications functionality despite the number of node failure. Two algorithms are proposed to achieve a highly available applications; optimal and sub-optimal algorithms. The optimal algorithm is proposed when the load of each component type is known while the suboptimal algorithm is proposed in case the load is unknown. By taking advantage of the component based architecture and the application scaling property, a highly available applications is build. A solution for determining the best number of component types on each node is presented. In addition, each node has a threshold of component load that cannot exceed it and the application running cost need to be minimized. The performance of the suboptimal algorithm is tested accordance to the node load, closeness to the optimal solution and the success rate.

A priority based job scheduling algorithm in cloud computing (PJSC) is proposed in [32] that is based on multiple criteria decision making model. This algorithm is based on the theory of Analytical Hierarchy Process (AHP) which is considered as a suitable method for priority based problem such scheduling a task with multi-attributes. The efficiency of the proposed algorithm is tested interm of consistency, complexity, and makespan. The experimental results show that the proposed algorithm has an acceptable complexity while needs more improvement to gain less makespan.

A taxonomy for cloud computing research is revealed based on an intensive literature survey for 205 journal article in cloud computing field [42]. These articles are classified into four main categories: technological issue, business issues, domain and application issues, and conceptualizing cloud computing. This study shows that the current state of cloud computing is skewed on technological issues. On the other hand, a new research issue is emerging that focus on social and organizational implications. This descriptive review is considered as a good reference to guide the practitioner and researchers on cloud computing for future research.

Scheduling algorithm is considered from one of the important issues on the cloud computing environment that enhance the workflow of the job tasks and improve the user satisfaction from the service provider. Because of that, a comprehensive survey on the different types of scheduling algorithm that is used on a cloud computing environment is presented by Vijindra and Shenai [38]. This study provides a detailed survey on the existing scheduling algorithms in cloud, grid, and workflows environment. Increasing number of parameters for scheduling algorithms may improve the framework for resource allocation and scheduling in cloud computing environment. Execution time, deadline, energy efficiency, transmission cost, performance issues, and makespan can be taken as an input for scheduling algorithm. The nature of the job such as size, availability of the resources and the environment decide which of the mentioned parameter will consider in the scheduling algorithm since considering all of them in one algorithm it will enter into the complexity problem.

A novel job scheduling algorithm in cloud computing environment based on Berger model of distributive justice is proposed by Xu *et al.* [33] through the expansion of cloudSim platform. In this algorithm, the scheduling is performed based on the fairness in resource allocation. Therefore, two fairness constraints are proposed to classify the user tasks. In the first constraint, the user's tasks are classified based on QoS to create an expectation function that control the fairness in resource allocation. Then, in the second constraint, the fairness justice function is defined to assess the fairness of the resource allocation. The effectiveness of the proposed algorithm is tested on the extended simulation platform. The experimental results show that the proposed algorithm is effective in achieving the user task with better fairness.

A route scheduling algorithm for a cloud database is proposed by Yan-Hua *et al.* [39] based on a combination of genetic and ant colony algorithm. This is performed by taking the initial value of genetic algorithm as an input to the ant colony algorithm (after transforming it into the pheromone initial value) to find the optimal solution from ant colony algorithm. The experimental results show that the proposed algorithm improves the efficiency of cloud computing by finding the suitable application's database quickly and effectively. A new scheduling algorithm is proposed by Mezmaz *et al.* [40] that based on a parallel bi-objective hybrid genetic algorithm that takes into account the makespan and the energy consumption. To minimize the energy consumption this algorithm uses the dynamic voltage scaling (DVS). The experimental results show that the proposed algorithm outperforms the other scheduling algorithm interm of time completion and energy consumption.

6. Discussion

In the literature, many researches tried to solve the problem of job scheduling in cloud computing using artificial intelligence techniques such as genetic algorithm and ant colony. Unfortunately, the proposed techniques have some problems. Neural Networks designed to mimic the way that the human brain executes a specific task or function. Its most important feature is the adaptive nature, where "learning by example" is used to solve complex or ambiguous systems problems, pattern classification and recognition. ANNs are trained using different learning rates, parameters and propagation methods. It learn by changing the connections between the input and output layers. Networks performance is affected by number of layers, number of nodes and training algorithms. ANNs are trained by iterating the recombination, mutation and fitness selection until developing chromosomes with accurate ANN. After neural network has been trained in certain information collection, it can be used to predict new situation. The suitable output is generated at the output layer at the end of the learning or training process. Better results can be achieved by using neural network architecture with proper selection of input variable and training set. Neural networks are widely used for identifications, classification, and prediction when a vast amount of information is available. By examining hundreds, neural network detects important relationships and patterns in information. The advantages of using neural networks are: learn and adjust to new cases on their own, lend them to massive parallel processing, function without complete or well-structured information and cope with huge volume of information with many dependant variables. Finally, neural network can learn to classify new input instantly that has not been seen before while the genetic algorithm finds acceptable solution within the solution space. Thus, the job scheduling result would be optimized using neural network by finding new set of classifications based on the provided tasks. Therefore, solving and optimizing the scheduling problems in cloud computing environment can be achieved using artificial neural networks.

7. Conclusions and Future Works

Could computing is considered one of the most important research areas that helps to get the maximum benefit of distributed resources, and aggregates them to achieve higher throughput and be able to solve large scale computation problems. Job scheduling is considered one of the main issues and hottest research topic in cloud computing. The main task of job scheduling system is to find the best resources in a cloud for the cloud computing user's jobs. In this research, the researchers review the literature of cloud computing job scheduling. In the literature, many researches tried to solve the problem of job scheduling in cloud computing. Most of them use artificial intelligence techniques such as genetic algorithm and ant colony to solve the problem of job scheduling and to find the optimal distribution of resources. However, as there are still problems in this research area, the researchers propose a new technique that solves the issue of job scheduling in cloud computing environment. This technique is based on using neural network to classify the job queues that exist on any resource and to give priorities to different jobs. The artificial neural network is an artificial intelligence system that is able of finding and differentiating pattern. It can learn by example and can adjust to new concepts and knowledge. Using artificial neural networks will be highly potential to solve and optimize the scheduling problems in cloud computing environment.

References

[1] Xu, X. (2012) From Cloud Computing to Cloud Manufacturing. *Robotics and Computer-Integrated Manufacturing*, **28**, 75-86. http://dx.doi.org/10.1016/j.rcim.2011.07.002

[2] Svantesson, D. and Clarke, R. (2010) Privacy and Consumer Risks in Cloud Computing. *Computer Law & Security*

Review, **26**, 391-397. http://dx.doi.org/10.1016/j.clsr.2010.05.005

[3] Jadeja, Y. and Modi, K. (2012) Cloud Computing—Concepts, Architecture and Challenges. 2012 *International Conference on Computing, Electronics and Electrical Technologies* (*ICCEET*). http://dx.doi.org/10.1109/ICCEET.2012.6203873

[4] Malathi, M. (2011) Cloud Computing Concepts. *3rd International Conference on Electronics Computer Technology* (*ICECT*).

[5] Nexogy (2014) The Impact of Cloud Computing for VoIP. http://nexogy.wordpress.com/2011/09/27/the-impact-of-cloud-computing-for-voip/

[6] Arshad, J., Townend, P. and Xu, J. (2013) A Novel Intrusion Severity Analysis Approach for Clouds. *Future Generation Computer Systems*, **29**, 416-428. http://dx.doi.org/10.1016/j.future.2011.08.009

[7] Modi, C., *et al.* (2013) A Survey of Intrusion Detection Techniques in Cloud. *Journal of Network and Computer Applications*, **36**, 42-57. http://dx.doi.org/10.1016/j.jnca.2012.05.003

[8] Subashini, S. and Kavitha, V. (2011) A Survey on Security Issues in Service Delivery Models of Cloud Computing. *Journal of Network and Computer Applications*, **34**, 1-11. http://dx.doi.org/10.1016/j.jnca.2010.07.006

[9] Rabai, L.B.A., *et al.* (2013) A Cybersecurity Model in Cloud Computing Environments. *Journal of King Saud University—Computer and Information Sciences*, **25**, 63-75. http://dx.doi.org/10.1016/j.jksuci.2012.06.002

[10] Karajeh, H., Maqableh, M. and Masa'deh, R. (2014) Security of Cloud Computing Environment. *23rd IBIMA Conference on Vision* 2020: *Sustainable Growth, Economic Development, and Global Competitiveness.*

[11] Mathur, P. and Nishchal, N. (2010) Cloud Computing: New Challenge to the Entire Computer Industry. 2010 *1st International Conference on Parallel Distributed and Grid Computing* (*PDGC*), Solan, 28-30 October 2010, 223-228.

[12] Maqableh, M., Samsudin, A. and Alia, M. (2008) New Hash Function Based on Chaos Theory (CHA-1). 20-27.

[13] Maqableh, M.M. (2010) Secure Hash Functions Based on Chaotic Maps for E-Commerce Applications. *International Journal of Information Technology and Management Information System* (*IJITMIS*), **1**, 12-22.

[14] Maqableh, M.M. (2010) Fast Hash Function Based on BCCM Encryption Algorithm for E-Commerce (HFBCCM). *5th International Conference on E-Commerce in Developing Countries*: *With Focus on Export*, Le Havre, 15-16 September 2010, 55-64.

[15] Maqableh, M.M. (2011) Fast Parallel Keyed Hash Functions Based on Chaotic Maps (PKHC). *Western European Workshop on Research in Cryptology*, Lecture Notes in Computer Science, Weimar, 20-22 July 2011, 33-40.

[16] Maqableh, M.M. (2012) Analysis and Design Security Primitives Based on Chaotic Systems for E-Commerce. Durham University, Durham.

[17] Wikipedia (2014) Cloud Computing. Wikipedia Contributors. http://en.wikipedia.org/w/index.php?title=Cloud_computing&oldid=616939446

[18] Khorshed, M.T., Ali, A.B.M.S. and Wasimi, S.A. (2012) A Survey on Gaps, Threat Remediation Challenges and Some Thoughts for Proactive Attack Detection in Cloud Computing. *Future Generation Computer Systems*, **28**, 833-851. http://dx.doi.org/10.1016/j.future.2012.01.006

[19] Benny Karunakar, D. and Datta, G.L. (2007) Controlling Green Sand Mould Properties Using Artificial Neural Networks and Genetic Algorithms—A Comparison. *Applied Clay Science*, **37**, 58-66. http://dx.doi.org/10.1016/j.clay.2006.11.005

[20] Abdella, M. and Marwala, T. (2005) The Use of Genetic Algorithms and Neural Networks to Approximate Missing Data in Database. *Computing and Informatics*, **24**, 577-589.

[21] Fraile-Ardanuy, J. and Zufiria, P.J. (2007) Design and Comparison of Adaptive Power System Stabilizers Based on Neural Fuzzy Networks and Genetic Algorithms. *Neurocomputing*, **70**, 2902-2912. http://dx.doi.org/10.1016/j.neucom.2006.06.014

[22] Kumar, P. and Verma, A. (2012) Scheduling Using Improved Genetic Algorithm in Cloud Computing for Independent Tasks. *ICACCT*12, Chennai, 3 November 2012.

[23] Goñi, S.M., Oddone, S., Segura, J.A., Mascheroni, R.H. and Salvadori, V.O. (2008) Prediction of Foods Freezing and Thawing Times: Artificial Neural Networks and Genetic Algorithm Approach. *Journal of Food Engineering*, **84**, 164-178. http://dx.doi.org/10.1016/j.jfoodeng.2007.05.006

[24] Group, L. (2014) Optimisation of Collector Form and Response. http://www.engineering.lancs.ac.uk/lureg/group_research/wave_energy_research/Collector_Shape_Design.php.

[25] Heckerling, P.S., Gerber, B.S., Tape, T.G. and Wigton, R.S. (2004) Use of Genetic Algorithms for Neural Networks to Predict Community-Acquired Pneumonia. *Artificial Intelligence in Medicine*, **30**, 71-84.

[26] Varahrami, V. (2010) Application of Genetic Algorithm to Neural Network Forecasting of Short-Term Water Demand.

International Conference on Applied Economics—ICOAE, Athens, 26-28 August 2010, 783-787.

[27] Chen, C.R. and Ramaswamy, H.S. (2002) Modeling and Optimization of Variable Retort Temperature (VRT) Thermal Processing Using Coupled Neural Networks and Genetic Algorithms. *Journal of Food Engineering*, **53**, 209-220. http://dx.doi.org/10.1016/S0260-8774(01)00159-5

[28] Tadiou, K.M. (2014) The Future of Human Evolution. http://futurehumanevolution.com/artificial-intelligence-future-human-evolution/artificial-neural-networks

[29] Li, L.Q. (2009) An Optimistic Differentiated Service Job Scheduling System for Cloud Computing Service Users and Provider. *3rd International Conference on Multimedia and Ubiquitous Engineering*, Qingdao, 4-6 June 2009, 295-299.

[30] do Lago, D.G., Madeira, E.R.M. and Bittencourt, L.F. (2011) Power-Aware Virtual Machine Scheduling on Clouds Using Active Cooling Control and DVFS. *9th International Workshop on Middleware for Grids, Clouds and e-Science*, Lisboa, 12-16 December 2011.

[31] Dutta, D. and Joshi, R.C. (2011) A Genetic-Algorithm Approach to Cost-Based Multi-QoS Job Scheduling in Cloud Computing Environment. *International Conference and Workshop on Emerging Trends in Technology* (*ICWET* 2011)-TCET, Mumbai, 25-26 February 2011.

[32] Ghanbari, S. and Othman, M. (2012) A Priority Based Job Scheduling Algorithm in Cloud Computing. *Procedia Engineering*, **50**, 778-785.

[33] Xu, B.M., Zhao, C.Y., Hu, E.Z. and Hu, B. (2011) Job Scheduling Algorithm Based on Berger Model in Cloud Environment. *Advances in Engineering Software*, **42**, 419-425. http://dx.doi.org/10.1016/j.advengsoft.2011.03.007

[34] Li, K., *et al*. (2011) Cloud Task Scheduling Based on Load Balancing Ant Colony Optimization. *6th Annual China Grid Conference*, Dalian, 22-23 August 2011.

[35] Morariu, O., Morariu, C. and Borangiu, T. (2012) A Genetic Algorithm for Workload Scheduling in Cloud Based E-Learning. *Proceedings of the* 2*nd International Workshop on Cloud Computing Platforms* (*CloudCP* 12), Bern, 10 April 2012.

[36] Palmieri, F., Buonanno, L., Venticinque, S., Aversa, R. and Di Martino, B. (2013) A Distributed Scheduling Framework Based on Selfish Autonomous Agents for Federated Cloud Environments. *Future Generation Computer Systems*, **29**, 1461-1472. http://dx.doi.org/10.1016/j.future.2013.01.012

[37] Frîncu, M.E. (2014) Scheduling Highly Available Applications on Cloud Environments. *Future Generation Computer Systems*, **32**, 138-153. http://dx.doi.org/10.1016/j.future.2012.05.017

[38] Vijindra and Shenai, S. (2012) Survey on Scheduling Issues in Cloud Computing. *Procedia Engineering*, **38**, 2881-2888. http://dx.doi.org/10.1016/j.proeng.2012.06.337

[39] Zhang, Y.H., Feng, L. and Yang, Z. (2011) Optimization of Cloud Database Route Scheduling Based on Combination of Genetic Algorithm and Ant Colony Algorithm. *Procedia Engineering*, **15**, 3341-3345. http://dx.doi.org/10.1016/j.proeng.2011.08.626

[40] Mezmaz, M., Melab, N., Kessaci, Y., Lee, Y.C., Talbi, E.G., Zomaya, A.Y. and Tuyttens, D. (2011) A Parallel Bi-Objective Hybrid Metaheuristic for Energy-Aware Scheduling for Cloud Computing Systems. *Journal of Parallel and Distributed Computing*, **71**, 1497-1508. http://dx.doi.org/10.1016/j.jpdc.2011.04.007

[41] Tsai, J.T., Fang, J.C. and Chou, J.H. (2013) Optimized Task Scheduling and Resource Allocation on Cloud Computing Environment Using Improved Differential Evolution Algorithm. *Computers & Operations Research*, **40**, 3045-3055. http://dx.doi.org/10.1016/j.cor.2013.06.012

[42] Yang, H. and Tate, M. (2012) A Descriptive Literature Review and Classification of Cloud Computing Research. *Communications of the Association for Information Systems*, **31**, 35-60.

6

Implementation and Evaluation of Transport Layer Protocol Executing Error Correction (ECP)

Tomofumi Matsuzawa[1], Keisuke Shimazu[2]

[1]Department of Information Sciences, Tokyo University of Science, Tokyo, Japan
[2]Service Operation Division, Internet Initiative Japan Inc., Tokyo, Japan
Email: t-matsu@is.noda.tus.ac.jp

Abstract

Technologies for retransmission control and error correction are available for communications over the Internet to improve reliability of data. For communications that require the data reliability be ensured, TCP, which performs retransmission control, is often employed. However, for environments and services where response confirmation and retransmission are difficult, error correction technologies are employed. Error correction is generally implemented on UDP, but the existing framework implemented on UDP frequently does not consider the maximum frame size of the data link layer and relegates data division to the IP module. The IP module divides data according to the maximum size for the data link, and the receiving IP module reconstructs the divided data. For a data link layer typified by the current Ethernet with an error detection function, the frame is often destroyed upon error detection. At the IP module, the specification allows destruction of the entire dataset whenever divided data necessary for reconstruction is incomplete. Consequently, an error in a single bit results in a total loss of data handed to the IP module, and thus error correction performance declines with the increase in data size handed to the IP module. The present study considers the MTU of the data link layer and proposes error correction protocol (ECP) over IP, which decreases the transfer data volume flowing to the data link layer by dividing data into blocks of appropriate size based on designated error correction code and its parameters (thus improving error correction performance) and assesses the performance of ECP. Experimental results demonstrate that performance is comparable or better than existing error correction frameworks. Results also show that when a specification not ensuring the reliability of the data link layer was employed, error correction was superior to existing frameworks on UDP.

Keywords

Component, FEC, Transport Protocol, MTU

1. Introduction

The spread of the Internet in recent years has come with increased communications in which real-time properties are emphasized, such as video teleconferences and voice communication. The transport layer for Web viewing and email services frequently employs the Transmission Control Protocol (TCP), which is embedded with functionality that the sending and receiving sides mutually handshake, conduct retransmission control for data errors or losses that occur in the communication pathway, and thus secure data reliability. TCP, however, is unsuited to communications that emphasize real-time properties. Instead, the User Datagram Protocol (UDP) is generally used. UDP is a connection-less protocol designed to be lightweight and high-speed. Unlike TCP, no handshaking occurs at the start and the end for communications and no mechanism is implemented to improve reliability, such as retransmission or congestion control. Techniques to improve reliability on UDP include retransmission control such as Automatic Repeat reQuest (ARQ) at the application and recovery of errors and losses with Forward Error Correction (FEC). Applications that emphasize real-time use, however, frequently employ FEC to avoid decline in real-time performance. UDP can handle data in lengths of at most 65,535 bytes, but when the data are handed to the IP module, the latter subdivides the received data according to the maximum transmission unit (MTU) size of the data link layer. The receiving IP module reconstructs the subdivided data to the size designated by the sending side and hands the data off to the UDP module. If even one frame of the subdivided data is destroyed in this process, then the IP module cannot reconstruct the data, and so the remaining subdivided data are all destroyed. Where the data size handled differs greatly between the MTU of the data link layer and UDP, an error of a single bit means loss of 1 UDP datagram. Thus, for technology such as FEC Framework [1] that performs error correction on UDP, in some cases the error correction performance declines. In recent years, technology that increases the MTU size of the data link layer has appeared, such as Ethernet Jumbo Frame [2], with the aim of attaining high speeds at the data link layer. Increasing the MTU size of the data link layer, however, tends to lower the error correction performance.

In response, this paper proposes a transport layer protocol with error correction capacity that considers the data size at the data link layer. The proposed protocol is implemented and its performance assessed.

2. Related Technologies

2.1. Ethernet Jumbo Frame

The standard specification for Ethernet defines the data size that can be transmitted in 1 frame as a minimum of 46 bytes and a maximum of 1500 bytes; however, Ethernet Jumbo Frame [2] expands the maximum size to between 9000 and 16,000 bytes while retaining the minimum size at 46 bytes. Since Ethernet Jumbo Frame has no standard specification, the maximum size differs by vendor or product. The specification simply expands the data size, which makes system-based support easy, improves network performance by several tens of percent, and achieves load reductions for network equipment.

2.2. Forward Error Correction (FEC)

FEC refers to technology that corrects errors and losses arising during communications. The sending side adds repair data (repair symbols) computed by specific calculations based on original data desired for sending, and the receiving side can restore the original data by specific calculations made on original data containing errors or repair data. The sum of the lengths for the original data and repair data is called the code length, and the ratio of original data length to code length is called the code rate. Generally, transmission efficiency is better for higher code rates but correction performance declines.

Algorithms existing for FEC include Reed-Solomon coding [3], LDPC coding [4], and BCH coding [5] [6].

For Reed-Solomon coding, multiple bits are considered one block (symbol), and code words are described by collections of symbols. Error correction is performed for each symbol. On the computer, 1 symbol is frequently implemented as 8 bits (1 byte), and N symbols constitute 1 code word. Within the word, K symbols represent the original data, and $N-K$ symbols represent the repair data. Correction of up to $(N-K)/2$ symbols is allowed.

2.3. FEC Framework

FEC Framework [1] is defined in RFC6363. It is defined between the transport layer protocol and the application layer protocol with the aim of making the application and FEC independent. The encoding algorithm actually used and the generation procedure for the decoder that performs decoding are defined under the assumption that sender and receiver agree in advance. These agreements must be made separately using SDP as defined in RFC6364 [7].

3. FEC Framework Issues

Error correction on UDP using FEC Framework suffers a decline in error correction performance when the data size handled by the UDP module and MTU of the data link layer differ greatly.

The maximum size in the specifications for UDP is 65,535, which includes that for the UDP header. Thus, when IPv6 [8] is used, the effective maximum size is 65,527 bytes after subtracting 8 bytes for the UDP header (65,507 bytes for IPv4, since the IPv4 header size is included in the data size that the IP module can handle). For example, in order to stream multiple UDP datagrams in 8192-byte units over Ethernet, since subdivision is performed in terms of MTU of the data link layer by the IP module, frames streamed over Ethernet are sent in blocks of 1500 bytes. If an error in 1 bit occurs, then Ethernet has an error detection mechanism employing CRC coding for the header and data [9], and any frame with a detected error is destroyed. The receiving IP module attempts to reconfigure the UDP datagram of the size designated by the sender UDP module by using a frame other than the destroyed frame; however, unless all of the subdivided frames are complete, the datagram is destroyed. In this case, data lost in 8192-byte units must be restored. For sending the UDP datagram at a size less than the MTU of the data link layer, for example at 1024 bytes, the loss from the error of 1 bit is 1024 bytes, and error correction performance is improved. For too small a size, however, the ratio of repair data to original data increases, and thus the sending data volume also increases. FEC Framework has no data subdividing function based on the MTU of the data link layer. Instead, the application must manage this function. Neither the FEC Framework nor the Reed-Solomon FEC scheme [10] explicitly states specifications that determine data subdivision size based on the MTU. The present paper, therefore, proposes Error Correction Protocol (ECP) as a transport layer protocol, which includes functions of maximum size management, attachment of repair data necessary for error correction, a repair data algorithm, and parameter notification, and assesses the performance of ECP.

4. Error Correction Protocol (ECP) Specifications

4.1. Overview

Unlike UDP, ECP provides a mechanism that conducts error correction, not error detection. Unlike TCP, moreover, ECP is envisioned for an environment in which acknowledgements are not obtained, and features no function to control retransmission to improve reliability. The ECP datagram is divided between header and data, and, unlike FEC Framework, has no need for agreement in advance between sending and receiving on subdivided size, error correction encoding to be used, its parameters, etc. Such designations are made by entry to ECP header. Because it is defined as a transport layer protocol, ECP has a port number concept to enable distinguishing between upper layer services.

4.2. ECP Datagram Configuration

ECP header is configured as 28 bytes (octets) and is located immediately after IP header. The ECP datagram configuration is shown in **Figure 1**.

ECP header is stored immediately after IP header, and data or repair data (repair symbols) are stored immediately after ECP header. Original data input from the application (Application Data Unit: ADU) is divided into lengths no greater than the MTU of the data link layer after subtracting IP and ECP header lengths, and ECP header is attached to each data block. For example, if the MTU of Ethernet is 1500 bytes and ADU is 5000 bytes, then subdivision is into four blocks of 1250 bytes each, where 1250 is referred to as the block length. For an ordinary unicast communication, a path MTU discovery [11] is conduced in advance, and its value is calculated as the data link layer MTU. For cases in which the path MTU discovery cannot be conducted in advance, such as an IP multicast, the minimum MTU specified for IP (1280 for IPv6) is used. As shown in **Figure 1**, repair data are computed across multiple blocks. When subdivided block lengths differ, padding (filling with zeroes) is carried out to reach the maximum block length before generating the repair data. When Reed-Solomon code RS(255, 239) is used for error correction encoding, all blocks are first padded to make their lengths as long as

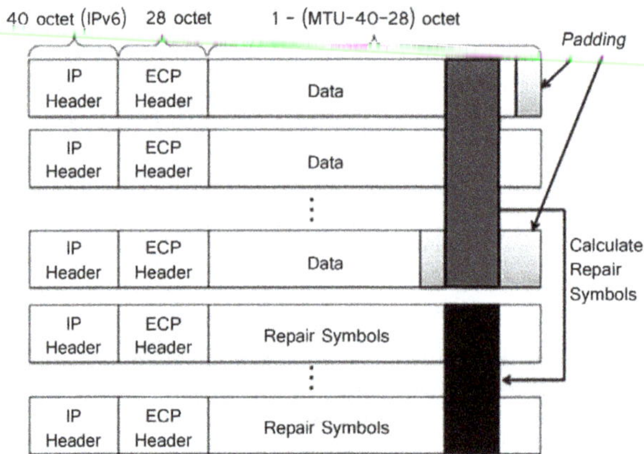

Figure 1. ECP datagram.

the largest of the 239 blocks, a 239 * (maximum block length) matrix is generated, and the respective values of the nth byte are used to calculate the values of the nth byte for 16 blocks of repair data. ADUs for 239 blocks do not need to be ready before data transmission can start. When ADUs are subdivided into blocks, one ECP header is attached to each block and handed to the IP module for sending. Repair data are generated when the 239th block is handed to the IP module, and then 16 blocks of repair data are additionally handed to the IP module. Thus, ECP must store 239 blocks worth of data in a buffer. Sent data do not need to include the padded portions. The receiving ECP module arranges block lengths of 239 + 16 blocks by padding to conduct error correction. If the number of blocks after the subdividing of input data exceeds 239, then repair data are generated and sent after sending the 239th block, the next block is handled as the first block of the next matrix, and this process is repeated. Thus, either storage of multiple ADUs in one matrix or storage of one ADU across two or more matrices by block subdivision may occur, but each block's ECP header, as described below, contains ADU number (Identification) and position within ADU (Data Offset) that enable reconfiguration of ADU at the receiving ECP module. Multiple ADUs are not stored within a single block, however.

4.3. Header Configuration

The ECP header format is shown in **Figure 2**.

Source Port stores the source port number and Destination Port stores the destination port number, both as 16 bits. ECP uses port number, such as TCP and UDP, to distinguish services.

Data Length stores the length of ADU handed from the application as 32 bits. Sequence Number is used to correctly rearrange received blocks when ECP conducts error correction. The first bit of Sequence Number is used as a flag (Repair Flag) to distinguish the data added after the header as either original data (0) or repair data (1) generated as a Repair Symbol. FEC Number stores the value representing the error correction code to be used and its parameters as 16 bits. For Reed-Solomon codes, for example, the RS(255, 239) code and RS(255, 249) code are defined by different numbers. If this field is 0, then error correction encoding is designated as unused. Block Length stores the lengths of each block. Identification stores the number attached per ADU. Data Offset stores the location (in byte units) from the beginning of ADU. Since ECP subdivides input ADUs into blocks and stores this in a matrix, the receiving ECP module must perform reconfiguration of the data, at which point the values stored in Identification and Data Offset are used. In fields other than header Sequence Number, FEC Number, and Header Repair Symbols of ECP header (fields shaded in grey in **Figure 2**), repair data calculated on the basis of the values of the same fields stored in the blocks of subdivided ADU, when used as a repair data header. As shown in **Figure 3**, the generation method of repair data is carried out in the same manner as the computation method for data by using the encoding method designated by FEC Number. In **Figure 3**, Sequence Number, Block Length, Identification, and Data Offset are provided as four examples, but the same generation method for repair data applies to all fields shaded in grey in **Figure 2**.

The example in **Figure 3** shows ADUs handed off from the application to the ECP module in sequences of 2000 bytes, 2800 bytes, and 1200 bytes, when the Ethernet MTU is 1500 bytes. If the three blocks in **Figure 3** whose Sequence Numbers are 2, 3, and 4 were lost, then even though the other blocks can be used to correct

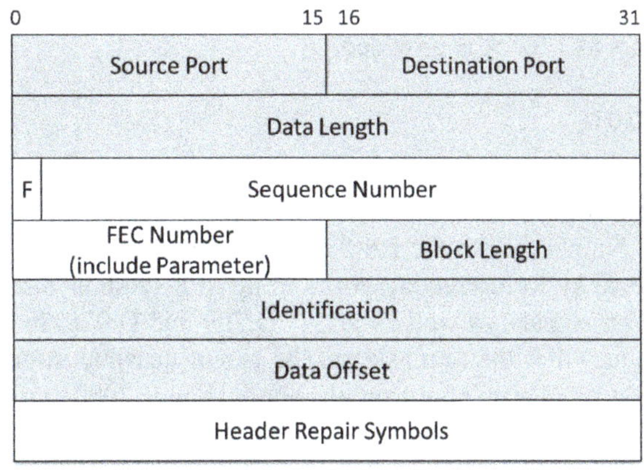

Figure 2. ECP header format.

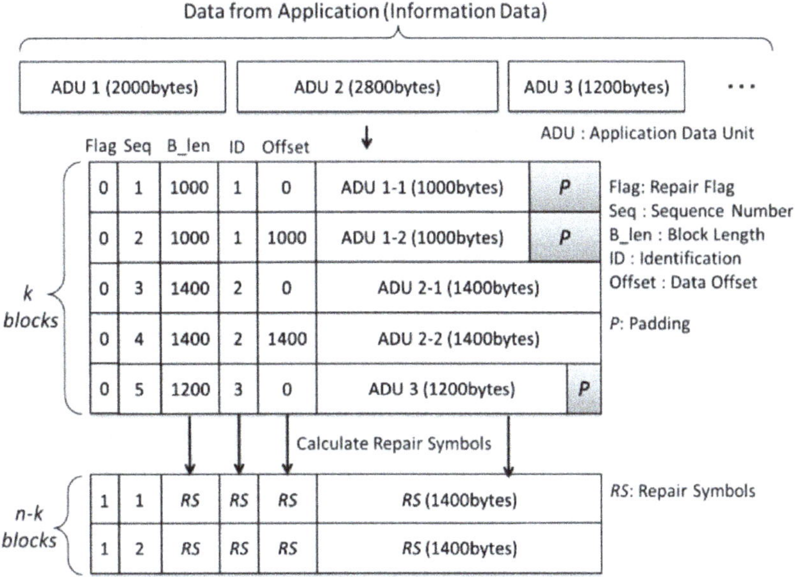

Figure 3. Flow of data division and repair data generation.

these three blocks and recover the data, unless the header information of Data Length, Block Length, Identification, and Data Offset is available, the ADU cannot be reconfigured from the restored blocks. Sequence Number and Repair Flag fields within the repair data header are used to position the blocks storing repair data in the matrix; therefore, flags are stored that indicate Sequence Number for a repair data block and status as a repair data block instead of Header Repair symbols. These two fields do not need to be restored from other blocks when ADU is reconfigured, since Sequence Number of any lost block can be restored from its position in the matrix. The FEC Number field is the value used to decide the matrix size, but this is unnecessary as header information of a restored block. Thus, correction symbols do not need to be stored in the header. In Header Repair Symbols shown in **Figure 2**, only 4 bytes of header repair data generated as RS(28, 24) code from a header of 24 bytes, excluding Header Repair Symbols, are stored. This field is used when header errors in block units are corrected. The type and parameters of error correction encoding used for header error correction and existence or absence of header error correction can be arbitrarily and dynamically changed by FEC Number in the same way as data error correction encoding and its parameters. The current FEC Framework has a strong nuance of "forward erasure correction", rather than "forward error correction". Single bit-unit errors remaining are not particularly envisioned. Instead, since data are lost in units of blocks for any error, FEC Framework has a function for restoring lost frames. Envisioned for ECP, however, are cases in which protocols are used that do not ensure reliability of the data link layer. Since an error of 1 bit at ECP header can cause errors during data reconfiguring, a function that can correct header errors as 1 block is provided.

5. ECP Implementation

In the performance assessment testing, the implementation of this paper employed Reed-Solomon codes RS(255, 239) for error correction encoding of data and RS(28, 24) for header error correction encoding, along with an assignment of "1" as FEC Number. With the RS(255, 239) code, 16 symbols of repair data are added for 239 symbols of data. With a capacity for correcting errors in up to 8 symbols out of 255 symbols, this encoding is also used in digital television signals, as well as ITU-T G.709 and G.975. In the implementation of this paper, ADU subdivision was conducted at the same size to the extent possible in order to avoid differences in block length. For example, the first ADU size input from the application is 2000 bytes and the second data size is 2800 bytes in **Figure 3**, where MTU is 1500 bytes. In this case, the first ADU would be halved to 1000 bytes each {1000, 1000} and the second ADU would be halved to 1400 bytes each {1400, 1400}, and then the blocks would be stored in a matrix. The number of blocks would be unchanged even when subdivided according to the maximum MTU size 1500 − 40 (IPv6 header) − 28 (ECP header) = 1432 bytes, and the sent data count of original data would also be unchanged. However, the largest block length among the blocks of stored original data would become the block length of repair data at the generation of repair data. Subdividing by the MTU size would always make the block length of repair data 1432, which would increase the sent data size of the repair data.

6. Performance Assessment and Discussion

6.1. Testing Environment

In order to assess ECP performance, the present study compared the error rates and data flow in the data link layer for the cases of using UDP, FEC Framework, and ECP. Because FEC Framework assumes that the types and parameters used for error correction encoding all agree in advance, the RS(255, 239) code was used for error correction encoding and the protocol used in Simple Reed-Solomon FEC Scheme for FECFRAME of RFC 6865 [12] was adopted. The data link layer used for testing was executed in a simulator implemented with Ethernet specifications that allowed MTU and bit error rate (BER) to be varied as desired. IPv6 was used for the network layer protocol. As shown in **Figure 4**, identical data were readied for sending and receiving in the test, data were sent from the sender, errors were entered at the data link layer based on BER, error correction was conducted at the receiver, and the ratio by which this differed from the data already on hand in bytes was calculated and compared as the error rate. The data sent from the sender with Ethernet header attached were measured and compared under the different methods in terms of data size.

6.2. Error Rates

The testing compared the various methods with BER set at 10^(−6). BER per hop under wired Ethernet is generally not greater than 10^(−9), versus 10^(−6) to 10^(−8) for wireless LAN. For environments that conduct path MTU discovery in advance, however, multiple hops across different data links frequently occur. Thus, the BER was set at 10^(−6) for testing. For ECP and FEC Framework, data encoding was conducted with RS(255, 239) and ECP header error correction with RS(28, 24). The Ethernet MTU used was 1500 bytes, and 6000 bytes was used for Ethernet Jumbo Frame. Results in terms of the sizes of ADU (1024, 2048, 4096, 8192, 16,384, and 32,768 bytes) handed from the application under each method sent 10,000 times each are shown in **Figure 5** and **Figure 6** for Ethernet MTUs of 1500 and 6000 bytes, respectively.

For both **Figure 5** and **Figure 6**, the errors are close to 0% for ADU sizes up to 2048 bytes for FEC Framework and ECP. Beyond 8192 bytes, however, correction under FEC Framework mostly failed to function and left errors comparable to UDP. Since an error of 1 bit in Ethernet destroys the frame, data loss is proportional to the size of MTU. Since UDP and FEC Framework delegate data division to IP, data reconstruction takes place under IP. Specifications call for destruction, however, unless all subdivided data are ready at reconstruction. Thus, an error of 1 bit becomes a loss of 1 ADU. Since ECP and FEC Framework employed RS(255, 239), a loss of up to 8 blocks out of 255 blocks could be restored. When ADU size increased, however, the probability of even 1 bit of error in 1 block declined for ECP when block length was no greater than MTU, compared to

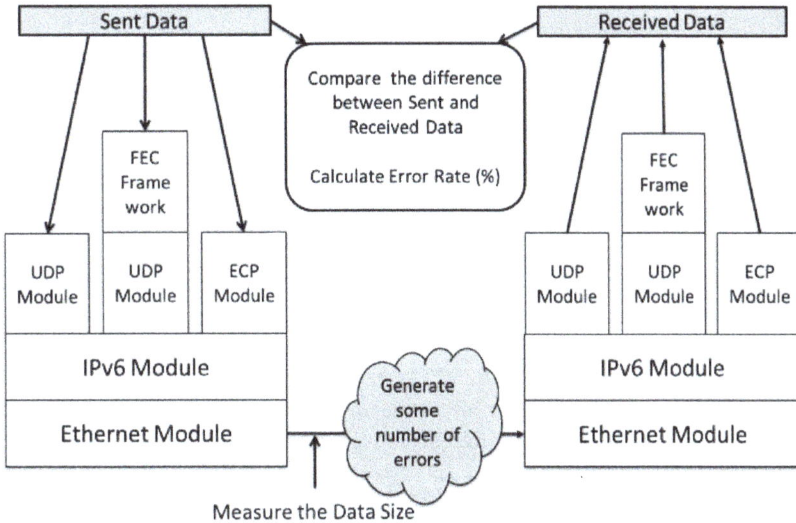

Figure 4. Experimental environment.

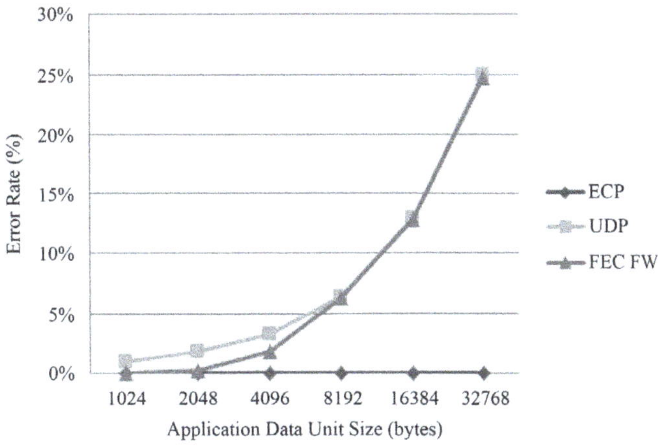

Figure 5. Error rate (Ethernet MTU, 1500 bytes; BER, 1.0 × 10^(−6)).

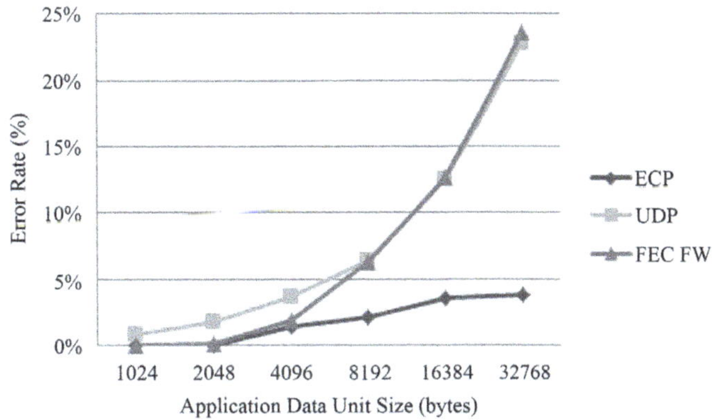

Figure 6. Error rate (Ethernet MTU, 6000 bytes; BER 1.0 × 10^(−6)).

FEC Framework where block length was the ADU size. Consequently, differences emerged in error correction performance. Correction performance between ECP and FEC Framework did not present any difference for ADU sizes no greater than MTU.

6.3. Sent Data Size

The data increase rates, determined by dividing the total sent data size under the respective methods, as measured by data link during testing of error rate measurement, by the total ADU size are shown in **Figure 7** and **Figure 8** for Ethernet MTUs of 1500 and 6000 bytes, respectively.

These test results do not include the sent data by FEC Framework for advance agreement and sent data generated from path MTU discovery conducted in advance by ECP. Since 16 symbols of repair data are generated for 239 symbols under RS(255, 239), the theoretical increase is approximately 7%. However, data measured on Ethernet have Ethernet header and Footer and IP header attached, ECP has one ECP header attached per Ethernet frame unit, FEC Framework has one FEC Framework header and UDP header attached per ADU, and UDP has one UDP header attached per ADU unit. Comparing the frame counts flowing on Ethernet, UDP < ECP—FEC Framework, but ECP header has 28 bytes, which increases the data size by 0.3% to just under 2% more than that with FEC Framework. Increasing the MTU size or the ADU size reduces the difference between ECP and FEC. Since the ADU sizes were fixed in this test, no padding process was carried out at matrix generation under FEC Framework. When ADU sizes differed between input, however, repair data were generated at a size to match the block with the largest block length, which caused data size to increase. For example, a test was conducted in which ADU size was varied from 8192 bytes for an Ethernet MTU of 1500 bytes. In this test, ADU was input 10,000 times, but a variable range x was set as the parameter such that input data between the sizes of $8192 - x$ and $8192 + x$ bytes were handed to ECP and FEC Framework. For example, for $x = 2000$, data took on an input data size randomly between 6192 and 10,191 bytes. The test results are shown in **Figure 9**.

Since sizes were subdivided at no greater than MTU for ECP, the block length for repair data was kept lower than that of FEC Framework. However, since repair data of the same block length as the maximum size among 239 ADUs were generated for FEC Framework, when the variations became large, the data volume flowing through Ethernet increased in comparison to ECP. Error rates in this test did not depend on the variable range and were nearly the same as the case for an ADU size of 8192 bytes in **Figure 4**: 0% for ECP and approximately 6% to 7% for FEC Framework. A similar trend was observed for results with an Ethernet MTU of 6000 bytes.

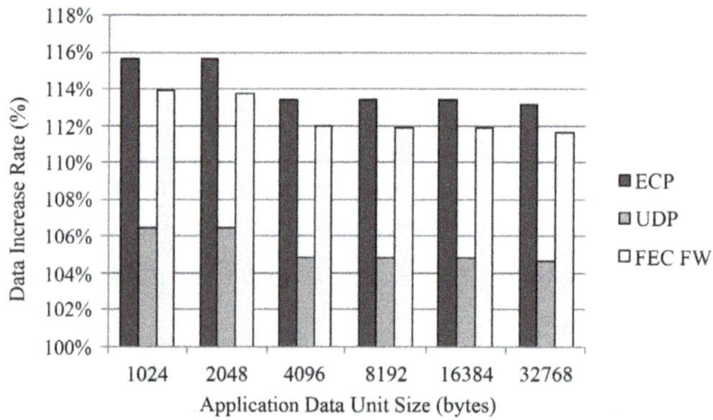

Figure 7. Data increase rate (Ethernet MTU, 1500 bytes).

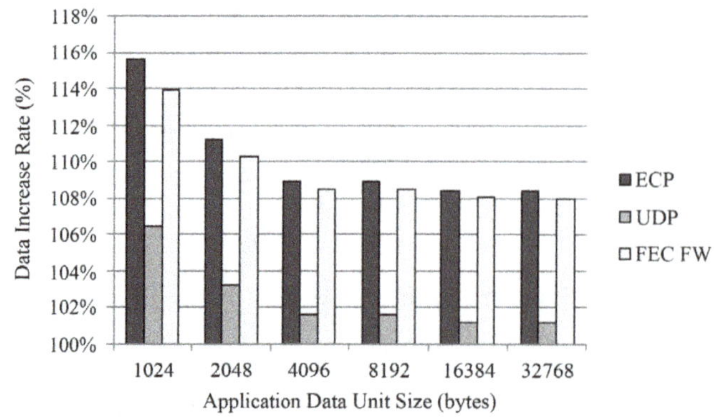

Figure 8. Data increase rate (Ethernet MTU, 6000 bytes).

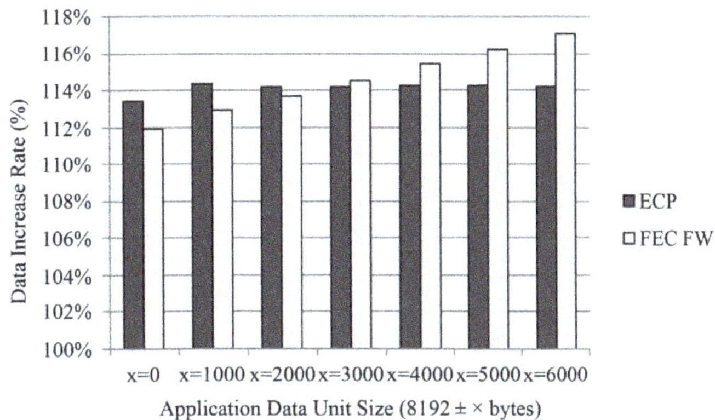

Figure 9. Data increase rate (Ethernet MTU, 1500 bytes; input data size fluctuations).

6.4. Assessment for Data Link Layer with No Error Detection Function

The following three types are defined as models for the data link layer [13].

- Unacknowledged Connectionless Service.
- Acknowledged Connectionless Service.
- Connection-Oriented Service.

Among these, the first type, Unacknowledged Connectionless Service, applies to a high-speed data link layer with relatively few errors, such as a LAN. Special processing is not performed in this type, even if errors occur. Under current Ethernet specifications, however, CRC error detection per frame is not conducted, and entire frames are destroyed when errors are detected. In the future, however, the appearance of Ethernet without an error detection function (error detection performed only on Ethernet frame header) can be envisioned with the emergence of transfer media with fewer errors and the demand for higher speeds to the current data link layers. In actuality, with the addition of the Jumbo Frame function even to today's Ethernet, efforts are underway to expand the MTU upper limit from 1500 bytes to a range of 6000 to 16,000 bytes. The current Ethernet Switch allows the upper limit to be increased beyond 16,000 bytes, but CRC error detection performance drops when around 9000 bytes is exceeded. Thus, the maximum upper limit is currently 16,000 bytes. Taking these conditions into account, it is not unreasonable to assume that application of error detection to just the Ethernet frame header could increase MTU size and advance attempts to improve transfer speeds tremendously. Current TCP and UDP error correction mechanisms, moreover, are simple addition checksums without very high precision, relying as they do on the performance of Ethernet CRC. Thus, disposing of or limiting the error detection function at the data link layer is difficult, but if the current dependence of TCP and UDP on lower layer error detection is inhibiting higher speeds of the data link, then a protocol different from TCP and UDP will be necessary. A test was performed in which Ethernet CRC checking was applied to just the Ethernet header for Ethernet MTUs of 9000 and 16,000 bytes. ADU sizes were varied between 4096, 8192, 16,384, and 32,768 bytes. The results showed no dependence on Ethernet MTU size; therefore, the test results for an Ethernet MTU of 16,000 bytes are shown in **Figure 10**.

Errors can be made to be nearly zero under ECP. For FEC Framework, however, error rates increase with ADU size. This is because, since FEC Framework is implemented on UDP, even when frames contaminated with errors in Ethernet are handed to the UDP, the errors are detected by UDP checksum and then the datagram is destroyed by UDP. Checksum checking can be designated arbitrarily for IPv4, but UDP-based checksum is mandatory for IPv6 in lieu of abandoning the error detection function in IP [8]. Thus, destruction by UDP cannot be avoided. Error correction can be conducted for ECP at a location of an error-contaminated frame in which no errors exist. Thus, error rates can be reduced to lower than when CRC checking is conducted for all data.

7. Future Work

Repair data size and error correction performance of ECP depend greatly on the types and parameters of error correction encoding employed. As such, by providing the function of deciding error correction encoding and pa-

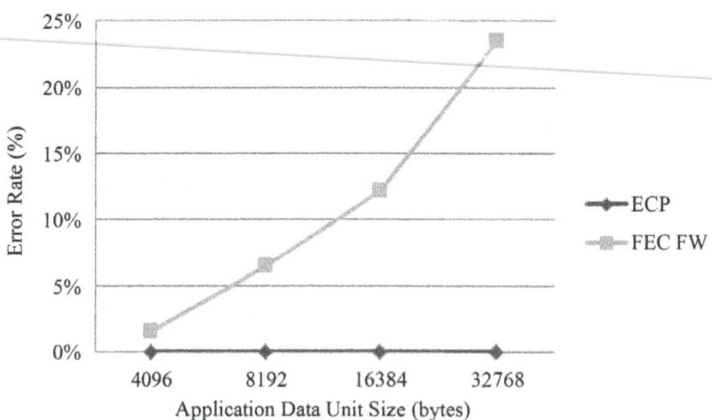

Figure 10. Error rate (Ethernet MTU, 16,000 bytes; BER, 1.0 × 10^(−6); no CRC error detection).

rameters used by ECP on the basis of the input data size and, where possible, transfer error rate, the application developer can engage in development work without considering error correction performance. The specifications presented in this paper did not include the costs of calculation and latency for encoding, decoding, or data division, which must be studied in the final determination of specifications. When cases of error contamination rather than data loss are envisioned for the data link layer, since ECP corrects across multiple blocks, ensuring data reliability requires waiting for the arrival of multiple blocks, which has an adverse impact on real-time communications. Considered as a countermeasure to this was the introduction of a mechanism to enable the detection of the necessity of error correction, by introducing a checksum such as CRC as an error detection mechanism for data contamination of each block; however, the introduction of a checksum causes further enlargement of the header and processing increases for the receiver. In the event of checksum introduction, the 2-byte checksum used for UDP offers low error detection performance. Instead, therefore, something along the lines of CRC32 should be introduced with the institution of a 4-byte field to the header. The necessity of checksum introduction and its algorithm need to be studied further, however. For the testing described here, errors were generated on the basis of BER, but losses in unpredictable bursts have been identified on actual networks. In the future, an ECP implementation involving the introduction of automatic decision-making for error correction encoding and parameters needs to be made on the IPv6 stack for assessment of calculation volume, latency, and performance against error bursts. A separate paper shall describe such assessments.

8. Conclusion

This paper proposed Transport Layer Protocol ECP, which introduces error correction mechanisms, and described performance assessments of the same. ECP holds block sizes for conducting error correction to be no greater than the MTU determined by path MTU discovery, which allows the data volume lost at errors of 1 bit at the data link layer to be suppressed, and attaches appropriately sized repair data. For a data link layer whose error correction function was restricted, ECP revealed error correction performance superior to that of even any error correction technology implemented on UDP.

References

[1] Watson, M., Began, A. and Roca, V. (2011) Forward Error Correction (FEC) Framework. *Request for Comments*, **6363** (Internet Engineering Task Force).

[2] Alteon Networks Extended Frame Sizes for Next Generation Ethernets, a White Paper. http://staff.psc.edu/mathis/MTU/AlteonExtendedFrames_W0601.pdf

[3] Reed, I.S. and Solomon, G. (1960) Polynomial Codes over Certain Finite Fields. *J.SIAM*, **8**, 300-304. http://dx.doi.org/10.1137/0108018

[4] Gallager, R.G. (1962) Low-Density Parity-Check Codes. MIT Press, Cambridge. (Preliminary Version in *IRE Trans on Inf. Theory*, **8**, 21-28.)

[5] Bose, R.C. and Ray-Chaudhuri, D.K. (1960) On a Class of Error-Correcting Binary Group Codes. *Inform. Control*, **3**, 68-79. http://dx.doi.org/10.1016/S0019-9958(60)90287-4

[6] Hocquenghem, A. (1959) Codes correcteursd'erreurs. *Chiffres*, **2**, 147-156.

[7] Begen, A. (2011) Session Description Protocol Elements for the Forward Error Correction (FEC) Framework. *Request for Comments*, **6364** (Internet Engineering Task Force).

[8] Deering, S. and Hinden, R. (1998) Internet Protocol, Version 6 (IPv6) Specification. *Request for Comments*, **2460** (Internet Engineering Task Force).

[9] IEEE 802.3-2002 (1985) IEEE Standard for Information Technology, Telecommunications and Information Exchange between Systems, Local and Metropolitan Area Networks, Specific Requirements Part: 3 Carrier Sense Multiple Access with Collision Detection (CSMA/CD) Access Method and Physical Layer Specifications Low-Density Parity-Check Codes.

[10] Lacan, J., Roca, V., Peltotalo, J. and Peltotalo, S. (2009) Reed-Solomon Forward Error Correction (FEC) Scheme. *Request for Comments*, **5510** (Internet Engineering Task Force).

[11] McCann, J., Deering, S. and Mogul, J. (1996) Path MTU Discovery for IP Version 6. *Request for Comments*, **1981** (Internet Engineering Task Force).

[12] Roca, V., Cunche, M., Lacan, J., Bouabdallah, A. and Matsuzono, K. (2013) Simple Reed-Solomon Forward Error Correction (FEC) Scheme for FECFRAME. *Request for Comments*, **6865** (Internet Engineering Task Force).

[13] IEEE 802.2. (1998) IEEE Standard for Information Technology, Telecommunications and Information Exchange between Systems, Local and Metropolitan Area Networks, Specific Requirements Part 2: Logical Link Control.

7

Stochastic Modeling and Power Control of Time-Varying Wireless Communication Networks

Mohammed M. Olama[1], Seddik M. Djouadi[2], Charalambos D. Charalambous[3]

[1]Computational Sciences and Engineering Division, Oak Ridge National Laboratory, Oak Ridge, USA
[2]Department of Electrical Engineering and Computer Science, University of Tennessee, Knoxville, USA
[3]Department of Electrical and Computer Engineering, University of Cyprus, Nicosia, Cyprus
Email: olamahussemm@ornl.gov, djouadi@eecs.utk.edu, chadcha@ucy.ac.cy

Abstract

Wireless networks are characterized by nodes mobility, which makes the propagation environment time-varying and subject to fading. As a consequence, the statistical characteristics of the received signal vary continuously, giving rise to a Doppler power spectral density (DPSD) that varies from one observation instant to the next. This paper is concerned with dynamical modeling of time-varying wireless fading channels, their estimation and parameter identification, and optimal power control from received signal measurement data. The wireless channel is characterized using a stochastic state-space form and derived by approximating the time-varying DPSD of the channel. The expected maximization and Kalman filter are employed to recursively identify and estimate the channel parameters and states, respectively, from online received signal strength measured data. Moreover, we investigate a centralized optimal power control algorithm based on predictable strategies and employing the estimated channel parameters and states. The proposed models together with the estimation and power control algorithms are tested using experimental measurement data and the results are presented.

Keywords

Wireless Networks, Time-Varying Wireless Fading Channel, Impulse Response, Doppler Power Spectral Density, Stochastic State-Space Model, Stochastic Modeling, Optimal Power Control, Expectation Maximization, Kalman Filter

1. Introduction

Time varying (TV) wireless channel models capture both the space and time variations of wireless systems, which are due to the relative mobility of the receiver, transmitter and/or scatterers [1]-[4]. The majority of research papers in this field such as in [5]-[7] use time-invariant (static) models for wireless channels. In these models, the speeds of mobile nodes are assumed to be constant and the statistical characteristics of the received signal are assumed to be fixed in time. But in reality, the propagation environment varies continuously due to mobility of the nodes causing network topology to dynamically change, the angle of arrival of the wave upon the receiver can vary continuously, and objects or scatters move in between the transmitter and the receiver resulting in appearance or disappearance of existing paths from one instant to the next. As a result, the current models that assume fixed statistics can no longer capture and track complex time variations in the propagation environment. These time variations compel us to introduce more advanced dynamical models based on stochastic state-space representation, in order to capture higher order dynamics of the wireless channel. In time-invariant models, channel parameters are random but do not depend on time, and remain constant throughout the observation and estimation phase. This contrasts with TV models, where the channel dynamics become stochastic processes [1]-[4]. This paper focuses on the development of stochastic short-term fading channel models based on system identification algorithms using received signal measurement data to extract various channel parameters and apply an optimal power control scheme.

In [4], the TV channel parameters are estimated from approximating the Doppler power spectral density (DPSD) of the wireless fading channel. However, in reality one does not have access to the TV DPSD at all times during the estimation process. Since these models are based on state-space forms, we propose to estimate the channel parameters as well as the inphase and quadrature components directly from received signal measurements, which are usually available or easy to obtain in any wireless network. A filter-based expectation maximization (EM) algorithm [8] and Kalman filter [9] are employed in estimating the channel parameters as well as the inphase and quadrature components, respectively. These recursive filters use only the first and second order statistics and therefore can be implemented online. These algorithms have been recently utilized in [1] and [2] to estimate the wireless channel parameters and states, and therefore the formulations of these algorithms are not presented in this paper. The proposed models and estimation algorithms are tested using received signal strength measurement data and the results are presented.

The developed stochastic channel models from received signal strength measurements are useful in most wireless applications. In this paper, these models are used to develop an optimal power control algorithm (PCA). Power control (PC) is important to improve performance of wireless communication systems. The benefits of power minimization are not just increased battery life, but also increased overall network capacity. The power allocation problem has been studied extensively as an eigenvalue problem for non-negative matrices [10] [11], resulting in iterative PCAs that converge each user's power to the minimum power [12] [13], and as optimization-based approaches [14]. Much of this previous work deals with static time-invariant channel models.

The proposed PCA is based on predictable power control strategies (PPCS) that were first introduced in [4]. PPCS basically means updating the transmitted powers at discrete times and maintaining them fixed until the next power update begins. The PPCS mechanism is proven to be effectively applicable to such dynamical models for an optimal PC. The outage probability (OP) is used as a performance measure. Since few TV dynamical channel models have so far been investigated with the application of any PCA, the suggested dynamical models and PCA thus provide a far more realistic and efficient optimal control of wireless networks.

The remainder of this paper is organized as follows. In Section 2, stochastic modeling and online estimation of TV wireless fading communication channels are presented. Section 3 introduces an optimal PCA based on predictable strategies and employing the estimated channel parameters and states. Section 4 presents numerical results that validate the proposed models together with the estimation and power control algorithms. Finally, Section 5 provides concluding remarks.

2. Stochastic Modeling and Estimation of Time-Varying Wireless Channels

The general TV model of a wireless channel is typically represented by the following multipath band-pass impulse response [6]

$$H(t;\tau)=\sum_{j=1}^{J(t)}\left(I_j(t,\tau)\cos(\omega_c t)-Q_j(t,\tau)\sin(\omega_c t)\right)\delta\left(\tau-\tau_j(t)\right) \tag{1}$$

where $H(t;\tau)$ is the band-pass response of the channel at time t, due to an impulse applied at time $t-\tau$,

$J(t)$ is the random number of multipath components, ω_c is the carrier frequency, $\delta(\cdot)$ is the Dirac delta function, and the set $\{I_j(t,\tau), Q_j(t,\tau), \tau_j(t)\}_{j=1}^{J(t)}$ describes the random TV inphase component, quadrature component, and arrival time of the different paths, respectively. Let $s_l(t)$ be the low-pass equivalent representation of the transmitted signal, then the band-pass representation of the received signal is given by

$$y(t) = \sum_{j=1}^{J(t)} \left(I_j(t,\tau)\cos(\omega_c t) - Q_j(t,\tau)\sin(\omega_c t) \right) s_l\left(t - \tau_j(t)\right) + v_I(t)\cos(\omega_c t) - v_Q(t)\sin(\omega_c t) \quad (2)$$

where $\{v_I(t)\}_{t\geq 0}$ and $\{v_Q(t)\}_{t\geq 0}$ are two independent and identically distributed (iid) white Gaussian noise processes.

It is shown in [4] and [15] that the DPSD of a wireless fading channel, denoted by $S(s)$, can be approximated by an even, stable, rational, and factorizable transfer function, $\tilde{S}(s) = H(s)H(-s)$, where $H(s)$ is given by

$$H(s) = \frac{b_{n-1}(t)s^{n-1} + \cdots + b_1(t)s + b_0(t)}{s^n + a_{n-1}(t)s^{n-1} + \cdots + a_1(t)s + a_0(t)} \quad (3)$$

Consequently, the inphase and quadrature components can be realized using the following stochastic state-space representation [16]

$$\begin{aligned} \mathrm{d}\boldsymbol{X}_{I,j}(t) &= A_I(t)\boldsymbol{X}_{I,j}(t)\mathrm{d}t + B_I(t)\mathrm{d}W_j^I(t) \\ I_j(t) &= C_I\boldsymbol{X}_{I,j}(t) + f_j^I(t) \\ \mathrm{d}\boldsymbol{X}_{Q,j}(t) &= A_Q(t)\boldsymbol{X}_{Q,j}(t)\mathrm{d}t + B_Q(t)\mathrm{d}W_j^Q(t) \\ Q_j(t) &= C_Q\boldsymbol{X}_{Q,j}(t) + f_j^Q(t) \end{aligned} \quad (4)$$

where

$$\begin{aligned} &\boldsymbol{X}_{I,j}(t) = \left[X_{I,j}^1(t), X_{I,j}^2(t), \cdots, X_{I,j}^n(t)\right]^{\mathrm{T}}, \\ &\boldsymbol{X}_{Q,j}(t) = \left[X_{Q,j}^1(t), X_{Q,j}^2(t), \cdots, X_{Q,j}^n(t)\right]^{\mathrm{T}}, \\ &\boldsymbol{A}_I(t) = \boldsymbol{A}_Q(t) = \begin{bmatrix} 0 & 1 & 0 & \cdots & 0 \\ 0 & 0 & 1 & \cdots & 0 \\ \vdots & \vdots & \vdots & \ddots & \vdots \\ 0 & 0 & 0 & \cdots & 1 \\ -a_0(t) & -a_1(t) & -a_2(t) & \cdots & -a_{n-1}(t) \end{bmatrix}, \\ &\boldsymbol{B}_I(t) = \boldsymbol{B}_Q(t) = \begin{bmatrix} b_{n-1}(t) \\ \vdots \\ b_2(t) \\ b_1(t) \\ b_0(t) \end{bmatrix}, \quad \boldsymbol{C}_I = \boldsymbol{C}_Q = [1 \ 0 \ 0 \ \cdots \ 0] \end{aligned} \quad (5)$$

$\boldsymbol{X}_{I,j}(t)$ and $\boldsymbol{X}_{Q,j}(t)$ are the state vectors corresponding to the inphase and quadrature components, respectively. $\{W_j^I(t)\}_{t\geq 0}, \{W_j^Q(t)\}_{t\geq 0}$ are independent standard Brownian motions, which correspond to the inphase and quadrature components of the jth path respectively, $f_j^I(t)$ and $f_j^Q(t)$ are arbitrary functions representing the presence or absence of line-of-sight (LOS) of the inphase and quadrature components for the jth path respectively, and T denotes matrix or vector transpose. Without loss of generality, we consider the case of flat fading, in which the fading channel has purely a multiplicative effect on the signal and the multipath components are not resolvable, and, thus, can be considered as a single path [6]. We also consider the non-line-of-sight (NLOS) case, *i.e.*, $f_j^I(t) = f_j^Q(t) = 0$, which represents an environment with large obstructions. The time-varying state space model described in (4) has a solution given by

$$X_L(t) = \Phi_L(t,t_0)X_L(t_0) + \int_{t_0}^{t} \Phi_L(t,u)B_L(u)\mathrm{d}W_L(u) \tag{6}$$

where $L = I$ or Q, $\Phi_L(t,t_0)$ is the state transition matrix associated to $A_L(t)$, and $\dot{\Phi}_L(t,t_0) = A_L(t)\Phi_L(t,t_0)$. Therefore, the mean of $X_L(t)$ is

$$E\left[X_L(t)\right] = \Phi_L(t,t_0)E\left[X_L(t_0)\right] \tag{7}$$

and the covariance matrix of $X_L(t)$ is

$$\Sigma_L(t) = \Phi_L(t,t_0)Var\left[X_L(t_0)\right]\Phi_L^{\mathrm{T}}(t,t_0) + \int_{t_0}^{t} \Phi_L(t,u)B_L(u)B_L^{\mathrm{T}}(u)\Phi_L^{\mathrm{T}}(t,t_0)\mathrm{d}u \tag{8}$$

For the time-invariant case, $\boldsymbol{A}_I(t) = \boldsymbol{A}_I, \boldsymbol{A}_Q(t) = \boldsymbol{A}_Q, \boldsymbol{B}_I(t) = \boldsymbol{B}_I$, and $\boldsymbol{B}_Q(t) = \boldsymbol{B}_Q$, then (6), (7), and (8) simplify to

$$\begin{aligned} X_L(t) &= \mathrm{e}^{A_L(t-t_0)}X_L(t_0) + \int_{t_0}^{t} \mathrm{e}^{A_L(t-u)}B_L\mathrm{d}W_L(u) \\ E\left[X_L(t)\right] &= \mathrm{e}^{A_L(t-t_0)}E\left[X_L(t_0)\right] \\ \Sigma_L(t) &= \mathrm{e}^{A_L(t-t_0)}Var\left[X_L(t_0)\right]\mathrm{e}^{A_L^{\mathrm{T}}(t-t_0)} + \int_{t_0}^{t} \mathrm{e}^{A_L(t-u)}B_LB_L^{\mathrm{T}}\mathrm{e}^{A_L^{\mathrm{T}}(t-u)}\mathrm{d}u \end{aligned} \tag{9}$$

It can be observed in (7) and (8) that the mean and variance of the inphase and quadrature components are functions of time. Thus, the statistics of the inphase and quadrature components, and therefore the statistics of the channel, are times varying.

Similarly, following the state space representation in (4) and the received signal in (2), the fading channel can be represented using general stochastic state-space representation of the form

$$\begin{aligned} \mathrm{d}\boldsymbol{X}(t) &= \boldsymbol{A}(t)\boldsymbol{X}(t)\mathrm{d}t + \boldsymbol{B}(t)\mathrm{d}\boldsymbol{W}(t) \\ y(t) &= \boldsymbol{C}(t)\boldsymbol{X}(t) + \boldsymbol{D}(t)\boldsymbol{v}(t) \end{aligned} \tag{10}$$

where

$$\begin{aligned} &\boldsymbol{X}(t) = \left[X_I(t) \quad X_Q(t)\right]^{\mathrm{T}}, \ \boldsymbol{A}(t) = \begin{bmatrix} A_I(t) & 0 \\ 0 & A_Q(t) \end{bmatrix}, \ \boldsymbol{B}(t) = \begin{bmatrix} B_I(t) & 0 \\ 0 & B_Q(t) \end{bmatrix}, \\ &\boldsymbol{C}(t) = \left[\cos(\omega_c t)C_I \quad -\sin(\omega_c t)C_Q\right], \ \boldsymbol{D}(t) = \left[\cos(\omega_c t) \quad -\sin(\omega_c t)\right] \\ &\boldsymbol{v}(t) = \left[v_I(t) \quad v_Q(t)\right]^{\mathrm{T}}, \ \mathrm{d}\boldsymbol{W}(t) = \left[\mathrm{d}W^I(t) \quad \mathrm{d}W^Q(t)\right]^{\mathrm{T}} \end{aligned} \tag{11}$$

In this case, $y(t)$ represents the received signal measurements, $\boldsymbol{X}(t)$ is the state variable of the inphase and quadrature components, and $\boldsymbol{v}(t)$ is the measurement Gaussian noise.

In [4] and [15], the channel parameters $\{a_{n-1}(t),\cdots,a_0(t),b_{n-1}(t),\cdots,b_0(t)\}$ are obtained from approximating the DPSD. However, in reality one does not have access to the DPSD at all times during the estimation process. Therefore, we propose estimating the channel parameters as well asinphase and quadrature components directly from received signal measurements, which are usually available or easy to obtain in any wireless network. A filter-based expectation maximization (EM) algorithm combined with the Kalman filter is employed to estimate the channel model parameters and states in (10). These filters use only the first and second order statistics and are also recursive, and therefore can be implemented online. These algorithms have been recently utilized in [1] and [2] to estimate wireless channel parameters and states, and therefore the formulations of these algorithms are not presented in this paper. Experimental results demonstrating the applicability of these algorithms in conjunction with the proposed stochastic wireless models are discussed in Section 4. In the next section, we introduce an important application based on the developed models; that is stochastic PC in wireless networks.

3. Optimal Power Control Based on the Stochastic Wireless Channel Models

In this section, an optimal PCA is investigated based on the estimated wireless channel models. Since the channel model parameters are estimated from received signal measurements, PC can be performed solely from having these measurements. The aim of the PCA described here is to minimize the total transmitted power of all users while maintaining acceptable quality-of-service (QoS) for each user. The measure of QoS is defined by the signal-to-interference ratio (SIR) for each link to be larger than a target SIR.

By generalizing the wireless channel model in (10), the state-space representation of a wireless network with M transmitters and N receivers can be described as

$$\begin{aligned} \mathrm{d}X_{ij}(t) &= A_{ij}X_{ij}(t)\mathrm{d}t + B_{ij}\mathrm{d}W_{ij}(t) \\ y_i(t) &= \sum_{k=1}^{M}\sqrt{p_k(t)}s_k(t)C_{ik}(t)X_{ik}(t) + v_i(t) \end{aligned} \tag{12}$$

where $y_i(t)$ is the received signal at the ith receiver at time t, $X_{ik}(t)$ is the states of the channel between transmitter k and the receiver assigned to transmitter i, $p_k(t)$ is the transmitted power of transmitter k at time t, which acts as a scaling on the information signal $s_k(t)$, $v_i(t)$ is the channel disturbance or noise at receiver i, and $1 \le i, j \le M$.

Consider the wireless network described in (12), the centralized PC problem for TV channels over a time interval $[0,T]$ can be stated as follows [4]

$$\begin{aligned} &\min_{(p_1\ge 0,\cdots,p_M\ge 0)} \left\{ \sum_{i=1}^{M}\int_0^T p_i(t)\mathrm{d}t \right\}, \\ &\text{subject to } \frac{\int_0^T p_i(t)s_i^2(t)\left[C_{ii}(t)X_{ii}(t)\right]^2 \mathrm{d}t}{\sum_{k\ne i}^{M}\int_0^T p_k(t)s_k^2(t)\left[C_{ik}(t)X_{ik}(t)\right]^2\mathrm{d}t + \int_0^T v_i^2(t)\mathrm{d}t} \ge \varepsilon_i \end{aligned} \tag{13}$$

where ε_i is the target SIR at receiver i and $i = 1,\cdots,M$. A solution to (13) is presented by first introducing the communication meaning of predictable power control strategies (PPCS). In wireless cellular networks, it is practical to observe and estimate channels at the base stations and then send the information back to the mobiles to adjust their power signals $\{p_i(t_k)\}_{i=1}^{M}$. Since channels experience delays, and power control is not feasible continuously in time but only at discrete time instants, the concept of predictable strategies is introduced [4]. Consider a set of discrete time strategies $\{p_i(t_k)\}_{i=1}^{M}$, $0 = t_0 < t_1 < \cdots < t_k < t_{k+1} < \cdots \le T$. At time t_{k-1}, the base stations estimate the channel information $\{I_{ij}(t_{k-1}), Q_{ij}(t_{k-1}), s_i(t_{k-1})\}_{i,j=1}^{M}$ as illustrated in Section 2. Using the concept of predictable strategy, the base stations determine the control strategy $\{p_i(t_k)\}_{i=1}^{M}$ for the next time instant t_k. The latter is communicated back to the mobiles, which hold these values during the time interval $[t_{k-1}, t_k)$. At time t_k, a new set of channel information $\{I_{ij}(t_k), Q_{ij}(t_k), s_i(t_k)\}_{i,j=1}^{M}$ is estimated at base stations and the time t_{k+1} control strategies $\{p_i(t_{k+1})\}_{i=1}^{M}$ are computed and communicated back to the mobiles which hold them constant during the time interval $[t_k, t_{k+1})$. Such decision strategies are called predictable. Using the concept of PPCS over any time interval $[t_k, t_{k+1}]$, Expression (13) is equivalent to

$$\min_{p(t_{k+1})>0} \sum_{i=1}^{M} p_i(t_{k+1}), \text{ subject to } \boldsymbol{p}(t_{k+1}) \ge \boldsymbol{\Gamma G}_I^{-1}(t_k, t_{k+1}) \times \left(\boldsymbol{G}(t_k, t_{k+1})\boldsymbol{p}(t_{k+1}) + \boldsymbol{\eta}(t_{k+1})\right) \tag{14}$$

where

$$g_{ij}(t_k,t_{k+1}) := \int_{t_k}^{t_{k+1}} s_j^2(t)\left[C_{ij}(t)X_{ij}(t)\right]^2 \mathrm{d}t,$$
$$\eta_i(t_k,t_{k+1}) := \int_{t_k}^{t_{k+1}} v_i^2(t)\mathrm{d}t,\ 1\le i,j\le M,$$
$$\boldsymbol{G}_I(t_k,t_{k+1}) := \mathrm{diag}\left(g_{11}(t_k,t_{k+1}),\cdots,g_{MM}(t_k,t_{k+1})\right),$$
$$\boldsymbol{G}(t_k,t_{k+1}) := \begin{Bmatrix} 0 & \text{if } i=j \\ g_{ij}(t_k,t_{k+1}) & \text{if } i\ne j \end{Bmatrix}, \tag{15}$$
$$\boldsymbol{\eta}(t_k,t_{k+1}) := \left(\eta_1(t_k,t_{k+1}),\cdots,\eta_M(t_k,t_{k+1})\right)^{\mathrm{T}},$$
$$\boldsymbol{p}(t_{k+1}) := \left(p_1(t_{k+1}),\cdots,p_M(t_{k+1})\right)^{\mathrm{T}},$$
$$\boldsymbol{\Gamma} := \mathrm{diag}(\varepsilon_1,\cdots,\varepsilon_M),$$

and $\mathrm{diag}(\cdot)$ denotes a diagonal matrix with its argument as diagonal entries. The optimization in (14) is a linear programming problem in $M\times 1$ vector of unknowns $\boldsymbol{p}(t_{k+1})$. Here $[t_k,t_{k+1}]$ is a time interval such that the channel model does not change significantly, *i.e.*, $[t_k,t_{k+1}]$ should be smaller than the coherence time of the channel. Throughout this section, we assume that the PC problem is feasible, *i.e.*, there exists a power vector $\boldsymbol{p}(t_k)$ that satisfies the inequality in (14) for all $[t_k,t_{k+1}]$ in $[0,T]$.

In the next section, a numerical example is presented to determine the performance of the proposed PCA under the estimated wireless channel models.

4. Numerical Results

In this section, two numerical examples are presented. In Example 1, the EM algorithm combined with Kalman filtering is performed to estimate the channel parameters as well as inphase and quadrature components from received signal measurements. In Example 2, the performance of the proposed PCA based on the estimated channel models in Example 1 is determined and compared with fixed transmitted powers.

4.1. Example 1: Wireless Channel Estimation

In this numerical example, a 4th order channel model as described in (10) and (11) is considered. Therefore, the system parameters $\theta_t = \{\boldsymbol{A}_t,\boldsymbol{B}_t,\boldsymbol{C}_t,\boldsymbol{D}_t\}$ can be represented as

$$\boldsymbol{A}_t = \begin{bmatrix} 0 & 1 & 0 & 0 \\ a_1 & a_2 & 0 & 0 \\ 0 & 0 & 0 & 1 \\ 0 & 0 & a_3 & a_4 \end{bmatrix},\ \boldsymbol{B}_t = \begin{bmatrix} b_1 & \delta_{12} & \delta_{13} & \delta_{14} \\ b_2 & \delta_{22} & \delta_{23} & \delta_{24} \\ \delta_{31} & \delta_{32} & b_3 & \delta_{34} \\ \delta_{41} & \delta_{42} & b_4 & \delta_{44} \end{bmatrix}, \tag{16}$$
$$\boldsymbol{C}_t = \left[\cos(\omega_c t)\ \ 0\ \ -\sin(\omega_c t)\ \ 0\right],\ \boldsymbol{D}_t = \left[d_1\ \ d_2\right]$$

Experimental data for a cellular network is provided by the Canadian Communication Research Center (CRC) and include measurement samples for the inphase and quadrature components and received signal strength.

As previously mentioned, the estimation of a flat fading wireless channel from received signal measurement data is considered. In particular, the estimation includes the channel parameters, inphase and quadrature components, and the received signal, which are then compared to the ones obtained from the provided measurement data. It is also assumed that the received signal measurement data are corrupted by white Gaussian noise sequences.

Figure 1 shows the measured and estimated inphase and quadrature components as well as the received signal using the EM algorithm together with the Kalman filter for 400 sampled data taken from the measurements of one channel chosen at random. At a certain time instant, the system parameters are estimated as

$$\hat{A} = \begin{bmatrix} 0 & 1 & 0 & 0 \\ -0.0756 & -0.0474 & 0 & 0 \\ 0 & 0 & 0 & 1 \\ 0 & 0 & -0.6638 & 0.0717 \end{bmatrix},$$

$$\hat{B}^2 = \begin{bmatrix} 0.0484 & -0.0029 & -0.0453 & 4.0686\times10^{-4} \\ -0.0029 & 0.0462 & 0.0013 & -0.0438 \\ -0.0453 & 0.0013 & 0.0573 & 0.0047 \\ 4.0686\times10^{-4} & -0.0438 & 0.0047 & 0.0564 \end{bmatrix}, \quad (17)$$

$$\hat{C} = \left[\cos(\omega_c t) \quad 0 \quad -\sin(\omega_c t) \quad 0\right], \ \hat{D}^2 = [0.0119].$$

From **Figure 1**, it can be observed that the inphase and quadrature components of the wireless fading channel as well as the received signal have been estimated with very high accuracy. It can also be noticed that the estimation error decreases as the number of samples increases; this is because the algorithm is recursive and the channel parameters converge to the actual values as more samples are being estimated. **Figure 2** shows the received signal estimates root mean square error (RMSE) for 100 runs. It can be noticed that it takes just few iterations (less than 15) for the filter to converge, and the steady state performance of the proposed channel estimation algorithm is excellent. Since we consider 4th order channel model, the computational cost of the proposed estimation algorithm is moderate and can be implemented in real time. Moreover, the filters of the expectation step are recursive and decoupled and hence are easy to implement in parallel on a multi-processor system [17].

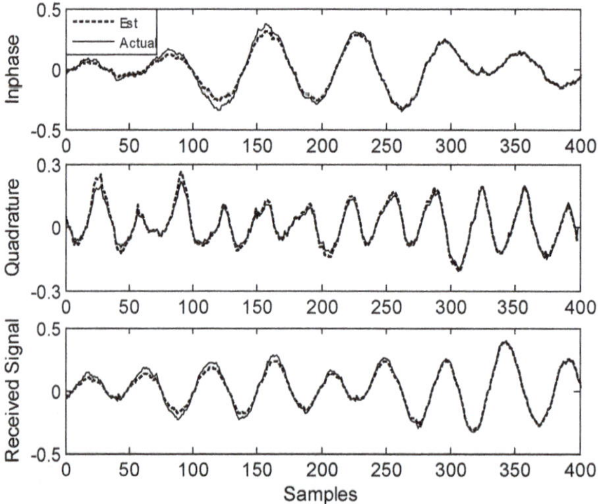

Figure 1. The measured and estimated inphase and quadrature components, and received signal for the 4th order channel model in Example 1, using the EM algorithm combined with the Kalman filter.

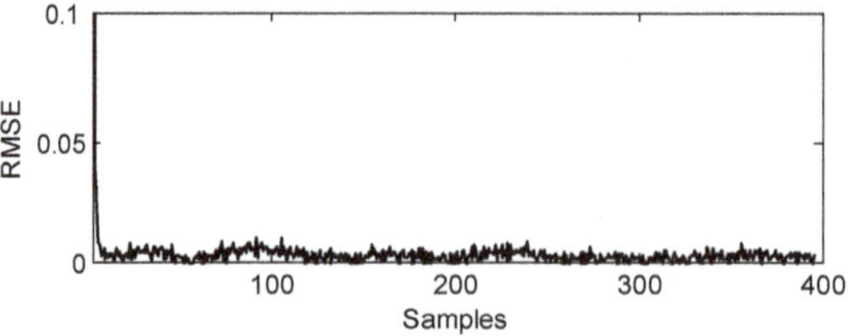

Figure 2. Received signal estimates RMSE for 100 runs using the EM algorithm combined with the Kalman filter.

4.2. Example 2: Optimal Power Control

In this numerical example, the received signal measurement data for 24 users are collected experimentally. They represent flat Rayleigh fading environment where the signal envelope at the receivers exhibit Rayleigh distributed density. The channel model parameters as well as the inphase and quadrature components for all users are estimated online from the measurement data using the EM algorithm together with the Kalman filter as illustrated in Example 1. The PCA described in (14) is performed using the estimated channel parameters and states. The outage probability (OP) is used as a performance measure for the PCA. A link with a received SIR, R_i, less than or equal to a target SIR, ε_i, is considered a communication failure. The OP, $O(\varepsilon_i)$, is expressed as $O(\varepsilon_i) = \text{Prob}\{R_i \le \varepsilon_i\}$, where R_i is the received SIR at receiver i.

It is assumed that the targets SIR, ε_i for all users are the same, and varied from 5 dB to 25 dB with step 5 dB. For each value of ε_i the OP is computed every 15 millisecond, *i.e.*, $[t_k, t_{k+1}] = 15$ millisecond. The simulation is performed for 4.5 seconds, *i.e.*, $[0, T] = 4.5$ seconds. The OP is computed using Monte-Carlo simulations. The performance of the proposed PCA is compared with the one of constant transmitted powers (CTP).

The OP for both the CTP and the proposed PCA based on PPCS are demonstrated in **Figure 3(a)** and **Figure 3(b)**, respectively. **Figure 3** shows how the OP changes with respect to target SIR and time. As the target SIR increases the OP increases. This is obvious since we expect more users to fail. The OP also changes as a function of time, since users move in different directions and velocities while gathering the measurements.

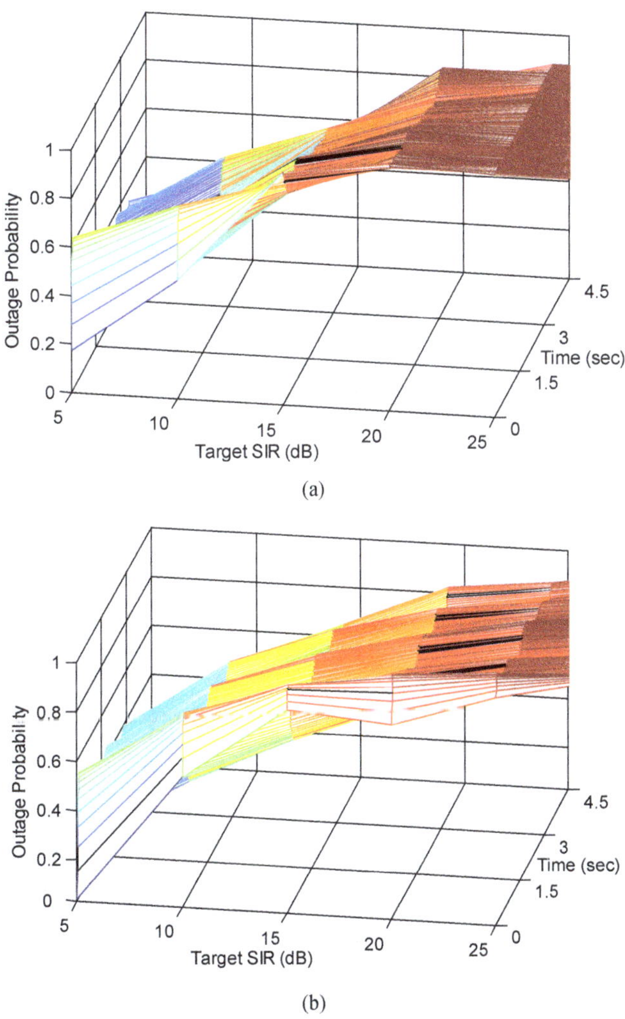

Figure 3. The outage probability for the dynamical flat Rayleigh wireless network in Example 2. (a) Using CTP; (b) Using PC based on PPCS.

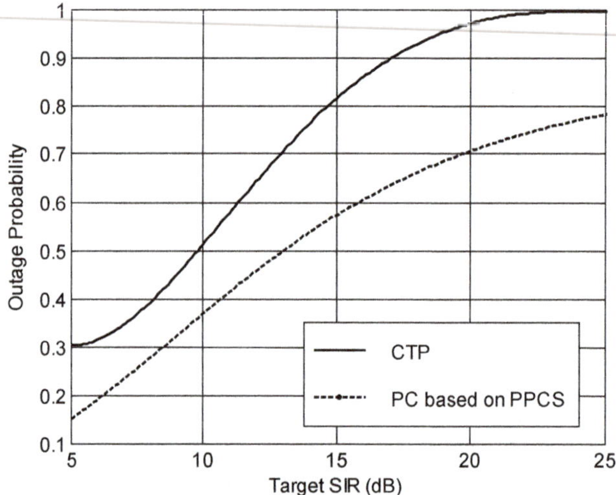

Figure 4. Average outage probability of the PC case based on PPCS and the CTP case in Example 2. Performance comparison.

The average OP versus target SIR for both cases over the whole simulation time (4.5 seconds) is demonstrated in **Figure 4**, which shows that the performance of the proposed PCA based on PPCS is on average much better than that of the CTP. For example, at 10 dB target SIR, the OP is reduced from 0.51 for the CTP algorithm to 0.37 for the PCA; this represents an improvement of over 27%.

5. Conclusion

This paper describes a general scheme for extracting mathematical wireless channel models from noisy received signal measurements, and performing power control based on the estimated channel parameters. The channel models are represented in stochastic state-space form. The proposed estimation algorithm consists of filtering based on the Kalman filter to remove noise from data, and identification based on the EM algorithm to determine the parameters of the model which best describe the measurements. Numerical results indicate that the measured data can be regenerated through a simple 4th order discrete-time stochastic state-space model. Moreover, a stochastic PCA based on the estimated parameters and channel states is investigated. Numerical results indicate that there is potentially large gain to be achieved by using the proposed PCA, which can be used as long as the channel model does not change significantly; that is $[t_k, t_{k+1}]$ is a subset of the coherence time of the channel.

Acknowledgements

This paper has been authored by employees of UT-Battelle, LLC, under contract DE-AC05-00OR22725 with the US Department of Energy. This work was also supported in part by NSF grant CMMI-1334094.

References

[1] Olama, M.M., Djouadi, S.M. and Charalambous, C.D. (2009) Stochastic Differential Equations for Modeling, Estimation and Identification of Mobile-to-Mobile Communication Channels. *IEEE Transactions on Wireless Communications*, **8**, 1754-1763. http://dx.doi.org/10.1109/TWC.2009.071068

[2] Olama, M.M., Li, Y., Djouadi, S.M. and Charalambous, C.D. (2007) Time-Varying Wireless Channel Modeling, Estimation, Identification, and Power Control from Measurements. *Proceedings of the American Control Conference (ACC'07)*, New York, 9-13 July 2007, 3100-3105.

[3] Olama, M.M., Djouadi, S.M. and Charalambous, C.D. (2006) Stochastic Power Control for Time-Varying Long-Term Fading Wireless Networks. *EURASIP Journal on Applied Signal Processing*, **2006**, Article ID: 89864.

[4] Charalambous, C.D., Djouadi, S.M. and Denic, S.Z. (2005) Stochastic Power Control for Wireless Networks via SDE's: Probabilistic QoS Measures. *IEEE Transactions on Information Theory*, **51**, 4396-4401. http://dx.doi.org/10.1109/TIT.2005.858984

[5] Jakes, W. (1974) Microwave Mobile Communications. IEEE Inc., New York.

[6] Proakis, J.G. (2000) Digital Communications. 4th Edition, McGraw Hill, New York.

[7] Rappaport, T.S. (2002) Wireless Communications: Principles and Practice. 2nd Edition, Prentice Hall, Upper Saddle River.

[8] Charalambous, C.D. and Logothetis, A. (2000) Maximum-Likelihood Parameter Estimation from Incomplete Data via the Sensitivity Equations: The Continuous-Time Case. *IEEE Transactions on Automatic Control*, **45**, 928-934. http://dx.doi.org/10.1109/9.855553

[9] Bishop, G. and Welch, G. (2001) An Introduction to the Kalman Filters. University of North Carolina, North Carolina.

[10] Zander, J. (1992) Performance of Optimum Transmitter Power Control in Cellular Radio Systems. *IEEE Transactions on Vehicular Technology*, **41**, 57-62. http://dx.doi.org/10.1109/25.120145

[11] Aein, J. (1973) Power Balancing in Systems Employing Frequency Reuse. *COMSAT Technical Review*, **3**, 277-299.

[12] Bambos, N. and Kandukuri, S. (2002) Power-Controlled Multiple Access Schemes for Next-Generation Wireless Packet Networks. *IEEE Wireless Communications*, **9**, 58-64. http://dx.doi.org/10.1109/MWC.2002.1016712

[13] Foschini, G.J. and Miljanic, Z. (1993) A Simple Distributed Autonomous Power Control Algorithm and Its Convergence. *IEEE Transactions on Vehicular Technology*, **42**, 641-646. http://dx.doi.org/10.1109/25.260747

[14] Kandukuri, S. and Boyd, S. (2002) Optimal Power Control in Interference-Limited Fading Wireless Channels with Outage-Probability Specifications. *IEEE Transactions on Wireless Communications*, **1**, 46-55. http://dx.doi.org/10.1109/7693.975444

[15] Olama, M.M., Djouadi, S.M. and Charalambous, C.D. (2006) Stochastic Channel Modeling for Ad-Hoc Wireless Networks. *Proceedings of the American Control Conference*, Minneapolis, 14-16 June 2006, 6075-6080.

[16] Oksendal, B. (1998) Stochastic Differential Equations: An Introduction with Applications. Springer, Berlin. http://dx.doi.org/10.1007/978-3-662-03620-4

[17] Elliott, R.J. and Krishnamurthy, V. (1999) New Finite-Dimensional Filters for Parameter Estimation of Discrete-Time Linear Guassian Models. *IEEE Transactions on Automatic Control*, **44**, 938-951. http://dx.doi.org/10.1109/9.763210

8

Analysis of Different Call Admission Control Strategies and Its Impact on the Performance of LTE-Advanced Networks

Mahammad A. Safwat[1], Hesham M. El-Badawy[1], Ahmad Yehya[2], Hosni El-Motaafy[3]

[1]Network Planning Department, National Telecommunication Institute, Cairo, Egypt
[2]Department of Electrical Engineering, Al-Azhar University, Cairo, Egypt
[3]Department of Electronics and Computer Engineering, HIT, Tenth of Ramdan City, Egypt
Email: mahammad.safwat@gmail.com, heshamelbadawy@ieee.org, ahmed_yahya_1@yahoo.com, hosni.motaafy@gmail.com

Abstract

The call admission control (CAC) optimizes the use of allocated channels against offered traffic maintaining the required *quality of service* (QoS). Provisioning QoS to user at cell-edge is a challenge where there is limitation in cell resources due to inter-cell interference (ICI). Soft Frequency Reuse is ICI mitigation scheme that controls the distribution of resources between users. In this paper, the Impact of four CAC schemes (Cutoff Priority scheme (CP), Uniform Fractional Guard Channel (UFGC), Limited Fractional Guard Channel (LFGC), New Call Bounding (NCB) scheme) at cell-edge have investigated using queuing analysis in a comparative manner. The comparison is based on two criteria. The first criterion guarantees a particular level of service to already admitted users while trying to optimize the revenue obtained. The second criterion determines the minimum of number of radio resources that provides hard constraints in both of blocking and dropping probabilities. The four schemes are compared at different scenarios of new and handover call arrival rates.

Keywords

Call Admission Control, Soft Frequency Reuse, QoS

1. Introduction

With the rapid development of wireless network communication, careful allocation of radio resources plays a significant role in achieving the required QoS in wireless cellular networks and considerable efforts have been focused on call admission control (CAC) in order to maintain the required QoS [1]. Cell-edge is an area at cell which is difficult to maintain suitable QoS for its users. This is because there is a limitation in radio resources due to inter-cell interference (ICI) [2] and sharing the radio resources with handover users.

Modern ICI mitigation schemes divide the cell into cell-edge and cell-core. In these schemes the users are divided according to their locations from the e-NodeB and the resources they can access to cell-edge and cell-core users. One of the most ICI mitigation schemes, which are used in LTE-Advanced network, is Soft Frequency Reuse (SFR). In SFR scheme, for each cell in the network; the cell is divided into two parts: cell-edge and cell-core. In addition, the available Resource Blocks (RBs) (basic resource element in LTE networks) are divided into cell-edge RBs and cell-core RBs. All of users within each cell are also divided into two groups which based on the SINR: cell-edge users and cell-core users. It is called SFR as the frequency partition only applies to the cell edge users, the cell edge users are restricted to use this frequency sub-band only and all frequency bands are available to the cell-core users. So the effective frequency reuse factor is still close to one [2] [3].

Cell-edge performance is such a subject that faces challenge in modeling and analysis. What makes the modeling difficult is that, to achieve some sense of accuracy, one need to consider ICI impact as well as handover traffic effect from adjacent cells.

Handoff priority-based CAC schemes have great impact on cell-edge users. First of all it occurs on cell-edges and secondly it depends on cell-edge radio resources. Therefore we cannot ignore its effect in cell-edge performance. The handover process depends on the policy that control handover access to the cell which in turn depends on the associated CAC scheme adopted.

Various handoff priority-based CAC schemes have been proposed. One of these schemes depends on reserving a portion of channel for handoff calls; whenever a channel is released, it is returned to the common poll of channels. This scheme is called the *cutoff priority scheme* [4]. On the other hand, the *fractional guard channels* schemes which depend on admitting a new call with certain probability. This scheme was first proposed by Ramjee *et al.* [5] and used extensively thereafter [6]-[9]; it is shown to be more general than the cutoff priority scheme. A CAC scheme named *New Call Bounding* (NCB) scheme [10]-[12] smoothly throttles the admission rates of calls according to their priorities as well as it aims to provide multiple prioritized traffic with a desired QoS. In the rigid division-based CAC scheme, all channels allocated to a cell are divided into two groups: one to be used by all calls and the other for handover calls only [13].

In [14], a traffic analysis is evolved for soft frequency reuse ICI mitigation scheme using queuing model and the performance metrics of blocking and outage probability is deduced in terms of resources availability. Handover users' traffic is not addressed in the analysis. In [15], the SFR is modeled in collaboration with UFGC schemes and blocking probability is deduced in cell-edge and cell-core separately.

The applied CAC scheme affects on the cell-edge performance. Each CAC scheme behaves differently according to network conditions and traffic load. Consequently by using the most suitable scheme according to network condition, the system performance can be optimized. So a comparison between the different CAC schemes is deeply needed. According to this comparison, the scheme that provides better performance will be preferred.

In this work, a comparison between different CAC schemes is evolved in collaboration with SFR in order to evaluate each scheme impact on cell-edge performance. The evaluation is based on the two QoS parameters: New call Blocking probability and handover dropping probability. Four schemes are addressed in this work; Cutoff priority scheme, Uniform Fractional Guard Channel, Limited Fractional Guard Channel and New Call Bounding scheme. This comparison is based on applying some restrictions on each scheme and compares the response against these restrictions. Two criteria will be adopted in this comparison [5] [16]:

- System behaviour with hard constraint on dropping probability: in this criterion, the parameters setting are adjusted for new call blocking optimality while the dropping probability meets its constraint at specific new call and handover arrival rate.
- System behaviour with hard constraint on dropping and blocking probabilities: in this criterion, in order to meet specific constraint in both of dropping and blocking probabilities (which are two contradictory performance metrics), the number of radio resources should be increased. So each scheme parameters are adjusted to meet the first metric and the number of radio resources is increased to meet the second.

The first criterion guarantees a particular level of service to already admitted users while trying to optimize the revenue obtained. The second criterion is more of network design problem where radio resources need to be allocated a priori based on, for example, traffic projection. In the second criterion, the preference will be the minimum number of radio resources required to fulfil blocking and dropping constraints.

Another important aspect of these criteria is the design of algorithms for determining the optimal setting for parameters associated with the criteria. It is important that the algorithms are simple and efficient.

2. Schemes Models

In the following, Four CAC schemes will be introduced and modeled using queuing theory. In order to investigate the traffic analysis at cell-edge, the CAC schemes are merged with SFR in the proposed models. Then their performance metrics will be deduced in order to compare between them according to the mentioned criteria.

The cutoff priority technique keeps a certain amount of channels to handover calls only while the rest of the channels can be shared by both new calls and handover calls. Hence, handover calls are given higher priority over new calls, and as a result the reduction in the handover probability comes at the expense of higher blocking rate. **Figure 1** illustrates the state diagram of Cutoff priority scheme with SFR.

In *Fractional Guard Channel* the new call is admitted by a certain probability which is a decreasing (or, more accurately, non-increasing) function of the number of occupied channels while a handover call is admitted as long as there is a free channel. Uniform FGC is special case of FGC where the acceptance probability has a constant probability that is independent of number of occupied channel. While Limited FGC is another type of FGC in which the acceptance probability varies between three values (1, β_l, 0) according to channels occupation. **Figure 2** depicts the state diagram of Fractional Guard Channel with SFR for UFGC and Limited FGC.

In *New Call Bounding* scheme, a threshold is used to limit the number of new calls in the cells. Handover calls are only blocked if all channels are occupied. The scheme works as follows: if the number of new calls in a cell exceeds a threshold when a new call arrives, the new call will be blocked; otherwise it will be admitted. The handover call is rejected only when all channels in the cell are used up. The idea behind this scheme is that we would rather accept fewer customers than drop the ongoing calls in the future, because customers are more sensitive to call dropping than to call blocking. **Figure 3** illustrates the state diagram of New Call Bounding scheme with SFR.

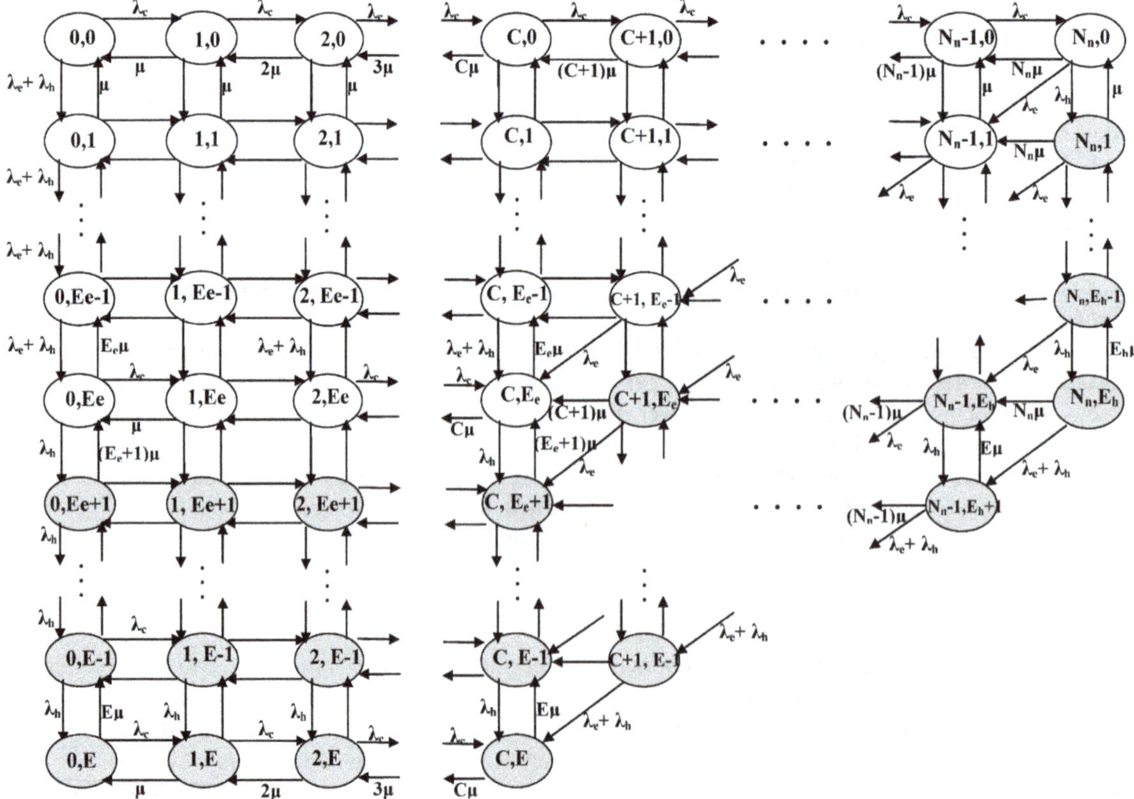

Figure 1. The state diagram of Cutoff priority scheme with SFR.

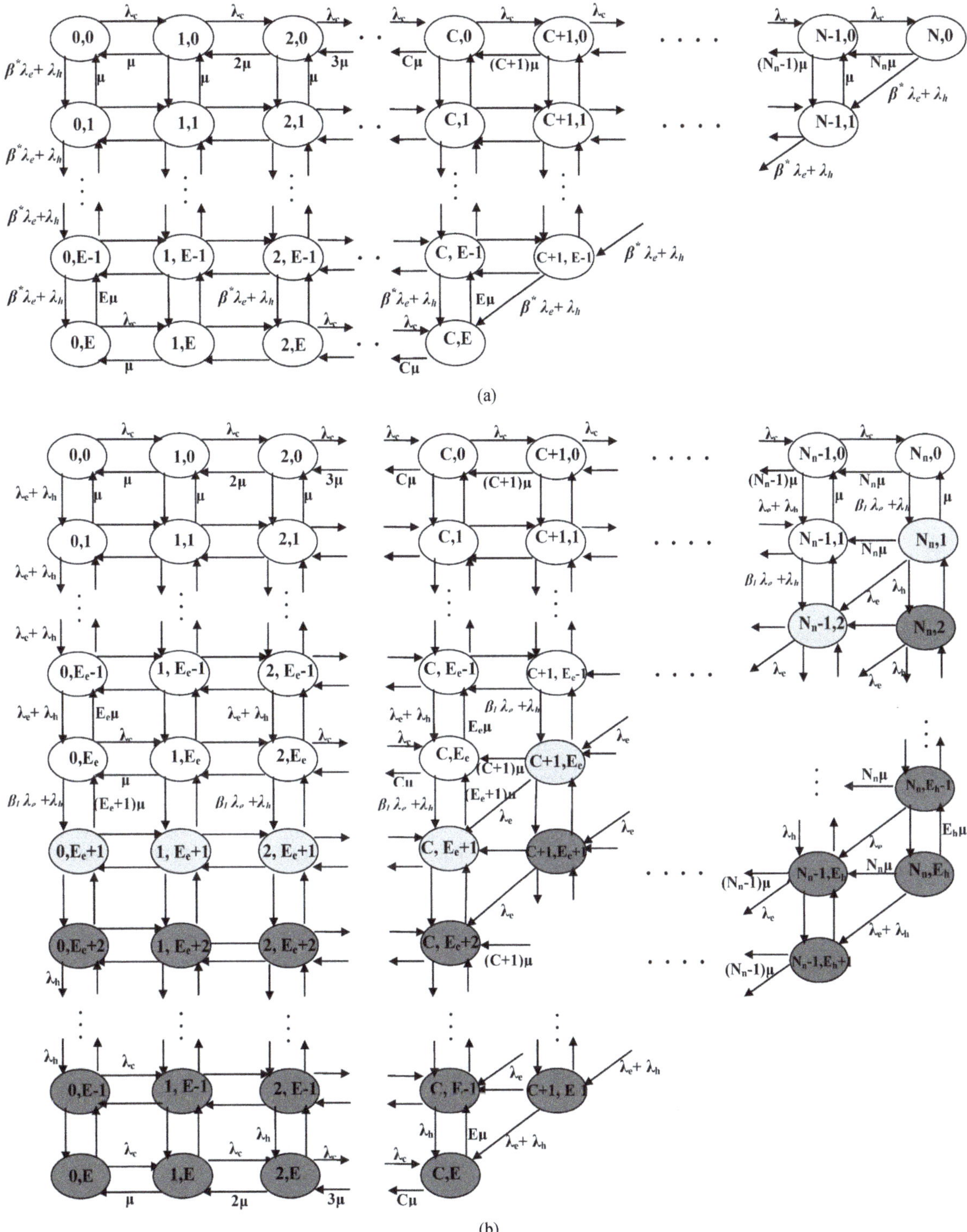

Figure 2. The state diagram of Fractional Guard Channel with SFR (a) UFGC (b) LFGC.

The four schemes are modelled using queuing analysis. For the sake of more inspection of traffic analysis at cell-edge, the resources at cell-edge are modelled in a separate dimension. The resources are distributed between cell-edge and cell-core according to SFR policies.

In the Markov representation of the four schemes which shown in **Figures 1-3**, Cutoff priority scheme, LFGC and UFGC scheme are modelled using two dimension Markov chains while NCB is modelled using three di-

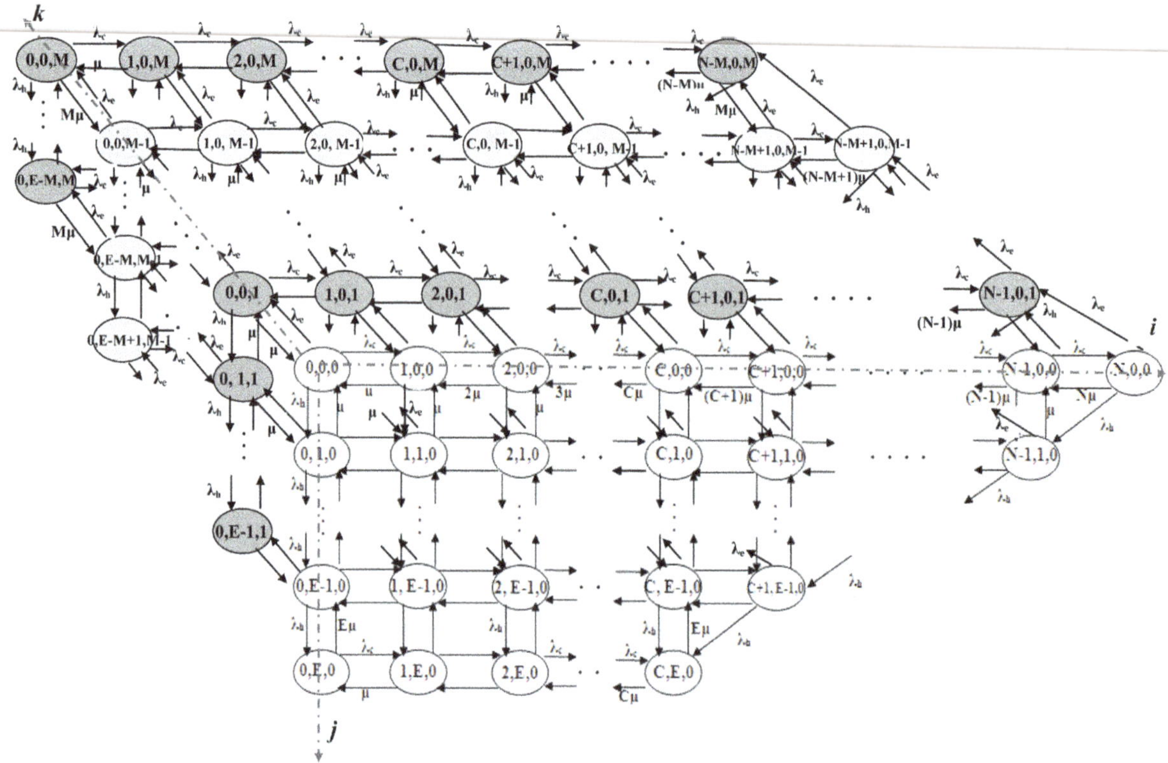

Figure 3. The state diagram of new call bounding scheme with SFR.

mensions Markov chain to separate between new and handover call at cell-edge. In each scheme, the state space Γ is defined with i representing the number of RBs used by cell-core users, j representing the number of RBs used by cell-edge users and k representing the number of RBs used by handover users (for NCB only). Let N denotes the number of available RBs that can be used for transmission in each transmission time interval (TTI) in the cell. The maximum number of RBs that can be assigned to the edge-users and core-users is E and C respectively; so we get $E + C = N$. Let λ denotes the arrival rate of new call and λ_c, λ_e, denote the arrival rate for new calls in cell-core, cell-edge respectively. So $\lambda = \lambda_c + \lambda_e$. λ_h is the arrival rate for handover calls. The current work considers that the arrival process for new calls and the arrival process for handoff calls are all Poisson. The channel holding times (μ) for new calls and handoff calls are exponentially distributed, and for simplicity it is assumed to be equal for three users. Each scheme has its own parameter setting as follow:

Cutoff priority scheme: the number of guard channels reserved for handover users T as shown in **Figure 1**. Let E_e denotes the number of cell-edge radio resources which are available for new call and handover users.

Uniform Fractional Guard Channel: the percent at which the new call is admitted at cell-edge which is called "acceptance probability β^*" as shown in **Figure 2(a)**.

Limited Fractional Guard Channel: The number of guard channels T and the new call acceptance probability β_l as shown in **Figure 2(b)**.

New Call Bounding scheme: the maximum number of RB that can be assigned for new call M as shown in **Figure 3**.

The proposed queuing model can be solved using Successive over Relaxation (SOR) method to get steady state probability for each state. The SOR is an iterative method [14] [15] that used to solve the set of linear equation. Most models under investigation are irreversible Markov process. So, the SOR may be one of the most suited techniques to obtain steady state probabilities and the required performance metrics. SOR not only supports the feasible solution but also, it gives sufficient stability for the obtained results. Performance metrics can be obtained from this steady state probability.

In this method, a new set of equations, called SOR equations, are deduced from balance equations, the left hand side of these equations is a new value of steady state probability which is obtained iteratively using previous value for steady state probability on the right hand side. The speed of convergence is determined by relaxation factor ω, the choice of relaxation factor is not necessarily easy, and depends upon the properties of the coefficient matrix. For symmetric, positive-definite matrices it can be proven that $0 < \omega < 2$ will lead to convergence, but we are generally interested in faster convergence rather than just convergence. **Figure 4** depicts the flow chart of SOR algorithm.

3. System Performance Metrics

In this work, we will use blocking and dropping probabilities to evaluate system performance. Cell blocking probability is the probability that a new arriving cell-core user and a cell-edge user are blocked. In this work, only cell-edge part will be evaluated. Let ψ_b be the subsets of states where a new arriving cell-edge user are blocked.

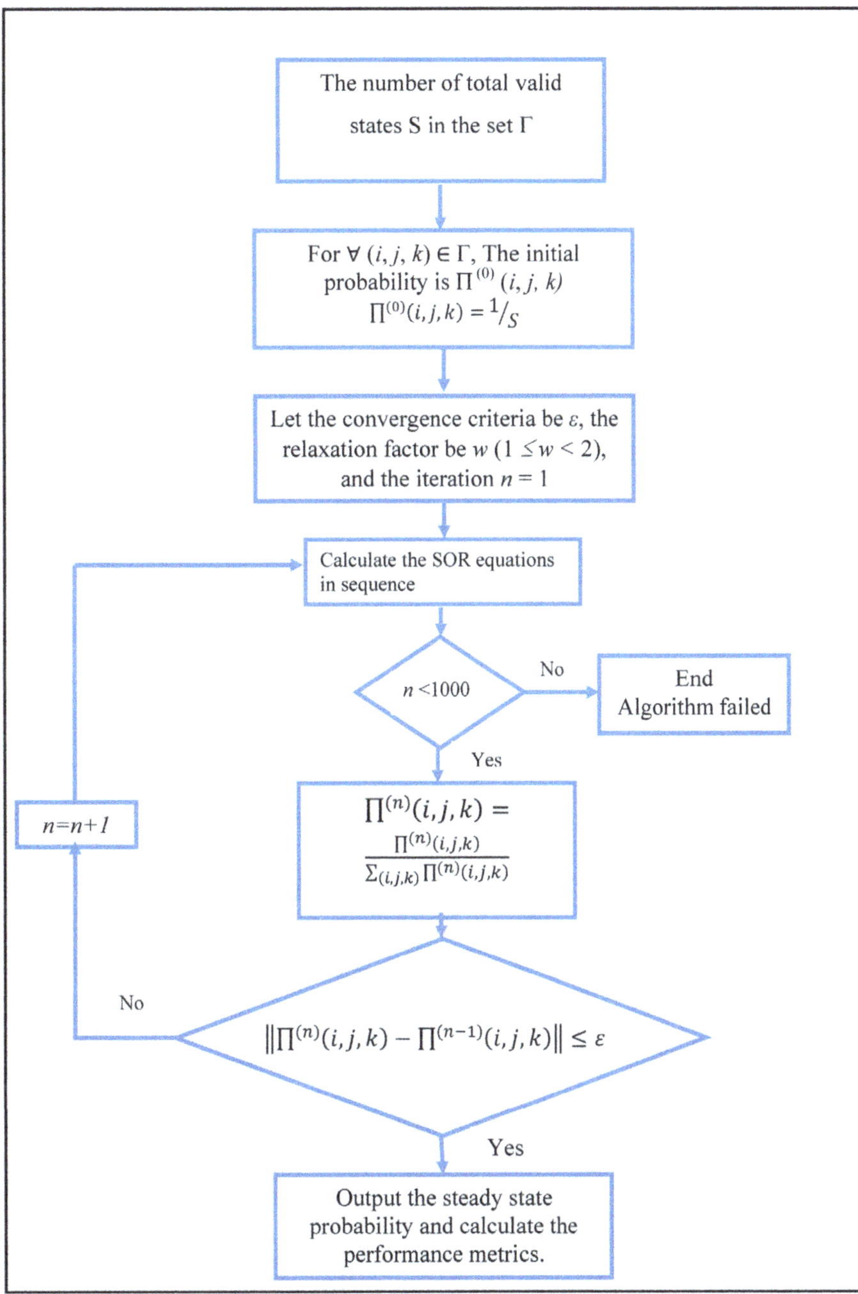

Figure 4. Flow chart for SFR algorithm.

$$\Psi_b\big|_{CP} = \left\{\prod(i,j) \in \Psi_b; E_e \le i \le E, 1 \le j \le C\right\} \tag{1}$$

$$\Psi_b\big|_{UFGC} = \left\{\prod(i,j) \in \Psi_{b1}; 1 \le i < E, 1 \le j \le N\right\} \cup \left\{\prod(i,j) \in \Psi_{b2}; i = E, 1 \le j \le C\right\} \tag{2}$$

$$\begin{aligned}\Psi_b\big|_{LFGC} = \Big\{&\prod(i,j) \in \Psi_{be1}; i = E_e, 1 \le j \le C \cup \prod(i,j) \in \Psi_{be2}; E_e < i \le E, 1 \le j \le C \\ &\cup \prod(i,j) \in \Psi_{be3}; C \le i \le N_n, 0 \le j \le E_h, i + j = N_n\Big\}\end{aligned} \tag{3}$$

$$\begin{aligned}\Psi_b\big|_{NCB} = &\left\{\pi(i,j,k) \in \Psi_{b1}; 1 \le i < N - M, 1 \le j \le E - M, k = M, i + j < N - M\right\} \\ &\cup \left\{\pi(i,j,k) \in \Psi_{b2}; 1 \le i < C, E - M < j \le E, k = E - j, i + j < N - M\right\}\end{aligned} \tag{4}$$

Then the blocking probability is calculated as [10]:

$$P_B\big|_{CP} = \sum\nolimits_{(i,j)\in\Psi_b} \xi_e \prod(i,j) \tag{5}$$

$$P_B\big|_{UFGC} = \sum\nolimits_{(i,j)\in\Psi_{b1}} \left(1-\beta^*\right) \xi_e \prod(i,j) + \sum\nolimits_{(i,j)\in\Psi_{b2}} \xi_e \prod(i,j) \tag{6}$$

$$\begin{aligned}P_B\big|_{LFGC} = &\sum\nolimits_{(i,j)\in\Psi_{bc}} \xi_c \prod(i,j) + (1-\beta_l) \sum\nolimits_{(i,j)\in\Psi_{be1}} \xi_e \prod(i,j) \\ &+ \sum\nolimits_{(i,j)\in\Psi_{be2}} \xi_e \prod(i,j) + (1-\beta_l) \sum\nolimits_{(i,j)\in\Psi_{be3}} \xi_e \prod(i,j)\end{aligned} \tag{7}$$

$$P_B\big|_{NCB} = \sum\nolimits_{\pi(i,j,k)\in\Psi_{bc}} \xi_c \pi(i,j,k) + \sum\nolimits_{\pi(i,j,k)\in\Psi_{b1}} \xi_e \pi(i,j,k) + \sum\nolimits_{\pi(i,j,k)\in\Psi_{b2}} \xi_e \pi(i,j,k) \tag{8}$$

where ξ_e is the probability that there are users at the edge of the cell.

Finally let ψ_d be the subsets of states where the system forces to terminate the ongoing handover call.

$$\Psi_d\big|_{CP} = \Psi_d\big|_{UFGC} = \Psi_d\big|_{LFGC} = \left\{\prod(i,j) \in \Psi_d; i = E, 1 \le j \le C\right\} \tag{9}$$

$$\Psi_d\big|_{NCB} = \left\{\pi(i,j,k) \in \Psi_d; 1 \le i < C, E - M < j \le E, k = E - j, i + j < N - M\right\} \tag{10}$$

Then the cell dropping probability is calculated as:

$$P_D\big|_{CP} = P_D\big|_{UFGC} = P_D\big|_{LFGC} = \sum\nolimits_{(i,j)\in\Psi_d} \xi_e \prod(i,j) \tag{11}$$

$$P_D\big|_{NCB} = \sum\nolimits_{\pi(i,j,k)\in\Psi_d} \xi_e \prod(i,j,k) \tag{12}$$

4. Comparison Criteria

4.1. Schemes Behavior with Hard Constraint on Dropping Probability

For a given number of channels, algorithms for determining the optimal parameters setting for each scheme will be deduced.

4.1.1. Cutoff Priority Scheme

The objective of this part is to minimize the number of guard channels T_{opt} which meets a hard constraint on dropping probability D_h. Algorithm 1 is used to find T_{opt}. It can be described from **Figure 5** as follows: At the beginning, the algorithm starts with minimum number of guard channels ($T_{opt} = 1$) then increase T_{opt} until the handover dropping probability P_D meets its constraint. If $T_{opt} = E$ while handover dropping probability P_D does not meet its constraint then the available radio resources for handover calls does not satisfy the level of QoS and the number of allocated channels to the cell is not sufficient and the algorithm terminates.

4.1.2. Uniform Fractional Guard Channel

The optimal value of new call acceptance probability β^*_{opt} that fulfils the first criteria in UFGC will be deduced. This will minimize the new call blocking probability subjected to a hard constraint on handover dropping probability or minimize P_B such that $P_D \le D_h$.

The presented algorithm in [5] will be used to find β^*_{opt}. This algorithm is given in **Figure 6** and can be described as follows: At the beginning, the algorithm considers two cases: the first the case when all channels are

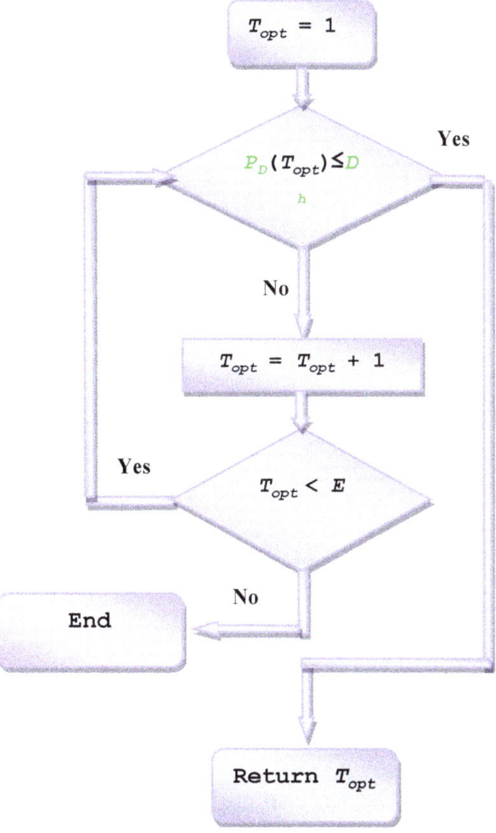

Figure 5. Algorithm 1 for adjust T to meet first criteria constraint.

exclusively used for handover calls. If the exclusive use of channels for handover calls does not satisfy the level of QoS, then the number of allocated channels to the cell is not sufficient and the algorithm terminates; the second the case when all channels are shared between handover and new calls. If the complete sharing satisfies the level of QoS, then the algorithm considers that $\beta^*_{opt} = 1$ and the algorithm terminates; otherwise the algorithm searches for the optimal value of β^*_{opt}. The search method used in this algorithm is binary search.

4.1.3. Limited Fractional Guard Channel

In LFGC, there are two parameters to adjust; the number of guard channels T and the new call acceptance probability β_l. The presence of two parameters instead of one as other schemes leads to more flexibility and more capability for scheme control but this is in price of scheme complexity.

The proposed algorithm should control the two parameters. So, it is more complex and depends on merging Algorithm 1 & 2. **Figure 7** illustrates Algorithm 3 to adjust both of T and β_l to minimize blocking probability while maintain dropping probability under specific level. The algorithm starts by $T = 1$ and $\beta_l = 0$, then the number of guard channels increases one by one till satisfy the level of QoS. While the required level is met, the new call acceptance probability β_l is adjusted using Algorithm 2 to fine tuning the scheme parameters with required constraints and obtaining minimum blocking probability.

4.1.4. New Call Bounding Scheme

The objective of this part is to obtain the maximum number of allowable new calls that maintain the dropping probability less than D_h. **Figure 8** illustrates Algorithm 4 for optimizing M_{opt} which can be described as follows: At the beginning, the algorithm starts with maximum number of allowable new call ($M_{opt} = E$) then decrease M_{opt} until the handover dropping probability P_D meets its constraint. If $M_{opt} = 0$ while handover dropping probability P_D does not meet its constraint then the available resources for handover calls does not satisfy the level of QoS and the number of allocated channels to the cell is not sufficient and the algorithm terminates.

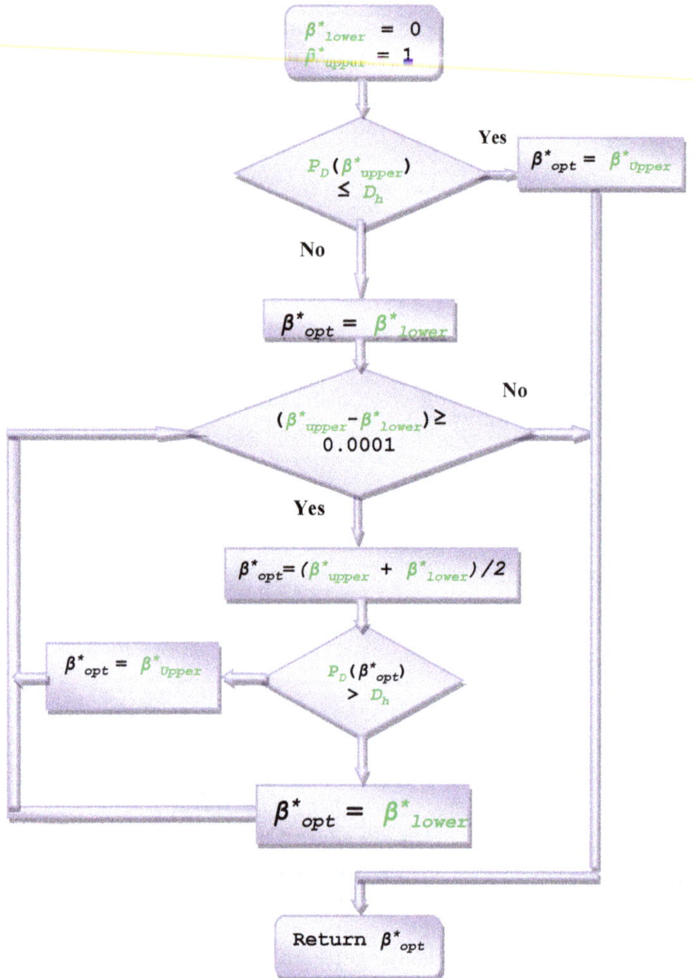

Figure 6. Algorithm 2 for adjust β^* to meet first criteria constraint.

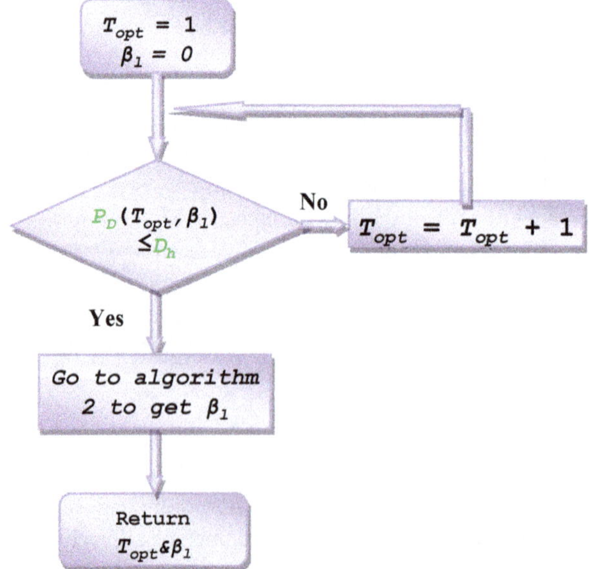

Figure 7. Algorithm 3 for adjust T & β_l to meet first criteria constraint.

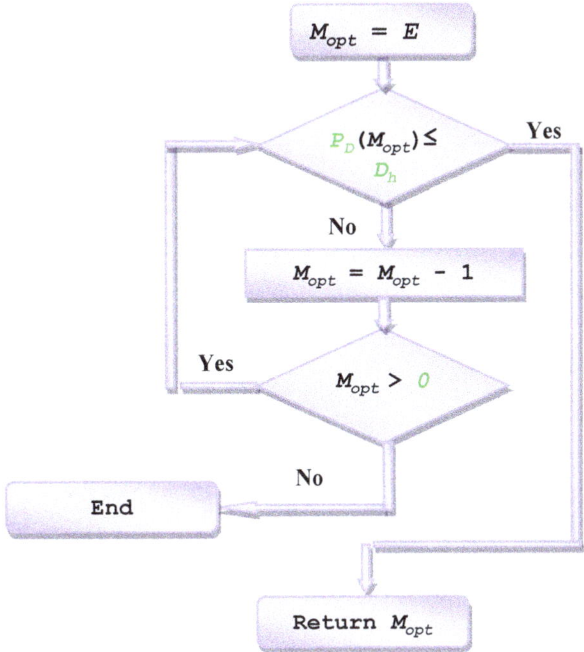

Figure 8. Algorithm 4 for adjust M to meet first criteria constraint.

4.2. Schemes Behavior with Hard Constraint on Dropping Probability and Blocking Probability

The blocking probability and the dropping probability are two contradictory performance metrics. It is very difficult to meet specific constraint in both of them by only adjust one parameter. So to solve this trade-off; the number of resources should be increased at each scheme. The objective of algorithms in this section is to minimizing the number of additive channels while satisfying the constraint in dropping and blocking probability.

The algorithms in the last section can be used to meet the first metric constraint and then we can add more resources till we meet second metric constraint. In the following, the algorithms to apply the second criteria at each scheme are concluded. Then the scheme with least resources will be elected.

4.2.1. Cutoff Priority Scheme

The objective of this algorithm is obtaining the minimum number of channels that meet the two constraints. The algorithm starts as illustrated in **Figure 9** with minimum number of channels ($E_c = 1$) and minimum number of guard channels ($T_{opt} = 1$) then increase E_c until the new call blocking probability meets its constraint. Then the algorithm starts to satisfy the dropping probability constraint without loose the adjusted blocking level. This is done by increasing T_{opt} until the handover dropping probability P_D meets its constraint without any increase in radio resources. After that a final check for blocking probability is executed and if it does not satisfy the required level, the radio resources should increase one by one again. Any tuning in T_{opt} should starts with $T_{opt} = 0$ to maintain optimality in blocking probability.

4.2.2. Uniform Fractional Guard Channel

Figure 10 depicts the algorithm to meet second criteria in UFGC. At the beginning, E_c and β^* are proposed to be 1 for both then the radio resources increase one by one till The blocking probability constraint is met. Then β^* is tuned according to Algorithm 2 to satisfy the dropping probability constraint. Applying Algorithm 2 may deviate the blocking over its constraint so additional radio resources may be added. Algorithm 2 should be applied each time channel is added.

4.2.3. Limited Fractional Guard Channel

In LFGC, there are three parameters to adjust (T, β_l, E_c). So, it is more complex and depends on merging Algo-

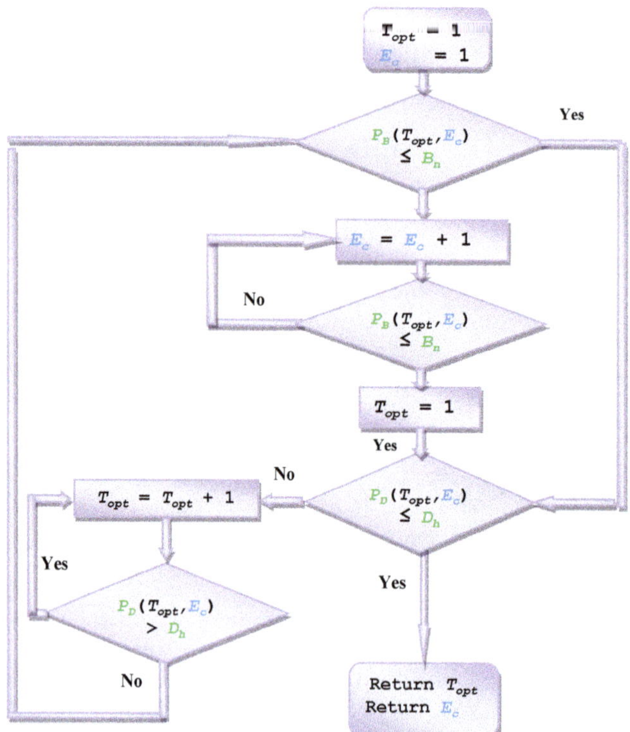

Figure 9. Algorithm 5 for obtaining minimum RBs in Cutoff priority scheme.

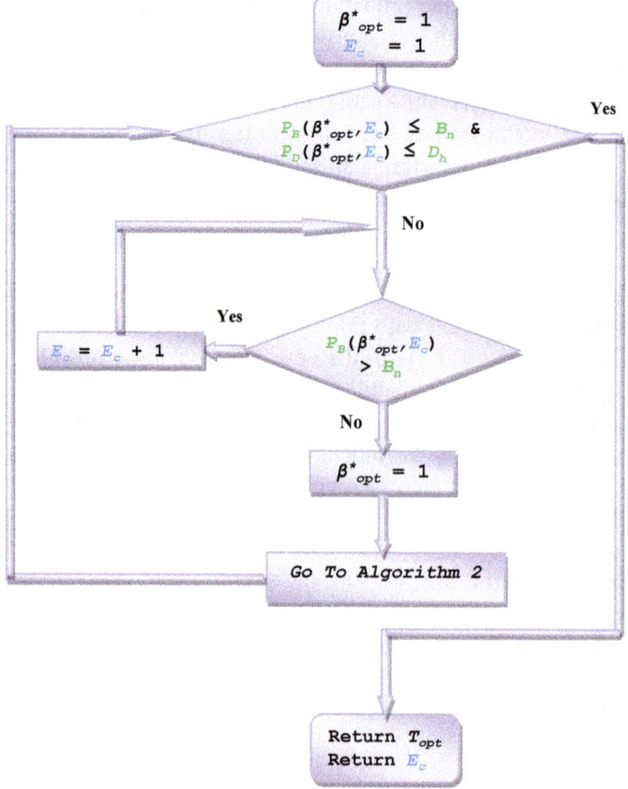

Figure 10. Algorithm 6 for obtaining minimum RBs in UFGC scheme.

rithm 3 & 4. **Figure 11** illustrates the algorithm to meet second criteria. At the beginning, the three parameters are adjusted to 1. The blocking level of QoS is satisfied by adding channel one by one. T & β_l are then adjusted using Algorithm 3 to meet the dropping level of QoS. A transition between blocking adjust and dropping adjust is executed till the two metrics are met.

4.2.4. New Call Bounding Scheme

Figure 12 illustrates applying the second criteria with new call bounding scheme. The algorithm starts with maximum number of allowable new call ($M = E$) and minimum number of radio resources ($E_c = 1$) then increase E_c until the new call blocking probability P_B meets its constraint. After that M decreases one by one till dropping probability meets its constraint. An important check should be done to M_{opt} after adding each channel in order to optimize M to be compatible with number of channels change.

5. Numerical Results and Analysis

In this section, the impact of four schemes of CAC in collaboration with SFR on cell-edge is analyzed and evaluated in comparative manner. The blocking probability P_B and dropping probability P_D are the metrics under consideration for QoS. The queuing model parameters for the presented results are as follow: the available RB in the cell (N) is 48, the ratio of cell-edge RBs to total cell RBs is 1/3; the probability that there are users at the edge of the cell ξ_e is 1/2, the mean service period (μ) is 90 seconds. The SOR parameters are $\omega = 1.05$, the convergence condition $\varepsilon = 10^{-5}$ and $n = 1000$.

5.1. Schemes Comparison Based on First Criteria

Figures 13-15 depict the blocking probability of each scheme at cell-edge with hard constraint of dropping probability D_h. The scheme parameters T_{opt}, β^*_{opt}, β_l & M_{opt} of Cutoff priority scheme, FGC and New Call Bounding scheme respectively are adjusted to meet the constraint of first criteria using Algorithms 1, 2, 3 & 4.

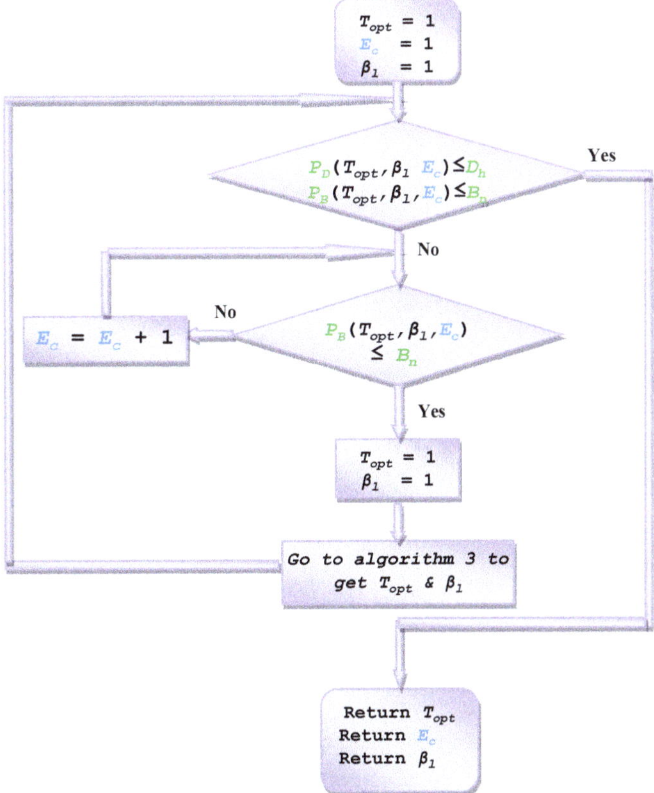

Figure 11. Algorithm 7 for obtaining minimum RBs in LFGC scheme.

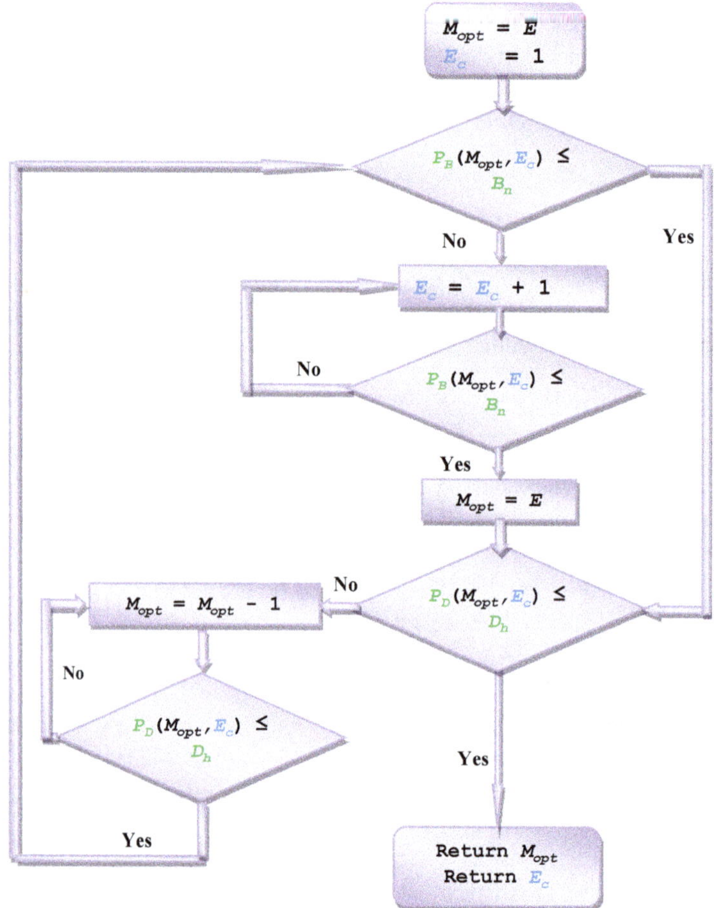

Figure 12. Algorithm 8 for obtaining minimum RBs in NCB scheme.

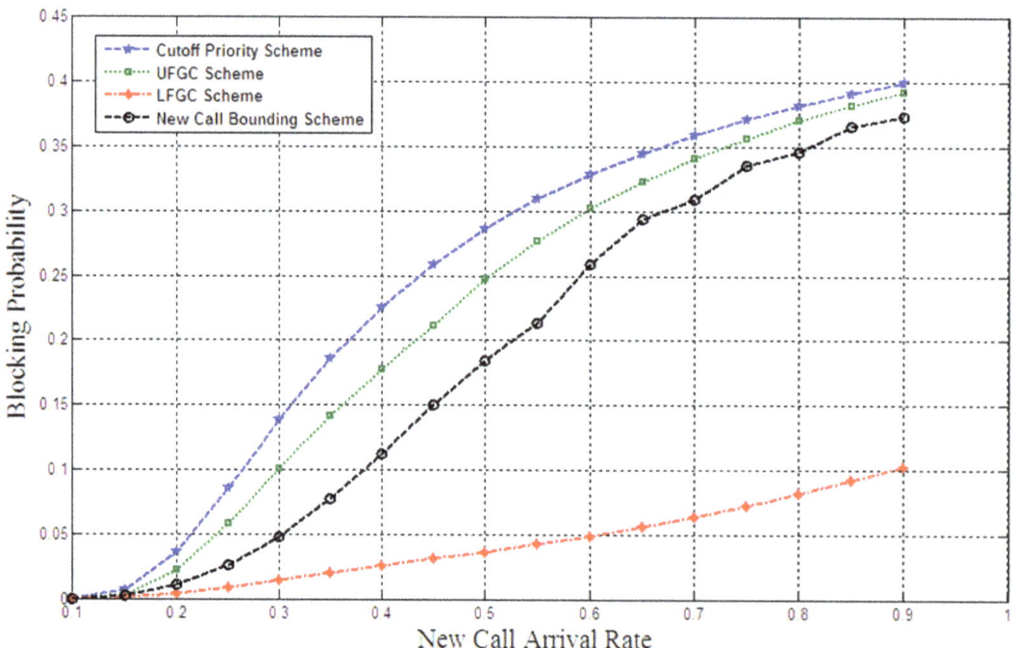

Figure 13. Comparison between the schemes based on first criteria at different new call arrival rate, λ_h = 0.25 λ_e call/sec.

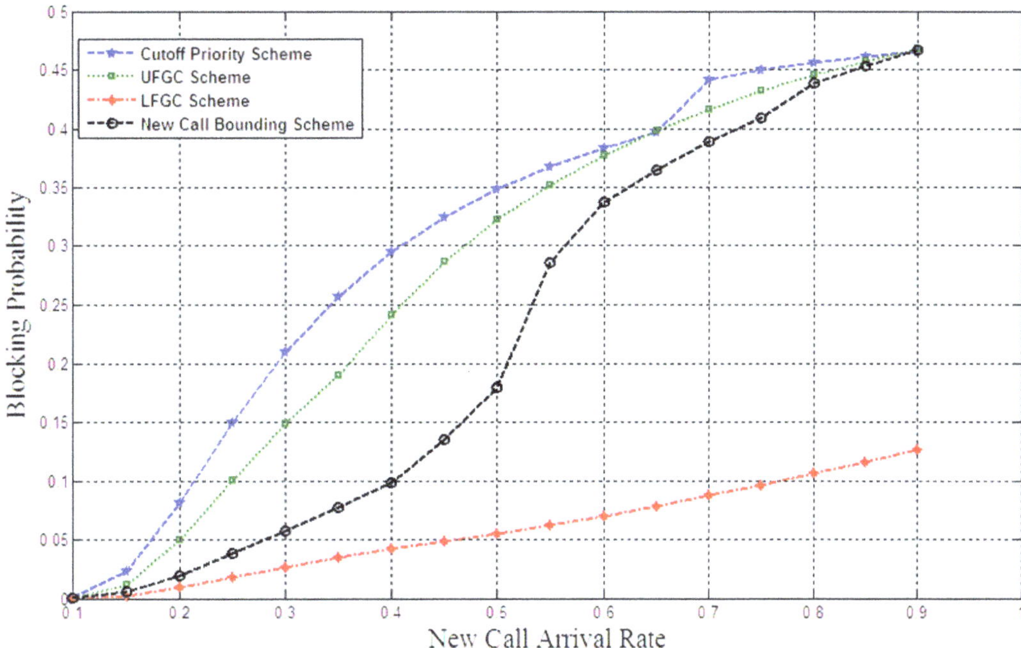

Figure 14. Comparison between the schemes based on first criteria at different new call arrival rate, $\lambda_h = 0.5 \lambda_e$ call/sec.

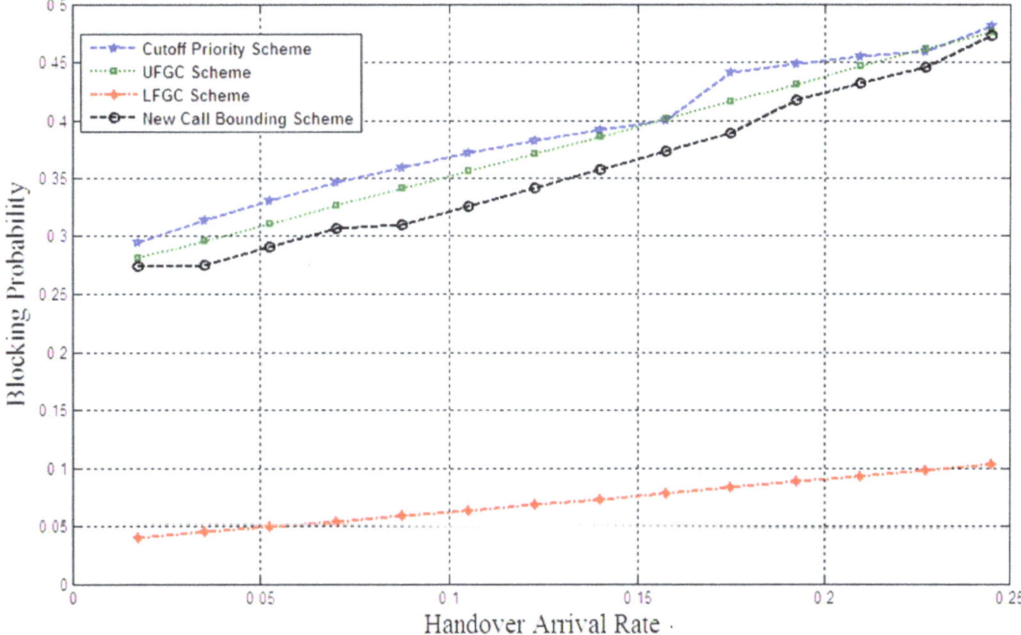

Figure 15. Comparison between the schemes based on first criteria at different handover arrival rate.

Figure 13 and **Figure 14** illustrate the blocking probability at cell-edge as a function of new call arrival rate at light and heavy handover traffic load when the maximum allowable dropping probability D_h is 0.2. The proposed handover arrival rate at **Figure 13** and **Figure 14** are 0.25 and 0.5 of new call arrival rate at cell-edge respectively. It is clear that the LFGC scheme provides minimum blocking probability than the other three schemes as illustrated from **Figure 13**.

Similar results are observed from **Figure 14** when the handover arrival rate is 0.5 of new call arrival rate. The LFGC provides best performance then NCB, UFGC and finally the cutoff priority scheme.

Figure 15 illustrates the obtained blocking probability as a function of handover arrival rate. The maximum allowable dropping probability at cell-edge in the obtained results D_h is 0.2 and The proposed new call arrival rate is 0.7 call/sec. It can be noticed from the results that the impact of increasing handover arrival rates on LFGC scheme dose not differ much than its impact in new call arrival rates. LFGC provides minimum blocking probability than the other schemes regardless of handover traffic. After LFGC with significantly great values, NCB scheme appears and then UFGC and finally cutoff priority scheme.

The comparison between the schemes according to **Figures 13-15** refers that the best network performance under hard constraint of dropping probability can be obtained using LFGC scheme at different new call and handover arrival rate. The interpretation of the obtained results as follow: each scheme has its own parameters that deeply affect the obtained performance metrics. In the cutoff priority scheme, the dominant parameter is the number of guard channels T. By adjusting T, the minimum blocking probability that maintain the dropping probability at certain level can be obtained. This can be adopted by increase (or decrease) T by unity step. Changing the number of guard channels by one channel has significant effect on both of blocking and dropping probabilities. So the unity step in cutoff priority scheme is significant large and will lead to have bad blocking probability. On the other hand, the dominant parameter in UFGC is the acceptance probability of new call β^*. Although the unity step of β^* is extremely low (fraction or less), but its effects covers all channels. So, any slight change in β^* has significant effect on blocking and dropping probabilities. The unity step in NCB is one channel and its effect is similar to cutoff priority scheme. On the contrary, LFGC is adjusted by using of the two parameters T & β_l. So, large steps can be adjusted by changing the number of guard channels while the acceptance probability can be used for fine tuning as its impact cover only one channel.

It can be concluded that to guarantee a particular level of service for admitted call while minimizing the blocking of new one, the best scheme to use is LFGC scheme then NCB and after that UFGC & cutoff priority scheme. This is valid at different traffic load of new call and handover arrival rate as illustrated in **Figures 13-15**.

5.2. Schemes Comparison Based on Second Criteria

In order to meet the contradictory constraints of dropping and blocking probability, we will follow Algorithms 4, 5 & 6 for the schemes under investigation. **Figures 14-16** depict the minimum number of required channels at cell-edge for each scheme to meet the second criteria constraint.

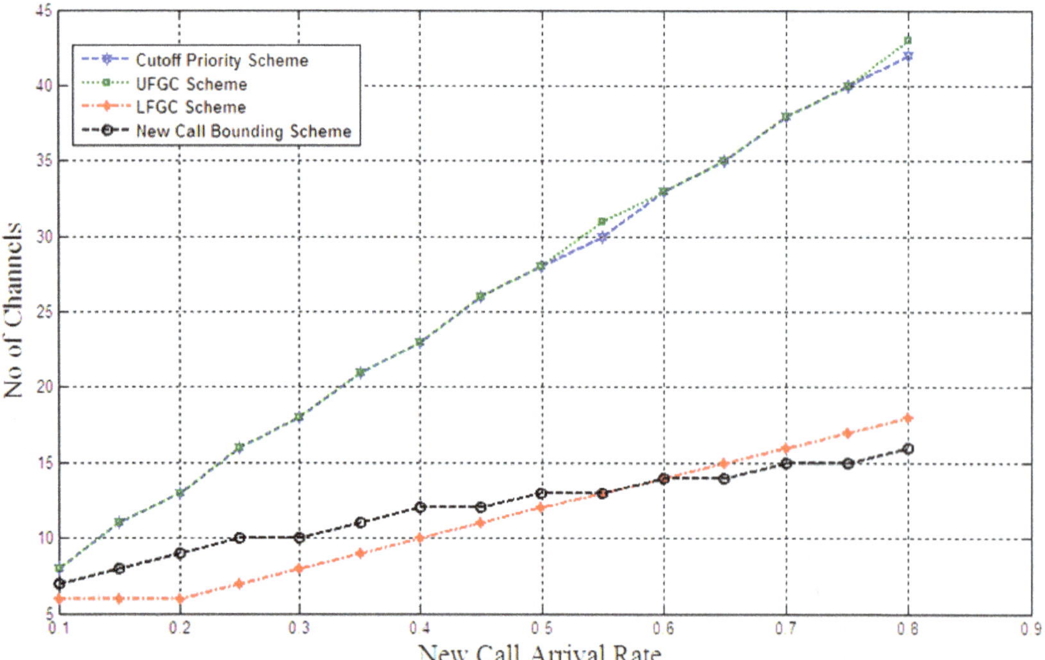

Figure 16. Comparison between the schemes based on second criteria at different new call arrival rate, $\lambda_h = 0.25\ \lambda_e$ call/sec.

Figure 16 illustrates the no of channels at different new call arrival rate. The required constraint of blocking (B_n) and dropping probability (D_h) at cell-edge in the obtained results are 0.1 & 0.2 respectively. The proposed handover arrival rate is 0.25 of new call arrival rate at cell-edge.

It can be observed from **Figure 16** that LFGC & NCB require less channels to satisfy the QoS levels with a comparative advantage to LFGC at low new call arrival rates and to NCB at large new call arrival rates. While the cutoff priority scheme and UFGC act similar at different new call arrival rates.

When the handover arrival rate increases to 0.5 of new call arrival rate, it does not affect on similarities between cutoff priority and UFGC scheme as shown in **Figure 17**. While LFGC requires the least channels regardless of the rates of new call arrival.

Figure 18 depicts the no of channels at different handover arrival rate. The proposed new call arrival rate in the obtained results is 0.7 call/sec. It can be noticed from the results that the LFGC scheme requires minimum channels, then NCB scheme. UFGC requires almost the same radio resources as the cutoff priority schemes.

Figures 16-18 illustrate that generally LFGC requires minimum radio resources to meets hard constraint of QoS. NCB provides better system performance than UFGC and Cutoff priority scheme. Finally UFGC and Cutoff priority scheme behave similar at different new call and handover rates.

Although LFGC provides best performance in the investigated criteria, but this is in price of scheme complexity. LFGC depends on two parameters as explained before. In addition, the associated algorithms to perform the criteria are more complex than other schemes as there are two parameters to adjust in case of first criteria and three parameters in case second criteria.

6. Conclusion

In this paper, the impact of call admission control schemes on cell-edge is discussed. A traffic analysis at the cell-edge is addressed by merging different CAC schemes with Soft Frequency Reuse. Four CAC schemes have investigated using queuing analysis in a comparative manner. The comparison is based on two criteria. The first criterion guarantees a particular level of service to already admitted users while trying to optimize the revenue obtained. The second criterion provides hard constraints in blocking new call and dropping handover one by control the number of radio resources. Algorithms for optimal parameters setting to meet criteria constraint for the four schemes are developed. From numerical analysis and results we can conclude that each scheme has its condi-

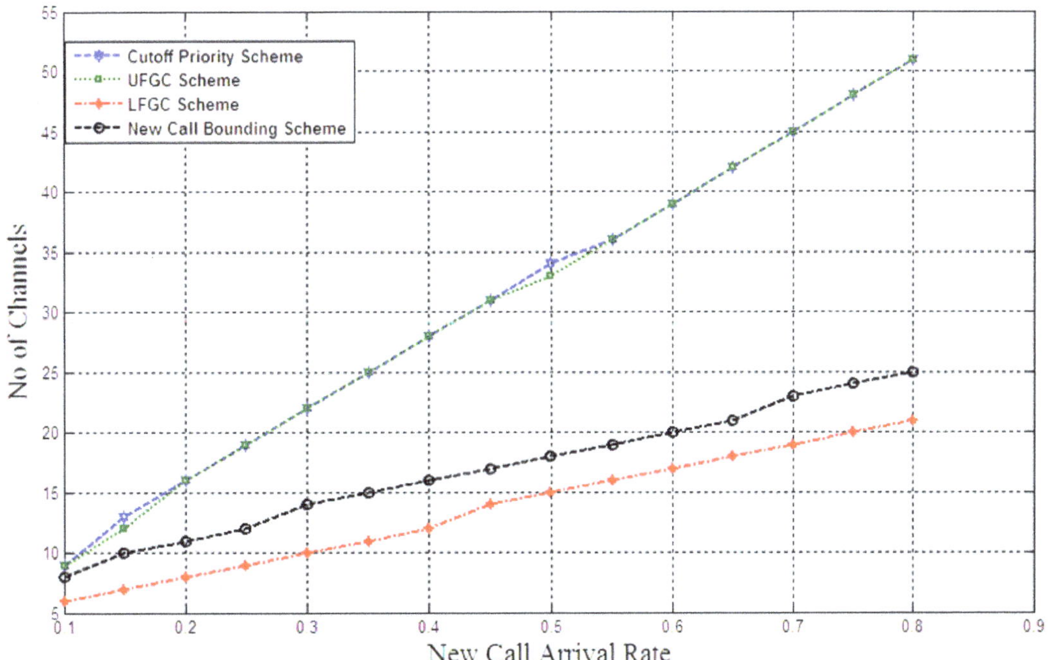

Figure 17. Comparison between the schemes based on second criteria at different new call arrival rate, $\lambda_h = 0.5\ \lambda_e$ call/sec.

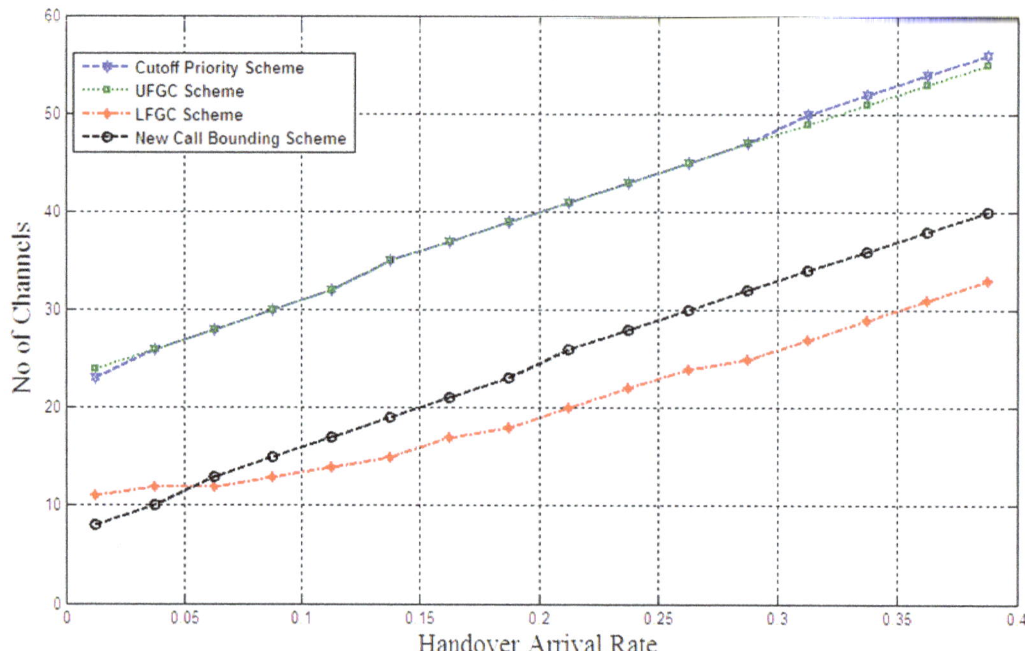

Figure 18. Comparison between the schemes based on second criteria at different handover arrival rate.

tions to provide best performance. The LFGC scheme behaves better even under heavy load of dropping and blocking rates but this is in price of the complexity. The performance of the other three schemes is close with comparative advantage to NCB.

References

[1] Huang, G.-S., *et al.* (2010) A Novel Dynamic Call Admission Control Policy for Wireless Network. *Journal of Central South University of Technology*, **17**, 110-116.

[2] Qian, M.L., *et al.* (2012) Inter-Cell Interference Coordination through Adaptive Soft Frequency Reuse in LTE Networks. *Wireless Communications and Networking Conference* (*WCNC*), 2012 *IEEE*.

[3] R1-050507: Soft Frequency Reuse Scheme for UTRAN LTE. *Huawei 3GPP TSG RAN WG1 Meeting no.*41 Athens, Greece, May 2005.

[4] Hong, D. and Rappaport, S. (1986) Traffic Model and Performance Analysis for Cellular Mobile Radio Telephone Systems with Prioritized and Nonprioritized Handoff Procedures. *IEEE Transactions on Vehicular Technology*, **35**, 77-92. http://dx.doi.org/10.1109/T-VT.1986.24076

[5] Ramjee, R., Nagarajan, R. and Towsley, D. (1996) On Optimal Call Admission Control in Cellular Networks. *Proc. 15th Annual Joint Conf. IEEE Comp. and Commun. Societies* (INFOCOM '96), **1**, 43-50.

[6] Vazquez-Avila, J., Cruz-Perez, F.A.C. and Orti-goza-Guerrero, L. (2006) Performance Analysis of Fractional Guard Channel Policies in Mobile Cellular Networks. *IEEE Transactions on Wireless Communications*, **5**, 301-305. http://dx.doi.org/10.1109/TWC.2006.1611053

[7] Goswami, V. and Swain. P.K. (2012) Analysis of Finite Population Limited Fractional Guard Channel Call Admission Scheme in Cellular Networks) *Procedia Engineering*, **30**, 759-766. http://dx.doi.org/10.1016/j.proeng.2012.01.925

[8] Kim, C., Klimenok, V.I. and Dudin. A.N. (2014) Analysis and Optimization of Guard Channel Policy in Cellular Mobile Networks with Account of Retrials. *Computers & Operations Research*, **43**, 181-190. http://dx.doi.org/10.1016/j.cor.2013.09.005

[9] Chau, T.-C., Michael Wong, K.Y. and Bo, L. (2006) Optimal Call Admission Control with QoS Guarantee in a Voice/Data Integrated Cellular Network. *IEEE Transactions on Wireless Communications*, **5**, 1133-1141.

[10] Fang, Y. (2003) Thinning Schemes for Call Admission Control in Wireless Networks. *IEEE Transactions on Computers*, **52**, 685-87. http://dx.doi.org/10.1109/TC.2003.1197135

[11] Fang, Y. and Yi, Z. (2002) Call Admission Control Schemes and Performance Analysis in Wireless Mobile Networks. *IEEE Transactions on Vehicular Technology*, **51**, 371-82.

[12] Firouzi, Z. and Hamid, B. (2009) A New Call Admission Control Scheme Based on New Call Bounding and Thinning II Schemes in Cellular Mobile Networks. *IEEE International Conference on Electro/Information Technology*, 2009. *eit'09.*

[13] Schneps-Schneppe, M. and Iversen, V.B. (2012) Call Admission Control in Cellular Networks. *Mobile Networks*, *In-Tech, Zagreb*, 111-136.

[14] Li, S., Wen, X.M., Liu, Z.J., Zheng, W. and Sun. Y. (2010) Queue Anlysis of Soft Frequency Reuse Scheme in LTE-Advanced. *Second International Conference on Computer Modeling and Simulation, 2010. ICCMS'10*, **1**, 248-252.

[15] Safwat, M.A., El-Badawy, H.M., Yehya, A. and El-Motaafy, H. (2013) Performance Assessment for LTE-Advanced Networks with Uniform Fractional Guard Channel over Soft Frequency Reuse Scheme. *Wireless Engineering and Technology*, **4**, 161. http://dx.doi.org/10.4236/wet.2013.44024

[16] Kolate, V.S., Patil, G.I. and Bhide. A.S. (2012) Call Admission Control Schemes and Handoff Prioritization in 3G Wireless Mobile Networks. *International Journal of Engineering and Innovative Technology* (*IJEIT*), **1**, 92-97.

9

Lyapunov Exponent Testing for AWGN Generator System

Hussein M. Hathal, Riyadh A. Abdulhussein, Sarmad K. Ibrahim

Electrical Engineering Department, College of Engineering, Al-Mustansiriya University, Baghdad, Iraq
Email: husssat@gmail.com, Riyadh_alhilali@yahoo.com, sarmad_8888@yahoo.com

Abstract

Additive White Gaussian Noise (AWGN) is common to every communication channel. It is statistically random radio noise characterized by a wide frequency range with regards to a signal in communication channels. In this paper, AWGN signal is generated through design an analogue circuit method, and then the multiple recursive method is also used to generate random data signal that is used for testing by Lyapunov exponent. Furthermore an algorithm for software generating of Additive White Gaussian Noise is presented. Lyapunov exponent test for chaos is used to distinguish between regular and chaotic dynamics of the generated data by the two methods. Simulation results are enhanced with the use of Microcontroller chip, since the hardware of the application is implemented by microcontroller-embedded system to obtain computerized noise generator. The results show that the generated AWGN signal by the analogue method and the multiple recursive method is chaotic which implies the random like-noise behavior.

Keywords

AWGN, Chaos Technique, Lyapunov Exponent, Microcontroller-Based System

1. Introduction

White noise is a random signal (or process) with a flat power spectral density. In other words, it is the signal contains equal power within a fixed bandwidth at any center frequency. A wide band communication circuit can be measured and tested by using this noise. AWGN signal generators are hardware cost-effective, thus this article presents a simple and inexpensive way to build white noise generators. Continuous dynamical of AWGN equations are very common in many applied sciences and engineering. Regrettably, most of these equations are nonlinear whereas most the methods of solution are linear [1]. A dynamical system is defined as a mathematical description for time evolution of a system in a state space. State space is the set of all possible states of a dynamical system, and each state corresponds to a unique trajectory in the space. Dynamical systems can be presented by attractors in phase space, and chaotic behavior sometimes occurs in these systems. In this paper, a description of chaos is offered and a description of the most important indicator of chaos, the Lyapunov exponent, is offered also. The digital computer cannot generate random numbers, and it is generally not convenient to connect the computer to some external source of random events. For most applications in statistics, engineering, and the natural sciences, this is not a disadvantage if there is some source of pseudorandom numbers, samples of which seem to be randomly drawn from some known distribution. There are many methods that have been suggested for generating such pseudorandom numbers [2]. The multiple recursive generator (MRG), which is based on a kth-order linear recurrence relation with a large prime modulus p, has become increasingly popular in recent years [3]. Maximum-period Multiple Recursive Generators (MRGs) of order k have become popular pseudorandom number generators (PRNGs) in many areas of applications because of the great properties of equidistribution over spaces up to k dimensions, long periods, and excellent empirical performances [4]. The performance of an MRG depends on the associated order k, the prime modulus p, and the multipliers used in the recurrence equation [3]. For random numbers to be useful in general applications, their distribution must be known, and they usually must be identically and independently distributed (i.i.d.) [2]. Each sample is generated in response to the user's request, so the samples are unique. Physical processes on the user's computer can also be used to generate random data. There are many ways to do this. For example, Davis, Ihaka, and Fenstermacher (1994) describe a method of using randomness in the air turbulence of disk drives [2]. A basic and generally accepted model for thermal noise in communication channels is the set of assumptions that:

- The noise is additive, *i.e.*, the received signal equals the transmit signal plus some noise, where the noise is statistically independent of the signal.
- The noise is white, *i.e.*, the power spectral density is flat, so the autocorrelation of the noise in time domain is zero for any non-zero time offset.
- The noise samples have a Gaussian distribution. The operation of the analogue system is based on the noise generated by the Zener breakdown phenomenon in an inversely polarized diode as shown in **Figure 1** [1].

2. Generation of Random Numbers

Since many statistical methods rely on random samples, applied statisticians often need a source of "random numbers". The use of random numbers in statistics has expanded beyond random sampling or random assignment of treatments to experimental units [2]. The digital computer cannot generate random numbers, and it is generally not convenient to connect the computer to some external source of random events. There are many methods that have been suggested for generating pseudorandom numbers.

2.1. Multiple Recursion Method

It is the most useful type of generator of pseudorandom processes updates a current sequence of numbers in a manner that appears to be random. Such a deterministic generator, f, yields numbers recursively, in a fixed sequence. The previous k numbers (often just the single previous number) determine(s) the next number [5]:

$$x_i = f\left(x_{i-1}, \cdots, x_{i-k}\right)$$

The number of previous numbers used, k, is called the "order" of the generator. The set of values at the start of the recursion is called the seed. Each time the recursion is begun with the same seed, the same sequence is generated. The length of the sequence prior to beginning to repeat is called the period or cycle length. The standard methods of generating pseudorandom numbers use modular reduction in congruential relationships. There are currently two basic techniques in common use for generating uniform random numbers: congruential methods and feedback shift register methods. The basic relation of modular arithmetic is equivalence modulo m, where m is some integer. This is also called congruence modulo m. Two numbers are said to be equivalent, or

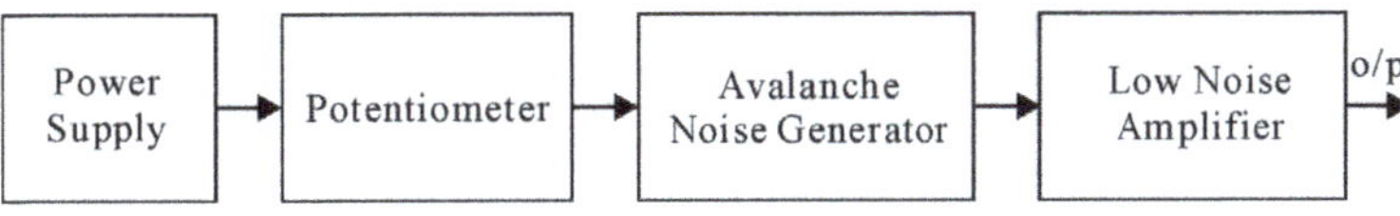

Figure 1. Additive white Gaussian noise generator system.

congruent, modulo m if their difference is an integer evenly divisible by m. For a and b, this relation is written as [4] [5]:

$$a \equiv b \bmod m$$

A simple extension of the multiplicative congruential generator is to use multiples of the previous k values to generate the next one:

$$x_i = f\left(a_1 x_{i-1} + a_2 x_{i-2} + \cdots + a_k x_{i-k}\right)(\bmod m) \tag{1}$$

When $k > 1$, this is sometimes called a "multiple recursive" multiplicative congruential generator. The number of previous numbers used, k, is called the "order" of the generator. (If $k = 1$, it is just a multiplicative congruential generator). The period of a multiple recursive generator can be much longer than that of a simple multiplicative generator.

2.2. Additive White Gaussian Noise Generator Method

In every communications channel noise is always present. Furthermore, other disturbances which cause the transmitted signal to change, such as fading, may also be present. In most cases, AWGN is used to evaluate the performance of a communication system in a noisy channel [6]. Additive white Gaussian noise (AWGN) is a channel model in which the only impairment to communication is a linear addition of wideband or white noise with a constant spectral density (expressed as watts per hertz of bandwidth) and a Gaussian distribution of amplitude. The model doesn't account for fading, frequency selectivity, interference, nonlinearity or dispersion. However, it produces simple and tractable mathematical models which are useful for gaining insight into the underlying behavior of a system before these other phenomena are considered. Wideband Gaussian noise comes from many natural sources, such as the thermal vibrations of atoms in conductors (referred to as thermal noise or Johnson-Nyquist noise), shot noise, black body radiation from the earth and other warm objects, and from celestial sources such as the Sun [1]. The AWGN channel is a good model for many satellite and deep space communication links. It is not a good model for most terrestrial links because of multipath, terrain blocking, interference, etc. However, for terrestrial path modeling, AWGN is commonly used to simulate background noise of the channel under study, in addition to multipath, terrain blocking, interference, ground clutter and self-interference that modern radio systems encounter in terrestrial operation.

3. Description of Chaos

The greatest power of science lies in its ability to relate causes to effects. The evolution of a dynamical system may occur either in continuous time or in discrete time. The former is called flow, and the latter is called map. For nonlinear systems, continuous flow and discrete map are also two mathematical concepts used to model chaotic behavior [7]. Chaos can be observed in a time series or in a phase space plot, but this is not very accurate. At present, there are a number of methods employed to detect chaos in dynamical systems, such as power spectrum analysis, the 0 - 1 test, calculating the Lyapunov exponents, and so on.

Lyapunov Exponent

The Lyapunov exponents of a system under consideration characterise the nature of that particular system. They are perhaps the most powerful diagnostic in determining whether the system is chaotic or not. Furthermore, Lyapunov exponents are not only used to determine whether the system is chaotic or not, but also to determine how chaotic it is. The Lyapunov exponents characterise the system in the following manner. Suppose that do is a measure of the distance among two initial conditions of the two structurally identical chaotic systems. Then, after some small amount of time the new distance is [8]:

$$\mathrm{d}(t) = do2^{\lambda t} \tag{2}$$

where λ denotes the Lyapunov exponent.

For chaotic maps, Equation (2), is rewritten in the form of Equation (3):

$$d_n = do2^{\Lambda t} \tag{3}$$

where Λ denotes the Lyapunov exponent and n a single iteration of a map. The choice of base 2 in Equations

(2) and (3) is arbitrary [8]. The Lyapunov exponents of Equations (2) and (3) are known as local Lyapunov exponents as they measure the divergence at one point on a trajectory (orbit). In order to obtain a global Lyapunov exponent the exponential growth at many points along a trajectory (orbit) must be measured and averaged. The largest Lyapunov exponents of a dynamical system have been developed to be the most important number of invariants, for characterizing chaos. Mathematically, the Lyapunov exponent of a dynamical system is a measure that used to characterize the rate of separation of infinitesimally close trajectories [9]. The Lyapunov exponent is used also to determine the level of chaos within it. In order to determine a global Lyapunov exponent, the exponential growth at many points along an orbit must be calculated and then averaged. The largest Lyapunov exponent is described as [2] [7] [9]:

$$\lambda = \frac{1}{t_{N-t_0}} \sum_{K=1}^{N} \log_2 \frac{\mathrm{d}(t_k)}{\mathrm{d}_0(t_k - 1)} \tag{4}$$

where $\left(\frac{\mathrm{d}(\ )}{\mathrm{d}_0(\ )}\right)$ means the growth rate of the distance between neighboring trajectories. And the largest lyapunov exponent Λ for chaotic maps is described by Equation (5):

$$\lambda = \lim_{n\to\infty} \frac{1}{N} \sum_{K\to\infty}^{N} \log_2 \frac{\mathrm{d}f(x_n)}{\mathrm{d}x} \tag{5}$$

where $(x_n) = x_{n+1}$.

A dynamical system is said to be chaotic if the Lyapunov exponent λ is positive (larger than zero). When the Lyapunov exponent λ is negative, this implies that the dynamical system is a fixed point or a periodic cycle. For a chaotic system, there are many Lyapunov exponent equal to its dimension. The logistic map (one dimensional map) given by:

$$X_{n+1} = 1 - X_n^2 \tag{6}$$

has a one positive Lyapunov exponent. The Hénon map (two dimensional maps) of has two Lyapunov exponents, one positive and the other negative. The Lorenz chaotic flow of (three dimensional maps) has three Lyapunov exponents, one positive, one negative and one equal to zero [2] [7].

4. Testing of Random Noise Generator Results Based on Chaos Technique

If all of the relevant information in the system is well known, the calculation of the theoretical Lyapunov exponents can be based on the equations of that system. This method includes repeatedly using equation linearization. In reality, the equations in a given system are not easy to obtain. However, time series data sets can easily be acquired. When only time series data are recorded, the calculation method introduced above is impossible to use. Alan Wolf [8] offers an algorithm to use in order to compute the LLE from time series data. Moreover, the dimension and origin of the dynamical system [3] and the form of the underlying equations are irrelevant. The input is the time series data and the output is 0 or 1, depending on whether the dynamics is non-chaotic or chaotic. The test is universally applicable to any deterministic dynamical system.

Description of the Lyapunov Exponent Test Algorithm for Chaos

This procedure can briefly be summarized as follows:

1) Choose and put the initial values of the multiple recursive method.
2) Calculate the next step of the multiple recursive method.
3) Calculate the derivative of the multiple recursive equation.
4) Calculate Lyapunov exponent value.
5) Calculate the Largest value of Lyapunov exponent, λ.
6) Test the λ value.
7) If > 0, then the system behavior is chaotic, and if <0, then the system behavior is periodic (*i.e.*, non-chaotic).

The flow chart for calculating Lyapunov exponent values is shown in **Figure 2**.

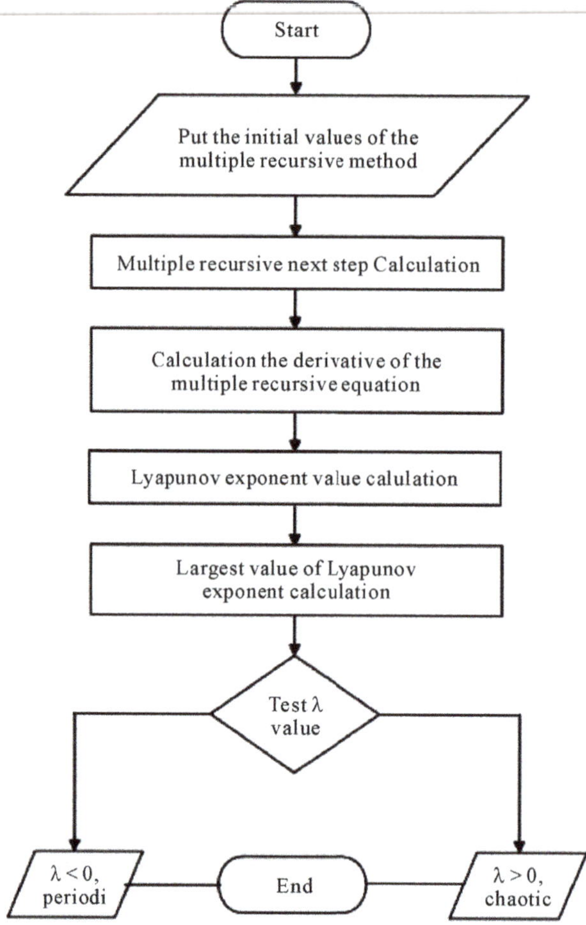

Figure 2. Flow chart of Lyapunov exponent test algorithm for chaos.

5. AWGN Generator Simulation Results

In this section, the simulation results are given to verify the theoretical results by implementing the AWGN generator system of **Figure 3**. The data that used for testing is generated by applying the multiple recursive method through the PIC microcontroller chip, and then it tested by using Lyapunov exponent chaos test.

The noise amplitude versus the number of samples is obtained by applying the multiple recursive method as shown in **Figure 4**, while **Figure 5** represents the amplitude of the power spectral density opposes the number of samples.

In order to test the AWGN signal that is generated by the PIC microcontroller, so Lyapunov exponent chaos test is applied. When the time series data, *i.e.*, AWGN signal, is entered as vector of values, then an embedding lag of state space reconstruction is initialized such that an embedding dimension must be calculated. If embedding dimension be selected correctly, then there would have smooth part (or fairly horizontal) on the Lyapunov exponent curve. So if there is no smooth section on the curve, it is better to try with other embedding dimensions. When there is not any information about proper value of embedding dimension, then should let it zero (0). In this case code automatically selects proper m by False Nearest Neighbors (or FNN) method, if this method fails due to high noise in data, the code will use another method named symplectic geometry. This method is a graphical in nature however it use test for selection of vector based on variance change of eigenvalues. The symplectic geometry method for determination embedded dimension is shown in **Figure 6**.

Figure 7 illustrates the nonlinear regression Layapunov exponents, for multiple recursive method the Lyapunov exponent λ is positive (*i.e.*, larger than zero) therefore, the generated data is chaotic. Also, for AWGN generator, the Layapunov exponents took positive values therefore; the generated data is chaotic too.

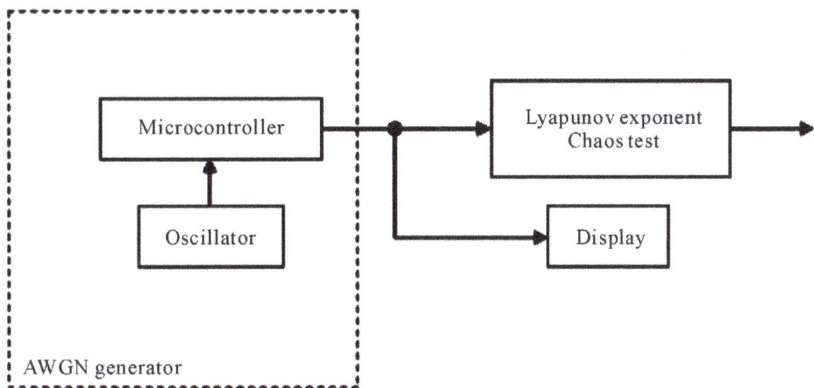

Figure 3. Block diagram of AWGN generator testing system.

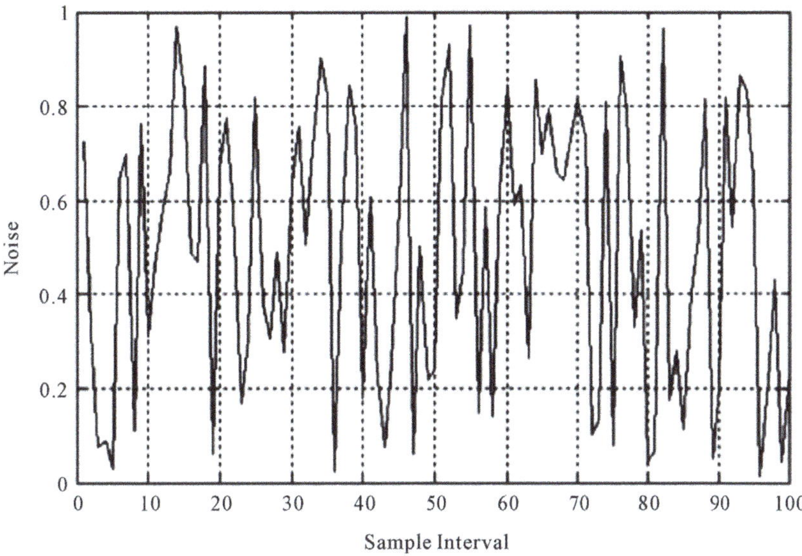

Figure 4. Noise amplitude versus number of samples.

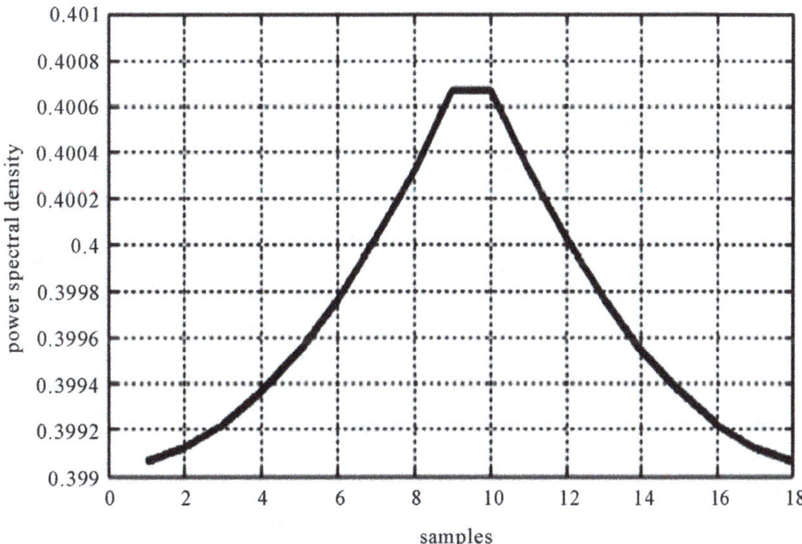

Figure 5. Power spectral density amplitude versus number of samples.

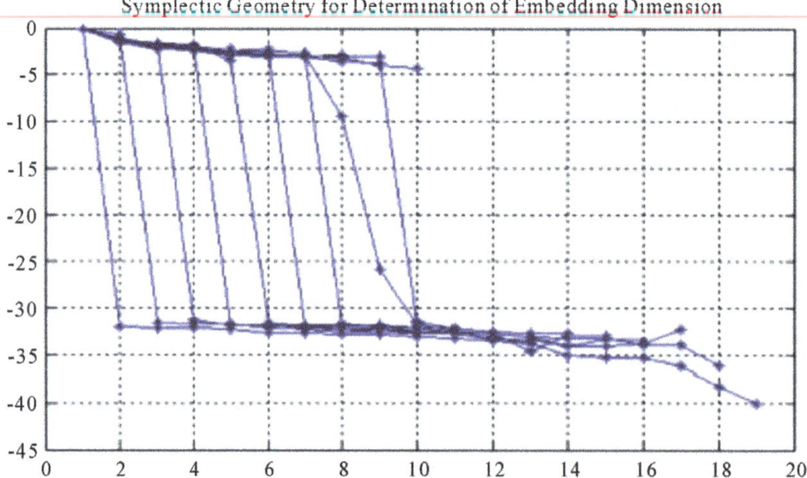

Figure 6. Symplectic geometry method for determination embedded dimension.

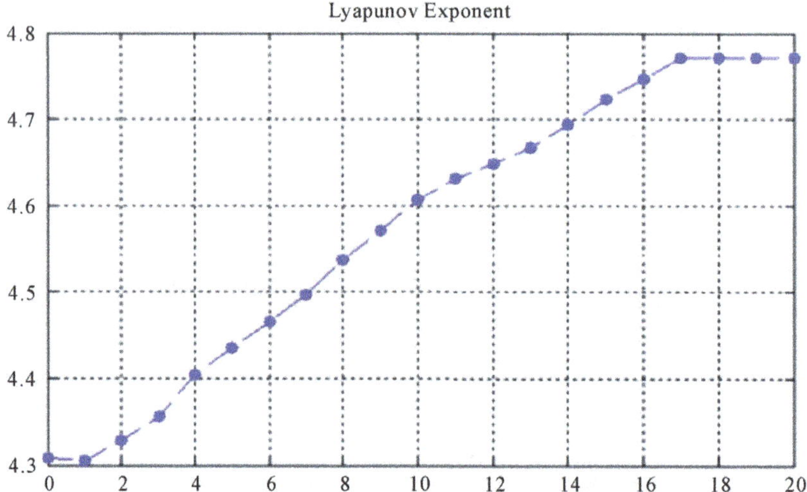

Figure 7. Large Lyapunov exponent for AWGN signal generated by multiple recursive method.

6. Conclusions

Chaos is a developing research topic. In this paper, Lyapunov exponent test for Multiple Recursive Method and additive white Gaussian noise generator is implemented. The application of Lyapunov exponent in real world dynamical systems is rarely developed. So, the method of testing by using Lyapunov exponent is proposed in this paper work. In order to obtain Lyapunov exponent from the experimental AWGN time series data, a method mainly based on phase space reconstruction is demonstrated. The phase space reconstruction requires a time delay and an embedding dimension. The digital software implementation of AWGN generators achieved; reliability and accuracy, where digital circuits have more reliability than analogue circuits, and wide number generation range, so in digital circuits, the range of random numbers is more wide than the range of random numbers obtained by analogue circuits, since there are various algebraic methods can be implemented by software.

Lyapunov exponent test for chaos is used for the analysis of nonlinear discrete and continues deterministic dynamical systems, the test distinguish between regular and chaotic dynamics. This distinction is extremely clear by means of the diagnostic variable λ, which has values either positive or negative. In particular the equations of the underlying dynamical system do not need to be known, and there is no practical restriction on the dimension of the underlying vector field. Simulation results show that the AWGN signal that is generated by the Multiple Recursive Method behaves randomly and can be used in simulating real world applications.

References

[1] Xiao, P. (2009) Effect of Additive White Gaussian Noise (AWGN) on the Transmitted Data. 1-15.

[2] Gentle, J.E. (2003) Random Number Generation and Monte Carlo Methods. 2nd Edition, Springer, Berlin.

[3] Deng, L.-Y., Shiau, J.-J.H. and Shing Lu, H.H. (2011) Large-Order Multiple Recursive Generators with Modulus 231-1. *INFORMS Journal on Computing*, 1-12.

[4] Deng, L.-Y., Shiau, J.-J.H. and Shing Lu, H.H. (2012) Efficient Computer Search of Large-Order Multiple Recursive Pseudo-Random Number Generators. *Journal of Computational and Applied Mathematics*, **236**, 3228-3237. http://dx.doi.org/10.1016/j.cam.2012.02.023

[5] Giacobazzi, R. and Ranzato, F. (1998) Some Properties of Complete Multiple Recursive Lattices. *Algebra Universe*, **40**, 189-200.

[6] Jovic, B. (2011) Synchronization Techniques for Chaotic Communication Systems. Springer, Berlin. http://dx.doi.org/10.1007/978-3-642-21849-1

[7] Aziz, M.M. and Faraj, M.N. (2012) Numerical and Chaotic Analysis of CHUA'S CIRCUT. *Journal of Emerging Trends in Computing and Information Sciences*, **3**.

[8] Alligood, K.T., Sauer, T.D. and Yorke, J.A. (1996) Chaos: An Introduction to Dynamical Systems. Springer, Berlin.

[9] Sun, Y. (2011) Fault Detection in Dynamic Systems Using the Largest Lyapunov Exponent. Thesis, Texas.

10

Implications of SSO Solutions on Cloud Applications

Mohamed Watfa*, Shakir Khan, Ali Radmehr

Faculty of Engineering and Information Sciences, University of Wollongong, Dubai, UAE
Email: *MohamedWatfa@uowdubai.ac.ae, ShakirKhan@uowdubai.ac.ae, ARadmehr@uowdubai.ac.ae

Abstract

The trend in businesses is moving towards a single browser tool on portable devices to access cloud applications which would increase portability but at the same time would introduce security vulnerabilities. This resulted in the need for several layers of password authentications for cloud applications access. Single Sign-On (SSO) is a tool of access control of multiple software systems. This research explores the effects and implications of SSO solutions on cloud applications. We utilize a new framework of different attributes developed by acquiring IT experts' opinions through extensive interviews to expand significant strategic parameters at the workplace. The framework was further tested using data collected from a sample of 400+ users in the UAE.

Keywords

Single Sign-On, Cloud Computing, Security, Value Added Services

1. Introduction

Cloud computing is a fast growing branch of information technology which is highly on demand. It is an opportunity to enhance capacity and capabilities based on hardware resources at distant locations with broader network access and reliable sources of data storage. It provides secure and quick access for applications to the cloud users on the multi-platform architectures in order to ease the use of dynamic request over the internet. Nowadays, various industries are exploiting cloud computing to facilitate business needs. The trend is moving towards a single browser tool on portable devices to access cloud applications and perform most of the business functions with the help of smart devices over the cloud through the internet. Security has been a major concern for cloud computing due to unavailability of an IT Infrastructure, lack of application manageability and control, and multiple accesses to the platform. The traditional client and server authentication process was adopted by

*Corresponding author.

different application vendors such as Microsoft, Oracle and Citrix. They increased the complexity to remember multiple user credentials for various applications which resulted in the need for having Single Sign-On authentication (SSO). Once client-server applications are transformed into web based applications, SSO was the only solution that can facilitate the broader access for these cloud applications. This research study is important due to the high usage of Single Sign-On features in cloud applications where the IT strategy demands the integration of this technique into business and organizational related applications.

2. Related Work

In recent years, it is becoming very common to use your single social login credential for logging into different websites. Although this is still increasing rapidly but privacy concerns and implications should also be considered as well. One of most successful single sign-on is OpenID [1] which provides a framework for deploying flexible centralized user authentication for web applications. In OpenID, user provides a variety of identity which may be any website or web-based application where user already has an user account (e.g. Google). Research studies on SSO suggest that while it greatly improves user experience by relieving them of the burden of remembering multiple user ids and passwords, it also noticeably reduces help desk calls, and improves security. However, it also cautions that an SSO product is not a cure-all. Without very careful planning, implementation and verification, SSO products can introduce new security holes [2]. There are several works discussing the implications of SSO on several factors. In [3] [4], four different methods were discussed in order to sort out issues associated with SSO and service continuity maintenance. Facebook was used as an example case to discuss all the undertaken privacy issues arising whenever you use your Facebook account to access many other websites. Several disadvantages including loss of anonymity, revealing of user's social cycle, loss of track, propagation of advertisements, disclosure of user's credentials and reverse Single Sign-On semantics were highlighted. Technically SSO appears like a simple solution; however, its implementation reveals hidden complexities. For example, a study by Josang *et al.* [5], analyzed some of the trust requirements resulting from various identity management models. The authors found that trust requirements for a particular authentication technology are directly correlated to the user's perceived risk exposure and that this trust is necessary for user acceptance of the technology. With respect to SSO technologies, it suggests that trust relationships between federated parties are harder to establish particularly if one party has a significantly higher risk exposure than the other.

Meniya *et al.* [6] bridged the gap between different cloud applications by introducing the federation of open cloud and invited different cloud service providers to be part of this body where only identical SSO is accepted across all cloud services providers and facilitated through interoperability. Zhu *et al.* [7] described the problems faced by web applications when different web services were offered to a viewer like: news management system, video on demand system, bulletin board system and the laboratory management system. As each system user has its own authentication system and verifying processing logic, this results in data inconsistency.

3. Research Objectives & Proposed Framework

In this paper, we focus on the effects and implications of Single Sign-On authentication in cloud applications from different viewpoints. We study the nature of current web applications and the benefits that can be utilized through single sign authentication. This research starts with the analysis of currently implemented single sign features across several companies. On the basis of this judgment, we would be able to evaluate whether companies can apply Single Sign-On as an effective solution to improve accessibility to their cloud applications and to increase productivity in the organizational processes. Our study is non-contrived and our primary source of data includes the people working in the IT industry in the UAE. We used a marketing database to find the IT companies utilizing Single Sign-On solutions. We conducted interviews with ten IT specialists in four UAE IT organizations. Our analysis is based on user preferences, productivity, efficiency, accessibility and some other key attributes related to our proposed framework as depicted in **Figure 1**.

Our proposed framework consists of three strategic dimensions summarized as follows.

3.1. IT Strategy—Data Protection

1) Security:

- Passive mode of authentication: It is difficult to impersonate the actual user credentials because the system is designed in such a way that accepts the response from the SSO assistant.

Figure 1. Proposed framework reflecting three different strategic dimensions.

- Dual factor authentication: Authentication through both the user and the SSO assistant where each session has a unique identity.
- Secure: Authenticated Single Sign-On access to the applications they need when they are outside the corporate firewall.

2) Privacy: To increase privacy control and access resources with privacy protection.

3) Reliability: Preferably zero down time including effective control of stolen Single Sign-On credentials.

3.2. Business Strategy—Critical Success Factor

1) Productivity: Continuous flow of tasks with increased number of assignments performed.

2) Efficiency: Time savings and more accurate results in a committed time frame.

3) Cost Effectiveness: Centralized authentication server for Single Sign-On has evident advantages over distributed authentication servers including user productivity enhancement resulting in higher revenues.

4) Usability: Back end plug-ins eases the access to web apps with greater user convenience.

3.3. Organizational Strategy—Effective Management

1) Manageability: Centralized management of user credentials with one time session identity utilized and two ways of credentials handling, one handled by the plug-in and the other by the user through the system.

2) Accessibility: Easily accessible from any web browser.

3) Availability: Unified authorization provides access to multiple web-apps with maintained service continuity in case of authentication server failure.

4. Data Collection and Analysis

In this section, we will introduce our descriptive and inferential analysis. Our sample included interviews of ten IT managers and a survey of 400 IT professionals utilizing SSO.

As summarized in **Figure 2**, the majority of our samples were extremely satisfied with the security of SSO. Also, it is evident that the majority witnessed no or rare effect on the interruption of the services through SSO whereas only about 10% believed in high frequency of failures. Also, about 70% of our sample believed in higher service availability after transitioning into SSO solution whereas the rest didn't notice any or noticed minimal impact on service availability. We also investigated the following hypotheses and performed inferential hypothesis testing using SPSS:

H1: There is an association between SSO mechanism or traditional mode of authentication and service availability.

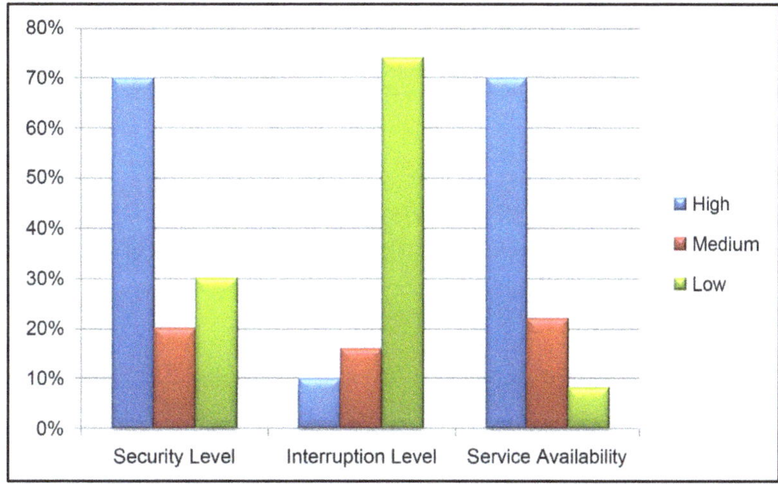

Figure 2. Results from the sample regarding security, interruption and service availability satisfaction levels of Single Sign-On solutions on cloud applications.

Result: Reject the null hypothesis and conclude that there is sufficient evidence to say that service availability is associated with SSO usage.

H2: Single Sign-On solution will be widely acceptable once all the cloud service providers come under the single federation of cloud computing.

Result: Reject the null hypothesis and conclude that there is sufficient evidence to say that the single federation of cloud computing will affect the acceptance of SSO solutions.

5. Conclusions and Recommendations

This study highlighted the effects and implications of Single Sign-On solutions in cloud applications using our proposed framework. More specifically, the following major points were concluded from the detailed interviews and surveys of our sample of 400+ IT professionals utilizing SSO authentication solutions.

5.1. Business Strategy—Critical Success Factors

The following major points were concluded regarding utilizing Single Sign-On solutions where the demand is high for productivity to speed up organizational processes.

- Reduced time to access and log on to IT systems.
- Reduced helpdesk contacts for password resets.
- Reduction in out of hours password "lock outs".
- Reduced time to switch between applications affecting positively on productivity.
- Support "terms and conditions" for access to critical business systems.

More specifically employees having more interactions with customers gain more benefits from Single Sign-On solutions to increase productivity. Moreover, Single Sign-On solutions reduce the pain for users to access their applications and data from different locations leading to higher performance and better usage. Also, the respondents agreed that SSO can lead to the following competitive advantages:

- Easing the process of job and duty transfer among the employees.
- Maintaining confidentiality of data on staff exit.
- Making the business environment more secure, manageable and credible.
- Leveraging the company productivity by minimizing the need of multiple accounts.
- Making new employee setup faster.
- Making remote assistance more effective and efficient.

5.2. IT Strategy—Data Protection

The below experiences were concluded from our selected sample in measuring the privacy control of the users

when accessing cloud services through SSO:
- The common feedback that users are always concerned about is privacy.
- Credentials are stored and encrypted within the central authentication server with no data leakage.
- Unified access for all.
- User details are controlled in one location.
- SSO by itself cannot guarantee the integrity of the data.
- SSO can remove the need to re-authenticate, by logging in user tickets.

The following recommendations were essential to build strong control over privacy with the SSO mechanism:
- Eliminate password sharing for individual applications by using SSO.
- Develop a strong SSO usage policy and then stick to that policy.
- Each user must have their allocated storage with encryption.
- Cloud providers need to implement multiple factor authentications to use all services seamlessly.

5.3. Organizational Strategy—Effective Management

The implementation of SSO for cloud applications can make manageability of access control more effective as follows:
- Less administrative overhead and configurations.
- Easy deployment of applications through SSO.
- User login can be monitored in real time.
- Once SSO is in place, organizations will have the King Key access.
- Easy implementation of governance policies including centralized audit and reporting.
- Increased efficiency and reduced efforts with more discipline in the attitude of the IT staff.
- Less chances and burdens for users to forget their passwords.

To conclude, with the increasing number of cloud applications in the business environment, the need to have a Single Sign-On to access all of those applications at once in order to accomplish different tasks in a shorter period of time will be growing. To make SSO as a portable and widely applicable solution for most of the cloud applications, it is suggested to reduce compatibility issues between different cloud vendors in order to build a uniform structure which is feasible to the needs of the organizations. Through SSO, an organization can obtain improved access with less complexity and increased productivity with better safeguard against any malicious activity. As technological advancements in central hardware authentication systems continue to grow with the flow of approvals in the organizational hierarchy, SSO solutions will be an added value for a faster and safer access.

References

[1] OpenID. www.openid.net

[2] Anchan, D. and Pegah, M. (2003) Regaining Single Sign-On Taming the Beast. *Proceedings of the 31st Annual ACM SIGUCCS Conference on User Services*, 166-171. http://dx.doi.org/10.1145/947469.947514

[3] Kakizaki, Y., Maeda, K. and Iwamura, K. (2011) Identity Continuance in Single Sign-On with Authentication Server Failure. *Proceedings of the 5th International Conference on Innovative Mobile and Internet Services in Ubiquitous Computing* (*IMIS*-2011), Seoul, 30 June-2 July 2011, 597-602.

[4] Kontaxis, G., Polychronakis, M. and Markatos, P. (2012) Minimizing Information Disclosure to Third Parties in Social Login Platforms. *International Journal of Information Security*, **11**, 321-332. http://dx.doi.org/10.1007/s10207-012-0173-6

[5] Jøsang, A., Fabre, J., *et al.* (2005) Trust Requirements in Identity Management. *Australasian Information Security Workshop*, Newcastle, 99-108.

[6] Meniya, A. and Jethva, H. (2012) Single-Sign-On (SSO) across Open Cloud Computing Federation. *International Journal of Engineering Research and Applications*, **2**, 891-895.

[7] Zhu, F. and Diao, H. (2010) Single Sign-On Assistant: An Authentication Broker for Web Applications. *3rd International Conference on Knowledge Discovery and Data Mining*, 2010, 146-149.

11

Optimizing Packet Generation Rate for Multiple Hops WBAN with CSMA/CA Based on IEEE802.15.6

Pham Thanh Hiep[1,2], Ryuji Kohno[1]

[1]School of Engineering, Yokohama National University, Yokohama, Japan
[2]Le Quy Don Technical University, Ha Noi, Viet Nam
Email: phamthanhhiep@gmail.com

Abstract

Wireless Body Area Network (WBAN) is considered to apply to both medical healthcare and entertainment applications. A requirement for each application is different, *i.e.* high reliability for medical healthcare whereas high throughput for entertainment application. However, for both applications, low energy consumption is requested. Multiple hops technics have been researching in many fields of wireless system, e.g., ad hod, mobile, ITS etc. and its energy-efficiency is reported to be high. We propose the multiple hops technic for WBAN, however, WBAN is different to another systems, almost sensors forward the vital data packet of another sensors while sensing and generating the data packet of itself. Therefore, according to a packet generation rate of all sensors, probabilities of successful transmission and packet loss because of collision, timeout and overflow, are changed. It means that the vital data is lost and the transmit power is wasted due to packet loss. In order to obtain the highest throughput and save the power, the successful transmission probability is analyzed and the packet generation rate is optimized for multiple hops WBAN that using CSMA/CA based on IEEE802.15.6. The numerical calculation result indicates that the optimized packet generation rate depends on the system model. Moreover, the relation between the system model, the optimized packet generation rate and the throughput is discussed in the paper.

Keywords

Multiple Hops Body Area Network, Optimal Packet Generation Rate, Successful Probability, Collision Probability, CSMA/CA of IEEE802.15.6

1. Introduction

Nowadays, elderly population in many countries are increasing and then in order to survey health situation of elderly peoples under the limited financial resources and current medical service, it is important to remotely monitor a body status and a surrounding environment. Moreover, doctors are hard to know what is really happening when each body function is monitored and separated by a considerable period of time. This is reason why the monitoring of movement and all body functions in daily life are essential. One of the monitoring systems is wireless body area network (WBAN). WBAN consists of wireless sensors attached on or inside human body for monitoring vital health related problems, *i.e.*, Electro Cardiogram (ECG), ElectroEncephalogram (EEG), Electronystagmogram (ENG) etc. These sensors continuously monitor data and send to a coordinator, the coordinator gathers data of all sensors and sends to Health care center through existing network. On the other hand, according to quick development of manufacturing industry, many wireless devices are developed, especially the devices that are using the vital data and/or be used around the body, e.g., wireless earphone, music/movie player, game and so on. Consequently, the high throughput is requested. Moreover, the long lifetime of battery meaning the low power consumption is important subject of WBAN. According to importance of WBAN, the standard IEEE802.15.6 was establish [1]-[4].

The transmission of sensors can be divided into 2 schemes; Scheme 1: all sensors transmit their data packet directly to the coordinator, Scheme 2: sensors transmit their data packet to coordinator via another sensor. At Scheme 1, the transmit power of sensors should use high because the coordinator isn't always close to. Therefore, the lifetime of batteries becomes shorter and each sensor causes an interference to almost all sensors in WBAN. Moreover, the connection between sensors and the coordinator maybe fails due to the interruption of body functions, especially when the human is moving. The research on physical (PHY) layer, media access control (MAC) layer and network layer of Scheme 1 are described in [5] [6] and the communication of implant sensors WBAN also was researched [7]. On the contrary, at Scheme 2, since each sensor transmits its data packet to neighbor sensors, the transmit power and the influenced area are small. Therefore, the number of interfered sensors decreases and the lifetime batteries increases. In additional, even the direct connection between a sensor to the coordinator is failed, the sensor can transmit to the coordinator via another sensors that connects to the coordinator. According to the advantage of multiple hops technic, in this paper, we focus on the multiple-hop WBAN system.

The multiple-hop system is being researched in many literatures of many fields, e.g. ad hoc network, mobile network, ITS system and so on [8]-[11]. The MAC layer, PHY layer, network layer and crosslayer of multiple hops scheme also are researched [12]-[15]. However, in these systems, senders send a data packet to receiver(s) via relays and relays just forward the received data packet. On the contrary, in WBANs, sensors forward the received data packet while monitoring a situation of body and generating the vital data by themselves. According to the number of generated packets at each sensor meaning the packet generation rate, probabilities of successful transmission and packet loss because of collision, timeout and overflow, are changed. It means that the vital data is lost and the transmit power is wasted due to packet loss. In order to obtain the highest throughput and save the power, the successful transmission probability is analyzed and the packet generation rate is optimized for multiple hops WBAN. The optimized packet generation rate is analyzed when factors of system model are changed. Since the standard IEEE802.15.6 was established for WBAN, the transmission scheme in this paper is indicated as carrier sense multiple access with collision avoidance (CSMA/CA) based on IEEE802.15.6.

The rest of the paper is organized as follows. We introduce a brief of standard IEEE802.15.6 in Section 2. Section 3 shows the system model and performance analysis of multiple hops WBAN. The numerical evaluation is expressed in Section 4. Finally, Section 5 concludes the paper.

2. Brief of Standard IEEE802.15.6

In this section, the standard IEEE802.15.6 is briefly described. The detail of this standard is represented in [1].

2.1. Physical Layer

The IEEE 802.15.6 defines three PHY layers, *i.e.*, Narrowband (NB), Ultra wideband (UWB), and Human Body Communications (HBC) frequency. The selection of each PHY depends on the application requirements. Since we focus on analysis performance of multiple hops WBAN based on CSMA/CA access scheme, any PHY can be applied, however, NB is considered as an example.

The NB PHY is responsible for activation/deactivation of the radio transceiver, Clear Channel Assessment (CCA) within the current channel and data transmission/reception. The Physical Protocol Data Unit (PPDU) frame of NB PHY contains a Physical Layer Convergence Procedure (PLCP) preamble, a PLCP header, and a PHY Service Data Unit (PSDU) as given in **Figure 1**. The PLCP preamble helps the receiver in the timing synchronization and carrier-offset recovery. It is the first component being transmitted at the given symbol rate. The PLCP header conveys information necessary for a successful decoding of a packet to the receiver. The PLCP header is transmitted after PLCP preamble using the given header data rate in the operating frequency band. The last component of PPDU is PSDU which consists of a MAC header, MAC frame body, Frame Check Sequence (FCS) and is transmitted after PLCP header using any of the available data rates in the operating frequency band. A WBAN device should be able to support transmission and reception in one of frequency bands summarized in **Table 1**. (Further detail for the modulation and the channel coding can be found in [1] [2]).

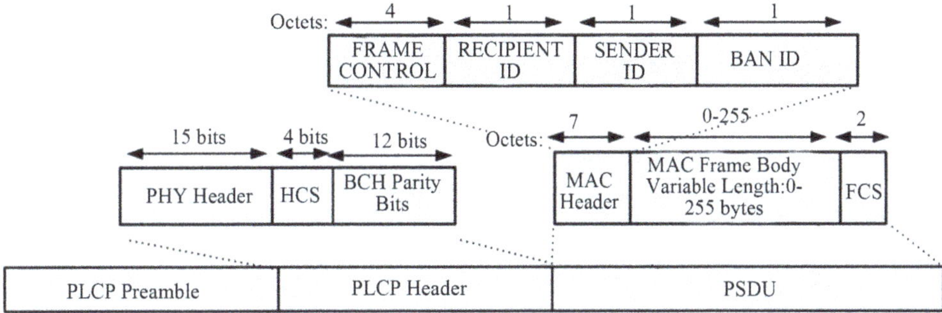

Figure 1. Structure of PPDU based on IEEE802.15.6.

Table 1. Main parameter for NB.

Frequency band	Packet component	Symbol rate (Ksps)	Data rate (Kbps)
420 - 450 MHz	PLCP header	187.5	57.5
	PSDU	187.5	75.9
	PSDU	187.5	151.8
	PSDU	187.5	187.5
863 - 870 MHz	PLCP header	250	76.6
950 - 956 MHz	PSDU	250	101.2
	PSDU	250	202.4
	PSDU	250	404.8
	PSDU	250	607.1
902 - 928 MHz	PLCP header	300	91.9
	PSDU	300	121.4
	PSDU	300	242.9
	PSDU	300	485.7
	PSDU	300	728.6
2360 - 2400 MHz	PLCP header	600	91.9
2400 - 2483.5 MHz	PSDU	600	121.4
	PSDU	600	242.9
	PSDU	600	485.7
	PSDU	600	971.4

2.2. CSMA/CA Based on IEEE802.15.6

In IEEE802.15.6, there are three access mechanisms that be comprehensively discussed in the standard. 1) Random access mechanism, which uses either CSMA/CA or a slotted Aloha procedure for resource allocation, 2) Improvised and unscheduled access (connectionless contention-free access), which uses unscheduled polling/posting for resource allocation, and 3) Scheduled access and variants (connection-oriented contention-free access), which schedules the allocation of slots in one or multiple upcoming superframes, also called 1-periodic or m-periodic allocations. Because of high flexibility and extensibility of CSMA/CA, it is considered in our analysis. The CSMA/CA procedure defined in the IEEE 802.15.6 standard is shown in **Figure 2** and its basic procedure is explained as follows.

A sensor sets its backoff counter to a random integer number within $[1,CW]$ where $CW \in (CW_{\min}, CW_{\max})$ is the contention window of this sensor. The values of $CW_{\min}$ and $CW_{\max}$ change depending on the user priority (UP) as given in **Table 2**. The sensor decreases the backoff counter by one for each idle CSMA slot of duration. Particularly, the sensor treats a CSMA slot to be idle if it determines that the channel has been idle between the start of the CSMA slot and clear channel assessment of duration time (*pCCATime*). If the backoff counter reaches zero, the sensor transmits a data packet. If the channel is busy because of transmission of another sensor, the sensor locks its backoff counter until the channel is idle. The CW is doubled for even number of failures until it reaches $CW_{\max}$. The failure means that the sensor fails to receive an acknowledgement from the coordinator. In random access period (RAP) 1, the sensor firstly waits for short interframe space ($SIFS$) = $pSIFS$ duration and then unlocks the backoff counter until it reaches zero where the transmission starts. But the sensor fails to receive an acknowledgement and the contention fails. As explained above, the CW is not doubled for odd number of failures and therefore the sensor sets its backoff counter to 5 and locks it. In contention access period (CAP), the sensor locks its backoff counter at 2 since the time between the end of the slot and the end of the CAP is not enough for completing the data transmission and the Nominal Guard Time represented by GT_n. The backoff counter is unlocked in the RAP2 period. Again the sensor fails to receive an acknowledgement and the contention fails. The CW gets doubled (for even number of failures) and the backoff counter is set to 8. When the data transmission is successful, the CW is set to $CW_{\min}$. Further details of CSMA/CA procedure can be found in the standard [1] [2].

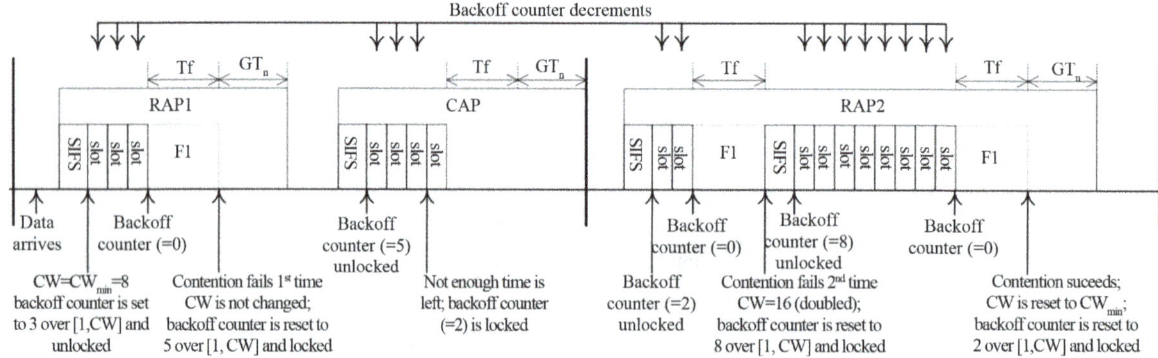

Figure 2. An example of IEEE802.15.6 CSMA/CA procedure.

Table 2. Contention window bound for CSMA/CA.

User priority	$CW_{\min}$	$CW_{\max}$
0	16	64
1	16	32
2	8	32
3	8	16
4	4	16
5	4	8
6	2	8
7	1	4

2.3. Calculation of Service Time

The service time (*T*) is defined as total time to transmit a data packet included the *backoff* time (T_{CW}), the time to transmit a data packet (T_{data}), interframe spacing (T_{pSIFS}), the time of acknowledgement packet (T_{ACK}) and delay time (α).

$$T = T_{CW} + T_{DATA} + T_{ACK} + 2T_{pSIFS} + 2\alpha. \tag{1}$$

Let's T_s denote a CSMA slot length, according to the standard, the average backoff time can be obtained as follows.

$$T_{CW} = \frac{CW_{\min} T_s}{2}. \tag{2}$$

As shown in **Figure 1**, since a data packet consists of a preamble, physical header, MAC header, MAC frame body and frame check sequence, the time to transmit a data packet becomes as

$$T_{DATA} = T_P + T_{PHY} + T_{MAC} + T_{BODY} + T_{FCS}, \tag{3}$$

here T_P, T_{PHY}, T_{MAC}, T_{BODY}, T_{FCS} represent the time to transmit a preamble, physical header, MAC header, MAC frame body and frame check sequence, respectively.

Since an immediate acknowledgement carries no payload, its transmission time is given by

$$T_{ACK} = T_P + T_{PHY} + T_{MAC} + T_{FCS}. \tag{4}$$

3. Multiple Hops Body Area Network System

3.1. System Model

Figure 3 shows an example of WBAN system. Many sensors are distributed around the body to monitor the health situation. Sensors transmit their vital data packet toward the coordinator. However, due to the interruption of body, some direct links between sensors and the coordinator is interrupted and the data packet of these sensors can't reach to the coordinator, especially, when the human is moving. Therefore, the multiple hops WBAN system is considered. According to multiple hops, a sensor that is out of transmission range of coordinator, can transmit its data packet to the coordinator via other sensors. We consider one link of multiple hops WBAN system consists of three sensors, A, B and C. The sensor A transmits its data packet to the sensor B, the sensor B transmits the received data packet as well as the data packet of itself to the sensor C and the sensor C forwards the received data packet and transmits its data packet to the coordinator (**Figure 4**). The WBAN system is constructed on or/and in the body, it means the space and the number of sensors are limited. Therefore, the multiple hops WBAN system has only few hops; it is different to other multiple hop systems, *i.e.* ad hoc, ITS and so on. This is the reason why the three hops of WBAN is considered.

The system model of multiple hops WBAN is described as follows. Since the packet loss due to collision is analyzed in this paper, the noise free is assumed. Therefore, packets are lost because of only collision of packets transmission in the same time. A link that consists of three sensors and one coordinator, is considered, and this link is assumed to be independent to another links and sensors. All sensors can transmit a packet to the neighbor sensor/coordinator only, however, all sensors in this link can sense the transmission of the others. The vital data packet is generated at each sensor by its packet generation rate. We assume that the system is started at time $t = -\infty$, hence it reaches its steady-state at the time $t = 0$. The buffer size of every sensors and the delay time of all packets are assumed to be limited, hence, if the throughput is smaller than the generated data meaning all the generated data aren't transmitted, the packets that aren't transmitted to the coordinator, will be lost. It is a reason of wasting transmit power and decreasing throughput of system. Consequently, in order to transmit all generated packets to the coordinator, the packet generation rate should be optimized.

3.2. Probabilities of Transmission Data

The transmission probability is defined as probability of sensor i in which the backoff counter is zero and denoted by Bf_i. A sensor can fail in transmission when more than one sensors send their data packet at the same time, namely collision of data frame. The backoff counter of a sensor counts down to zero when this sensor is in a transmission and the channel is idle (the channel is ready to transmit). If the channel is busy, the

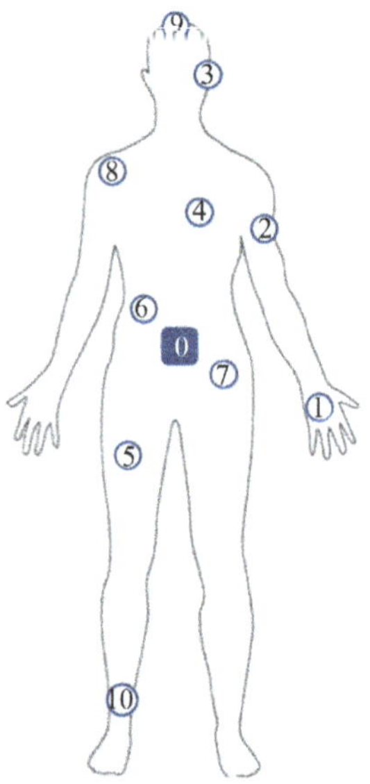

Figure 3. Body area network.

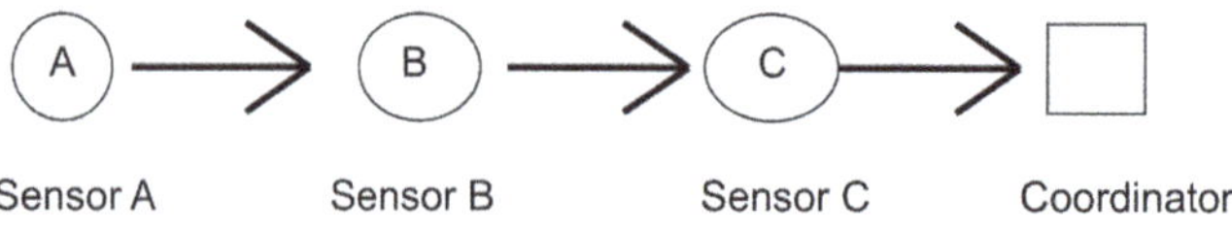

Figure 4. Multiple hops WBAN system.

sensor stops counting down its backoff counter. Therefore, the transmission probability is equal to the probability that the channel is ready to transmit $\left(\prod_{j\neq i}\left(1-Bf_j\right)\right)$. In this paper, we assume that all sensors follow the same mechanism. Therefore the transmission probability of all sensors is the same as Bf and can be expressed as

$$Bf=\left(1-Bf\right)^{n-1}, \tag{5}$$

where n is the number of sensors that is in transmission.

A sensor transmits a data packet successfully when only one sensor transmits the data packet. Let's P_{succ} denote the successful probability of transmission data of sensor i.

$$P_{succ_i}=Bf_i\prod_{j\neq i}(1-Bf_j). \tag{6}$$

The collision probability $\left(P_{coll}\right)$ is the probability that the data packet transmitted by sensor i collides with at least one of another data packets. It is the production of the transmission probability of sensor i $\left(Bf_i\right)$ and the probability at least one sensor transmits a data packet $\left(1-\prod_{j\neq i}\left(1-Bf_j\right)\right)$.

$$P_{coll_i}=Bf_i\left(1-\prod_{j\neq i}\left(1-Bf_j\right)\right). \tag{7}$$

The collision probability of sensor i also can be calculated from another view point. It is the probability of unsuccessful transmission data of sensor i within the transmission probability $\left(Bf_i\right)$.

$$P_{coll_i}=Bf_i-P_{succ_i}=Bf_i-Bf_i\prod_{j\neq i}\left(1-Bf_j\right). \tag{8}$$

Compare the P_{coll_i} in (7) and (8), it is the same.

3.3. Analysis Performance of Multiple Hops WBAN

The packet generation rate of sensors A, B and C is denoted by λ_A, λ_B and λ_C; furthermore, the number of packets in queue at sensors A, B and C is q_A, q_B and q_C, respectively. Let's l represent the maximal number of transmission packets in one unit time, therefore $l = \frac{1}{T}$. We analyze the performance of multiple hops in one unit time started from $t = 0$. For each duration of T, the number of packets that is successfully transmitted from sensor i is equal to the successful probability $\left(P_{succ_i}\right)$. The sensor i is assumed to successfully transmit one packet after k times of T duration, hence after duration [0,1], the remained packets is $q_i = \lambda_i \left(1 - P_{succ_i}\right)^k$. All generated packets at sensor i are successfully transmitted if $q_i < 1$. The number of T durations in which the sensor i has a packet to send whether the backoff counter of sensor i is zero or not, is denoted as $tran_i$.

$$tran_i = \lambda_i \sum_{j=0}^{k-1} \left(1 - P_{succ_i}\right)^j = \lambda_i \frac{1 - \left(1 - P_{succ_i}\right)^k}{1 - \left(1 - P_{succ_i}\right)} = \lambda_i \frac{1 - \left(1 - P_{succ_i}\right)^k}{P_{succ_i}}. \tag{9}$$

In the proposed system, the sensor A transmits its data packet to the sensor B, the sensor B transmits the data packet to the sensor C and then the sensor C transmits the data packet to the coordinator. Therefore if the sensor A has the packet to transmit, the sensor B also has the packet to transmit, and if the sensor B has the packet to transmit, the sensor C also has the packet to transmit. All sensors are assumed to be equal in priority, therefore the successful probability, the collision probability of all sensors are the same. In case the sensor A is in transmission, the number of sensors that is in transmission (n) is three, the system in this case is indicated for Scheme 1. Furthermore, the successful probability in Scheme 1 is denoted by $\dot{P}_{succ}$. At Scheme 2, the sensor A isn't in transmission, the sensors B and C transmit with the successful probability $\ddot{P}_{succ}$ and at Scheme 3, the sensors A and B aren't in transmission, the sensor C transmits with the successful probability $\dddot{P}_{succ} = 1$.

Let's $\dot{k}$, $\ddot{k}$ and $\dddot{k}$ denote the average number of T durations to successfully transmit one packet of the Schemes 1, 2 and 3, respectively. The $tran_A$, $tran_B$ and $tran_C$ are described as follows.

$$\begin{aligned} tran_A &= \lambda_A \frac{1 - \left(1 - \dot{P}_{succ}\right)^{\dot{k}}}{\dot{P}_{succ}}, \\ tran_B &= tran_A + \lambda_B \frac{1 - \left(1 - \ddot{P}_{succ}\right)^{\ddot{k}}}{\ddot{P}_{succ}}, \\ tran_C &= tran_B + \lambda_C. \end{aligned} \tag{10}$$

The performance of system is analyzed in an unit time. If $tran_A \le l$, all packets of the sensor A are transmitted to the sensor B. If $trans_A > l$, some packets are remained at the sensor A. Similar to the sensor A, if $tran_B \le l$, all packets of the sensor B (included the packet received from the sensor A) are transmitted to the sensor C; if $tran_B > l$, some packets of sensor B are remained at the sensor B. If $tran_C \le l$, all packets of sensor C (included the packet received from the sensor B) are transmitted to the coordinator. If $tran_C > l$, some packets are remained at the sensor C. Consequently, in case $tran_C \le l$, all packets of sensors A, B and C are transmitted to the coordinator.

3.4. Optimizing Packet Generation Rate

As the analysis in previous section, packets of all sensors are transmitted to the coordinator if the packet generation rate is low. The optimal packet generation rate is defined as the maximum packet generation rate with that all packets of sensors A, B and C can be transmitted to the coordinator. The optimal packet generation rate is denoted as λ_{opt}. The optimal packet generation rate should be found in order to save the energy consumption and obtain the highest throughput.

As mentioned above, all packets of sensors can be transmitted to the coordinator if $tran_C \le l$. Therefore, the optimal packet generation rate is the value that can satisfies $tran_C = l$.

$$l = tran_B + \lambda_{opt} = tran_A + \lambda_{opt} \frac{1 - \left(1 - \ddot{P}_{succ}\right)^{\ddot{k}}}{\ddot{P}_{succ}} + \lambda_{opt} = \lambda_{opt} \frac{1 - \left(1 - \dot{P}_{succ}\right)^{\dot{k}}}{\dot{P}_{succ}} + \lambda_{opt} \frac{1 - \left(1 - \ddot{P}_{succ}\right)^{\ddot{k}}}{\ddot{P}_{succ}} + \lambda_{opt}. \tag{11}$$

Hence,

$$\lambda_{opt} = \frac{l}{\frac{1-\left(1-\dot{P}_{succ}\right)^{\dot{k}}}{\dot{P}_{succ}} + \frac{1-\left(1-\ddot{P}_{succ}\right)^{\ddot{k}}}{\ddot{P}_{succ}} + 1}. \quad (12)$$

In (12), the variables $\dot{k}$ and $\ddot{k}$ are indeterminate. In order to determinate the variables $\dot{k}$ and $\ddot{k}$, the remained packets of sensors A and B after duration [0,1] is considered.

$$\begin{aligned} q_{A} &= \lambda_{opt}\left(1-\dot{P}_{succ}\right)^{\dot{k}}, \\ q_{B} &= \lambda_{opt}\left(1-\ddot{P}_{succ}\right)^{\ddot{k}}. \end{aligned} \quad (13)$$

All packets of sensors A and B are transmitted to the sensor C if $q_{A}, q_{B} < 1$. Therefore, the $\dot{k}$ and $\ddot{k}$ are the minimum number that satisfies $q_{A}, q_{B} < 1$. It means that $\dot{k}$ and $\ddot{k}$ are the number with that $q_{A}, q_{B} \to 1$. From (12) and (13), the λ_{opt} is expressed as

$$\lambda_{opt} = \frac{\frac{1}{\dot{P}_{succ}} + \frac{1}{\ddot{P}_{succ}} + l}{\frac{1}{\dot{P}_{succ}} + \frac{1}{\ddot{P}_{succ}} + 1}. \quad (14)$$

4. Numerical Evaluation

4.1. Successful Probability

According to (5) and (6), the successful probability of Schemes 1, 2 and 3 (corresponding to the number of sensors in transmission is 3, 2 and 1) can be calculated and shown in **Figure 5**. The successful probability decreases rapidly when the number of sensors in transmission increases.

4.2. Theoretical Result

In order to evaluate the theoretical analysis, the parameter that is summarized in **Table 3**, is used as an example. Hence, $T = 0.0099$ second and the maximal number of transmission packets l is 100.9 times per second. According to (14) and successful probability in **Figure 5**, the optimal packet generation rate is 9.3 packets per second. The number of successfully transmitted packets of all sensors is shown in **Figure 6**. The number of

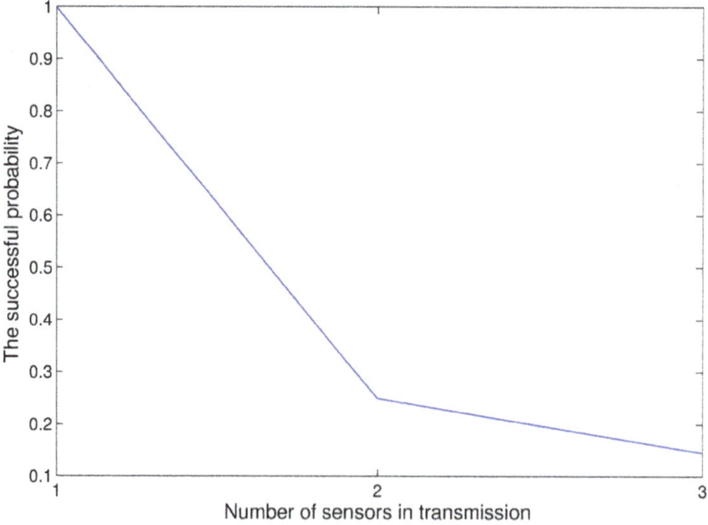

Figure 5. The successful probability of Schemes 1, 2 and 3.

Table 3. Numerical parameters.

Frequency band [MHz]	2400 - 2483.5
Packet component	PSDU
Modulation	$\pi/2$-DBPSK
Symbol rate R_s [Ksps]	600
Physical header rate R_{hdr}	242.9
Payload size [byte]	250
Minimum contention windows CW_{min}	16
Maximum contention windows CW_{max}	64
Clear channel assessment time	$63/R_s$
MAC header [byte]	56
MAC footer [byte]	16
Minimum interframe spacing time [μs]	20
Short interframe spacing time T_{sifs} [μs]	50
Transmission time of preamble [s]	88/Rs
Propagation delay α [μs]	1

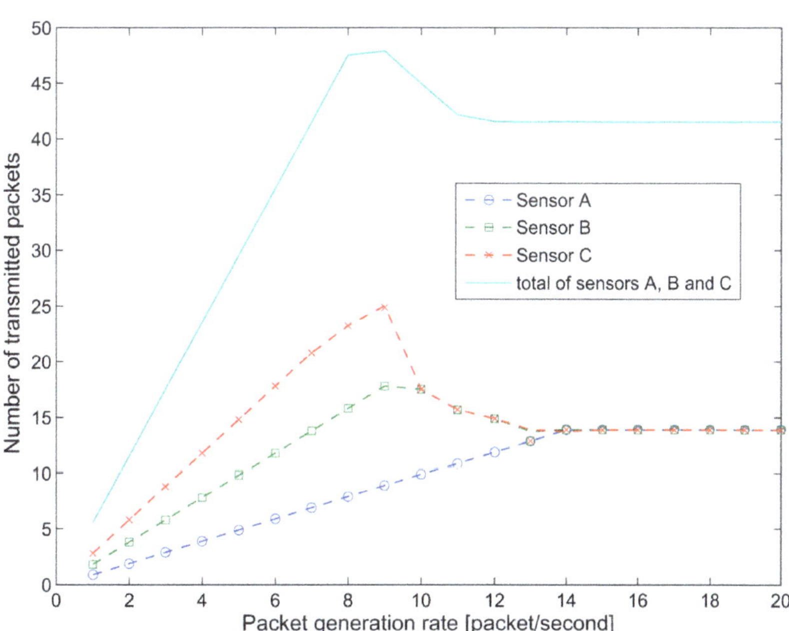

Figure 6. The number of successfully transmitted packets of all sensors.

successfully transmitted packets at sensors A, B and C increases when the packet generation rate (λ) increases. However, the number of successfully transmitted packets of sensor C starts decreasing when packet generation rate reaches 9 packets per second. The optimal packet generation rate has a slight difference between the theoretical and the simulation result. The reason can be explained that in simulation the packet generation rate is an integer value and increased by one. Furthermore, the reason why the number of successfully transmitted packets of sensors C and B start decreasing when the packet generation rate is respectively 9 and 10, can be explained as follows. When the packet generation rate increases, the number of generated packets at each sensor increases. Moreover, since all sensors transmit a packet to the next sensor/coordinator, the number of packets at the sensor C increases considerably and all packets can't be transmitted in an unit time when the packet generation rate increases. Similarly, the sensor B can't transmit the packet of itself and the packet received from the sensor A when the packet generation rate reaches 10. When the packet generation rate is over 15, the number of successfully transmitted packets of all sensors is the same. In this scenario, the sensor A also can't transmit all its packets, the successful probability of all sensors is the same in all over [0,1]. Therefore, the number of successfully transmitted packets of all sensors is the same.

Figure 7 shows the number of remained packets of all sensors that isn't transmitted to the neighbor sensor or the coordinator in an unit time. The sensor C has the remained packet when the packet generation rate is over the optimal value. it reconfirms that the optimal packet generation rate is the maximum of packet generation rate with that all packets of sensors can be transmitted to the coordinator. The sensor A has the remained packet when the packet generation rate is over 15, it means that three sensors are in transmission in all duration [0,1] and all sensors have the same number of successfully transmitted packets.

4.3. Optimal Packet Generation Rate Based on System Model

Any change in system model leads to the change in service time, however, in a system, the changeable factor is the payload. The optimal packet generation rate is calculated according to changing of payload and shown in **Figure 8**. The optimal packet generation rate decrease when the payload increases. The reason is that the service time increases when the payload increases meaning the maximal number of transmission packets decreases. However, the maximal throughput of system increases when the payload increases (**Figure 9**).

5. Conclusions

We have proposed the multiple hops scheme for WBAN and analyzed the performance of multiple hops WBAN with IEEE802.15.6 CSMA/CA protocol. The transmission probability, the successful probability as well as the number of successfully transmitted packets of all sensors were represented. Furthermore, the optimal packet generation rate with that all generated packets at sensors can be transmitted to the coordinator was obtained and the optimal packet generation rate based on system model was discussed. When the payload increases, the optimal packet generation rate decreases, whereas the throughput of system increases.

In this paper, due to the limited space on and/or in the body, we considered the WBAN with three sensors and one coordinator. However, this link was assumed to be independent to other sensors and links. The effect of other sensors that don't joint to this link, will be considered in the future work. Moreover, the noise was as-

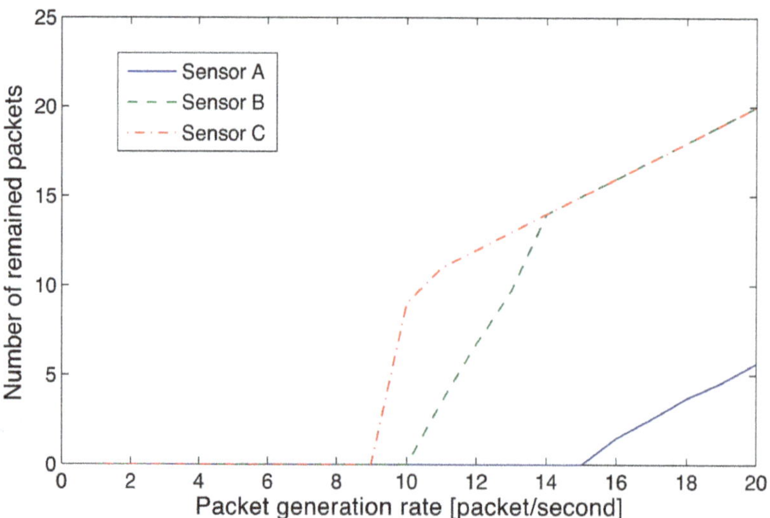

Figure 7. The number of remained packets.

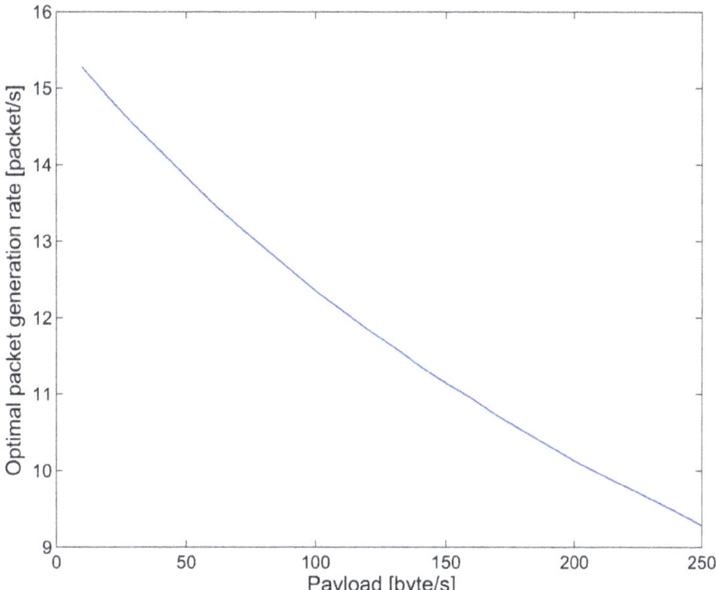

Figure 8. The optimal packet generation rate based on changing of payload.

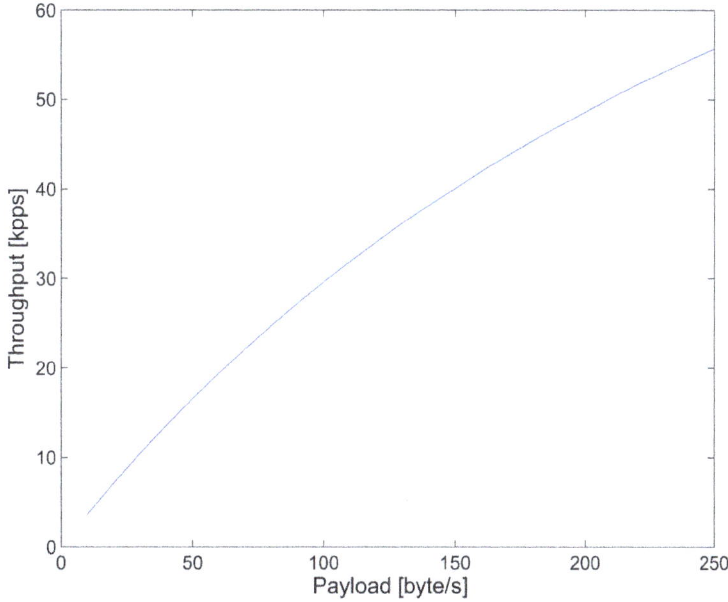

Figure 9. The throughput based on changing of payload.

sumed to be free and the distance between each sensors wasn't considered. We left them to the future work.

References

[1] Wireless Personal Area Network Working Group (2012) EEE Standard 802.15.6, Wireless Body Area Networks. *IEEE Standards*, 1-271.

[2] Kwak, K.S., Ullah, S. and Ullah, N. (2010) An Overview of IEEE 802.15.6 Standard. *3rd International Symposium on Applied Sciences in Biomedical and Communication Technologies* (ISABEL), Rome, November 2010.

[3] Martelli, F., Buratti, C. and Verdone, R. (2011) On the Performance of an IEEE 802.15.6 Wireless Body Area Network. *European Wireless* 2011, Vienna, 27-29 April 2011.

[4] Astrin, A.W., Li, H.-B. and Kohno, R. (2009) Standardization for Body Area Networks. *IEICE Transactions on Com-*

munications, **E92.B**, 366-372.

[5] Ullah, S., Higgins, H., Braem, B., Latre, B., Blondia, C., Moerman, I., Saleem, S., Rahman, Z. and Kwak, K.S. (2012) A Comprehensive Survey of Wireless Body Area Networks: On PHY, MAC, and Network Layers Solutions. *Journal of Medical Systems*, **36**, 1065-1094.

[6] Li, C.L., Geng, X.Y., Yuan, J.J. and Sun, T.T. (2013) Performance Analysis of IEEE 802.15.6 mac Protocol in Beacon Mode with Superframes. *KSII Transactions on Internet and Information Systems*, **7**, 1108-1130,

[7] Zhen, B., Li, H.B. and Kohno, R. (2008) IEEE Body Area Networks and Medical Implant Communications. *Proceedings of the ICST 3rd International Conference on Body Area Networks*, Tempe, Ariz.

[8] Hiep, P.T., Ryuji, K. and Ono, F. (2012) Optimizing Distance, Transmit Power and Allocation Time for Reliable Multi-Hop Relay System? *EURASIP Journal on Wireless Communications and Networking*. http://dx.doi.org/10.1186/1687-1499-2012-153

[9] Hiep, P.T. and Ryuji, K. (2010) Optimizing Position of Repeaters in Distributed MIMO Repeater System for Large Capacity? *IEICE-Transactions on Communications*, **E93-B**.

[10] Wang, J.-B., Wang, J.-Y., Chen, M., Zhao, X.B., Si, S.-B., Cui, L.R., Cao, L.-L. and Xu, R.H. (2013) Reliability Analysis for a Data Flow in Event-Driven Wireless Sensor Networks Using a Multiple Sending Transmission Approach. *EURASIP Journal on Wireless Communications and Networking*, **2013**.

[11] Zhao, D.B. and Chin, K.-W. (2013) Approximation Algorithm for Data Broadcasting in Duty Cycled Multi-Hop Wireless Networks. *EURASIP Journal on Wireless Communications and Networking*, **2013**. http://dx.doi.org/10.1186/1687-1499-2013-248

[12] Hiep, P.T., Hoang, N.H., Chika, S. and Ryuji K. (2013) End-to-End Channel Capacity of MAC-PHY Cross-Layer Multiple-Hop MIMO Relay System with Outdated CSI. *EURASIP Journal on Wireless Communications and Networking*, **2013**. http://dx.doi.org/10.1186/1687-1499-2013-144

[13] Li, Y.Y., Liu, K., Liu, F. and Xu, Z. (2013) A Rapid Cooperation-Differentiated Medium access Control Protocol with Packet Piggyback for Multihop Wireless Networks. *EURASIP Journal on Wireless Communications and Networking*, **2013**.

[14] Dromard, J., Khoukhi, L. and Khatoun, R. (2013) An Efficient Admission Control Model Based on Dynamic Link Scheduling in Wireless Mesh Networks. *EURASIP Journal on Wireless Communications and Networking*, **2013**. http://dx.doi.org/10.1186/1687-1499-2013-288

[15] Hiep, P.T., Sugimoto, C. and Kohno, R. (2012) MAC-PHY Cross-Layer for High Channel Capacity of Multiple-Hop MIMO Relay System. *Communications and Network*, **4**, 129-138.

12

Handover Time Delay Reduction and Its Effects in Cloud Computing

Qassim Bani Hani[1], Julius Dichter[1], Jamal Fathi[2]

[1]Department of Computer Science, University of Bridgeport, Bridgeport, USA
[2]Department of Electrical and Electronics Engineering, Near East University, Nicosia, Northern Cyprus
Email: qbanihan@my.bridgeport.edu, dichter@my.bridgeport.edu, jamalfathi2004@gmail.com

Abstract

Mobile devices connected by cellular service require a constant connection to a base station. As these devices move from place to place, they need to disconnect from one base station and connect to another. The process of transferring between base stations found in neighboring geographic areas is referred to as the handover course of action. During the handover course of action, the connection to the disconnected cellular device and the serving basic base station terminates. The quality of cell WiFi networks will suffer considerably from any handover latency as well as the supply decline percentage. In our work, we propose and implement in MATLAB a simple handover program applying mobility behavior pertaining to WiMAX networks. The ability to produce mobility pattern table is implemented to help in determining another available basic base station and as a consequence eliminate needless verification. Additionally, the serving basic base station forwards important computer data packets which it received throughout the entire handover course of action for the targeted base station giving a reduction of the supply decline percentage. Extensive simulation studies are executed to judge the efficiency inside suggested program using MATLAB. The outcome proves that our program can with certainty reduce the handover latency as compared with other solutions found in the literature.

Keywords

WiMAX, Handover, Signal Strength, Packet Drop, Base Station

1. Introduction

Looking at the development that is along these few lines that joining the target BS as incidentally rather than a mobile station system separating each of the pushed base station. This may give a diminishing which yields that the compass not with remaining on an extraordinarily key level minimized extending and cooperation works out so that the critical deferral time in reacting to the base station. The aching with the neighboring base station ideal for a specific handover change is poverty stricken upon these key parameters:

1) Direction of the mobile station system movement.
2) Average time interval in between each hop of the mobile station.
3) Current load of a neighboring base station.
4) Position and coverage of the neighboring base station with regards to the current select base station.

Selecting the right base station for the scanning activity is usually a joint decision on the select base station according to its signal strength and the response time taken, as well as the concerned mobile station system while using select base station utilizing the most decision-making responsibilities belong to the base station in order not to get disconnected with the visitor mobile station even for a while. This importance in connection belongs to the fast in respond of the base station and its capacity.

By topic of performance and the interference which depends on the BSs separation distances, that proportional to the overlap region and the antenna used in each BS, for this none of the appropriated effects were broad enough to unmistakably evaluate precisely, which depends on the different stages that are going to be taken as the maximum offer of the aggregate handover time. Thusly basic exertion was obliged to this study, reproduce and break down the execution of the sorted out WiMAX handover. This proliferation was completed by MATLAB programming.

The rest of the paper is organized as follows: In the next Section 2 presents the related work. In Section 3, performance and interferences of the proposed model explanations. The proposed handover detailed design scheme is presented in Section 4. The simulation results and performance evaluation is described in Section 5. Section 6 concludes this paper and presents future work.

2. Related Work

Many researchers submitted several works in the subject of data security and in particular in the subject of steganography. The following are some of the current works in the field of the handover, where handover means exchanging a progressing call or information sessions one phone to trade. Handovers happen as a consequence of the change of the adaptable client starting with one achieve then onto the accompanying range. Handovers are utilized to keep a progressing call to be separated as Hyeyeon *et al.* [1] demonstrated several handover longing numbers to decrease the handover latency by fast handover impelling. Moreover, broke down multipath transmission control protocol (MPTCP's) essentialness usage and handover execution in distinctive operational modes. Finally found that (MPTCP) engages smooth handovers offering sensible execution really for extraordinarily asking for procurements, for instance, voice over internet protocol (VOIP). To the degree that, proposed a low-flightiness received signal strength indicator (RSSI)-based computation and, then, an improved mixture RSSI/ extraordinary put version. Where, the proposed RSSI-based vertical handover (VHO) figuring guarantees a constraining extraordinary put increase at the mechanical tester sensor (MTS). Where, the estimations showed a possibly extensive change using universal mobile telecommunications system (UMTS) showing data with relationship to Global system for mobile communications (GSM) as to handover range precision.

Vasos *et al.* [2] softened down the idleness sections up Mobile Ipv6 handovers. What's more, gave genuine execution results for enormous parts of the handover handle through estimations in a veritable Mipv6 use on a remote proving ground centered on IEEE 802.11b. Khan [3] introduced a diagnostic work that improves the handover system. The creator talked about, gatekeeper channels, call induction and handover queuing focused around the covering scope ranges in the neighboring cells. Nishtith *et al.* [4] displayed diverse parts of handover to the extent that demonstrated handover usage, and the systems of handover and the assessment of handover and its execution. Wong *et al.* [5] explained quickly the high dangers of irregularity of tend to patients, and explored the vitality of clinical handover, to the extent that outlined the dissection accessible on distinguishing clinical handover process, gave a writing audit in regards to clinical handover and worldwide distributed meets expectations.

Pandey *et al.* [6] clarified issues inside handover handle, and proposed system to enhance handover time inactivity. Hsieh *et al.* [7] handled two imperative difficulties: 1) Enhancing handover execution in heterogeneous remote system, and 2) enhancing Transmission Control Protocol TCP execution in multi-jump remote system. In heterogeneous system, clients expect continuous administrations moving from a solitary system to another. Institute of Electrical and Electronics Engineers (IEEE) proposed media independent handover (MIH) to bring about a noticeable improvement handover execution. Fu *et al.* [8] essayed that at present mobile IPv4 (MIP) will be the overwhelming instrument for versatility administration and should persevere into the future. Mortaza *et al.* [9] subsequently presented the neighboring cells may experience the ill effects of inordinate impedance that is produced by this MS. Besides, a hazard that connection quality declines all of a sudden change an extensive part, *i.e.* consequently, remove handover needs to be begun up. Chao *et al.* [10] explained the high dangers of irregularity of nurture patients quickly, and audited the essentialness of clinical handover, to the extent that abridged the dissection accessible on recognizing clinical handover process.

Purnendu *et al.* [11] clarified issues inside handover handle, and proposed system to enhance handover time inertness. Abduloulaziz *et al.* [12] displayed another vertical handover choice to minimize the amount of disappointment and unnecessary handover in remote systems, their proposed calculation relies on upon the estimation time and figuring of limit time. To extent that the handovers that happening between mobile station (MS) and the remote neighborhood wireless local area networks (WLANs), where this strategy vanished the disappointments and the unnecessary handover time by 70% to 80%. Akki *et al.* [13] explored the properties of Asynchronous Transfer Mode ATM and its profits, to the extent that clarified how it manages its characteristics, necessities, convention architectures and the worldwide exercises. Hu *et al.* [14] introduced a strategy for taking care of the directing issues by overlaying static sensible topology over the physical star grouping by producing close ideal most limited ways. Mushtaq *et al.* [15] distinguished the execution of the handover over worldwide interoperability for microwave access WiMAX-WiMAX, WiMAX-UMTS and WiMAX-Wifi regarding the chose measurements. To decrease the handover time idleness for portable Ipv6 (Mipv6).

An *et al.* [16] proposed an instrument with extra primitives and parameters to the media free handover administrations characterized in the IEEE 802.21 in order to decrease the handover time delay in the FMipv6. To comprehend the impacts of Duplicating Address Detection on the handover time delay. [17] Vasos *et al.* analyzed the well-known methodology of Mobile Ipv6 in the genuine remote proving ground, which is focused around IEEE 802.11b and extricated the taken information by system elements throughout the development of the versatile endorser. Shin *et al.* [18] created another system to diminish the Media Access Control (MAC) layer handover inertness on account of Voice over IP (VOIP) gets consistent. The proposed model which is called Spmipv6 might be restricted to one Round-Trip Time (RTT) between the versatile endorsers and the target access switch to diminishing the handover.

3. Performance and Interference

By topic of performance and the interference which depends on the Base Stations (BSs) separation distances, that proportional to the overlap region and the antenna used in each BS, for this none of the appropriated effects were broad enough to unmistakably evaluate precisely, which depends on the different stages that are going to be taken as the maximum offer of the aggregate handover time. Thusly basic exertion was obliged to this study, reproduce and break down the execution of the sorted out WiMAX handover. This proliferation was completed by MATLAB programming. It introduced the outline of an improved cross-layer based handover calculation, which comprehends the delayed handover handling acquired when utilizing portable WiMAX by wiping out the checking stage performed by versatile supporter stations. The calculation used the presently associated vehicles to gather MAC and PHY layer data about target base stations, and afterward show the data to briefly disengaged ones. The separated vehicles then alter their WiMAX connectors and resume correspondence promptly in the wake of joining the transmission zone. It was exhibited by system test system 2 (NS2) reproductions that the Sehlabaka *et al.* [19] proposed calculation gave a lessened handover deferral, expanded system throughput and minimized number of lost parcels at different velocities of vehicles and bundle sizes.

3.1. Simulation Environment

In this study, the simulations parameters are taken natural as 15 base stations and 150 subscribers in a small area of 5000 m × 5000 m area in a circumstance was reenacted in MATLAB programming. The shifting time and the total handover operation time were focused on with the support of IEEE 802.16e OFDMA model realized using MATLAB. The pace of SSs was contrasted reliably from 0 - 100 m/sec as a maximum speed, which suggests that both traveler and vehicular advancements of SSs were perceived. The standard parameters are classified in **Table 1**.

Table 1. Standard simulation parameters.

Parameter	Value
No. of Base Stations	15
No. of Mobile Stations	150
Simulation Time	1000 second
Area Range	5000 m × 5000 m
Maximal Velocity	100 m/s
Overlap Range	200 m
Radio Range	1000 meter
Frequency	2.4 GHz

3.2. Analysis

Regardless, in 802.16e such evidently injured looking at is to a degree stayed far from with the SBS once in a while saving and radio information about the neighboring BSs. Similarly, the standard does not clearly indicate the measure of hindrances, which respects the increment in the handover postponement time. Endeavoring activities take in the wake of breaking down. Moreover, the standard does not clearly show the measure of disadvantages. In like way, since all around the checking between times, diverse sorts of transmissions between the MSS and the SBS are carried out; it prompts enormous throughput corruption and particularly hampers the QoS of deferral sensitive foreseeable traffics.

4. Proposed Model Explanation

The SS steers the possible BSS in a need based case, while using pass on proficiency case table and moreover the information of possible BSS furnished with the current BS. In total, the proposed method can minimize the handover grievous deficiency of change with low package hardship degree and subsequently an essentially handover. With this zone, it is showing the urging handover blueprint using versatility diagrams. For the base stations, adaptability specimen tables are well known and utilized with help the smaller stations suspect the checked base stations. The data recorded from the adaptability case table is overhauled all around every suitable handover skeleton depending upon the handover decisions got from additionally unassuming station. The reenactment study shows that our strategy can on sensationally key level decrease handover slowness.

Detailed Design

Within the scheme below, the SS uses the mobility pattern table to predict the mark BS. The mobility pattern table, where the pairs with the previous BS, as well as the target BS, are recorded, is maintained through the serving BS mounted in the center of the cell. Among the mobility pattern, table is shown in **Table 2**, and **Table 3** is produced by the first scan to all base stations by assuming all mobile stations are located in the first cell as shown in **Figure 1**.

The handover times relate to how now and again the adaptability arrangement zone shows up inside a certain period. In the occasion the pass on breaking point outline table is dealt with, the table is void, and table ranges are joined and updated in the handover process. The serving BS then requests if the pair exists in the flexibility sample table. In the event that the pair exists, the handover note worth is reached out by 1; if all else fails, a substitute table section holding the pair of the past base station ID moreover the target base station ID is cemented and in addition the handover respect for the new way is planned to a solitary.

For every one table portion in the adaptability representation table, the serving BS considers the system for past Bsprev and Bsprev; if the nature of previous base station in the table path is the same with the ID of past base station exemplified in the thickness request message, the target BS in such a table distribution is considered as the contender BS, and the concentrate on base station ID is solidified with the chipper BSS rundown. In all probability, examining each of the BSS inside the separating once-over takes truly a while, and that is not preceded in stillness sensitive enduring offers. Inside this response message, the BS encapsulates the exuberant BSS rundown from the lessening ask for on the handover times as demonstrated by the flexibility illustration table.

When the signal strength of the serving BS drops below the predefined signal strength threshold, the mobile station enters the scanning stage to find the next base station to associate with. In any case, the obliging station tries to synchronize with the entire BSS that has the most vital likelihood (centrality this kind of handover decision appeared with most astonishing repeat) and bits of taking in at change physical brilliant information in the certain BS as showed by its most shocking marker quality as indicated in **Figure 2**. Where, **Figure 2** shows the occasion the channel condition fits the need from the adaptable station, the width framework is completed without extra neighbor BS must be analyzed. If all else fails, the flexible station ought to yield the running as a laced unit with BS until the perfect BS is found. The last BS's ID regardless of the concentrate on BS's ID will furthermore be embodied in that message to keep up the adaptability outline table. Precisely when the serving BS perceives the mobile handover interrupt message, it upgrades its portability case table focused around the past base station ID and the concentrate on base station ID as shown in **Table 4**.

Table 2. Obtained table after first scan.

Base Station ID	Avg. Handover Time	Av. Signal Strength	Av. Load	Av. Load Ratio
2	195.90312500	−50.78717594	43.6250	0.218125
3	195.97125000	−50.78717594	43.6250	0.218125
4	197.01312500	−50.78717594	43.6250	0.218125
5	196.96812500	−50.78717594	43.6250	0.218125
6	196.67687500	−50.78717594	43.6250	0.218125
7	196.47000000	−50.78717594	43.6250	0.218125
8	155.14416667	−50.8569515	40.5000	0.2025
9	155.45166667	−50.8569515	40.5000	0.2025
10	155.61083333	−50.8569515	40.5000	0.2025
11	155.02583333	−50.8569515	40.5000	0.2025
12	155.04666667	−50.8569515	40.5000	0.2025
13	155.50916667	−50.8569515	40.5000	0.2025
14	155.61500000	−50.8569515	40.5000	0.2025
15	155.19416667	−50.8569515	40.5000	0.2025

Table 3. Mobility pattern table.

Previous Base Station	Target Base Station	Av. Signal Strength	Av. Load	Av. Load Ratio
8	9	−50.86	40.50	0.2025
8	10	−50.86	40.50	0.2025
8	11	−50.86	40.50	0.2025
8	12	−50.86	40.50	0.2025
8	13	−50.86	40.50	0.2025
8	14	−50.86	40.50	0.2025
8	15	−50.86	40.50	0.2025
8	2	−50.79	43.625	0.2181
8	3	−50.79	43.625	0.2181
8	4	−50.79	43.625	0.2181
8	5	−50.79	43.625	0.2181
8	6	−50.79	43.625	0.2181
8	7	−50.79	43.625	0.2181

Table 4. Obtained table after second scan.

Base Station ID	Avg. Handover Time	Av. Signal Strength	Av. Load	Av. Load Ratio
1	110.5314103	−51.45694641	48.97435897	0.108831911
2	110.5397436	−51.45694641	48.97435897	0.108831911
3	110.9121795	−51.45694641	48.97435897	0.108831911
4	111.1455128	−51.45694641	48.97435897	0.108831911
5	111.6628205	−51.45694641	48.97435897	0.108831911
6	112.2275641	−51.45694641	48.97435897	0.108831911
7	112.0929487	−51.45694641	48.97435897	0.108831911
9	111.4365385	−51.45694641	48.97435897	0.108831911
10	111.8356688	−51.3709758	48.98089172	0.108846428
11	111.0141935	−51.54402632	48.96774194	0.108817206
12	111.0474359	−51.45694641	48.97435897	0.108831911
13	109.9897436	−51.45694641	48.97435897	0.108831911
14	110.4237179	−51.45694641	48.97435897	0.108831911
15	110.5685897	−51.45694641	48.97435897	0.108831911

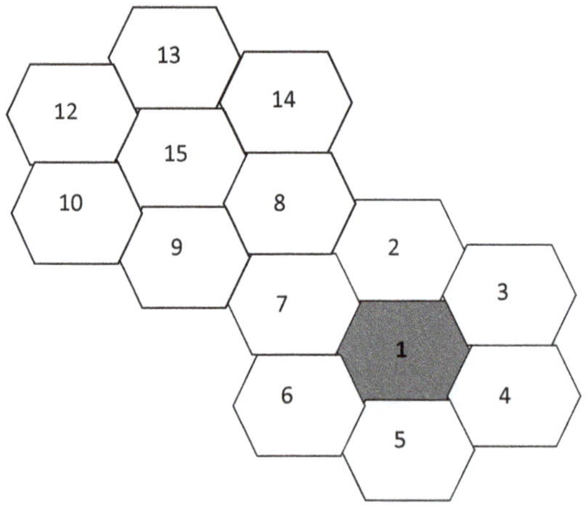

Figure 1. Cell distribution.

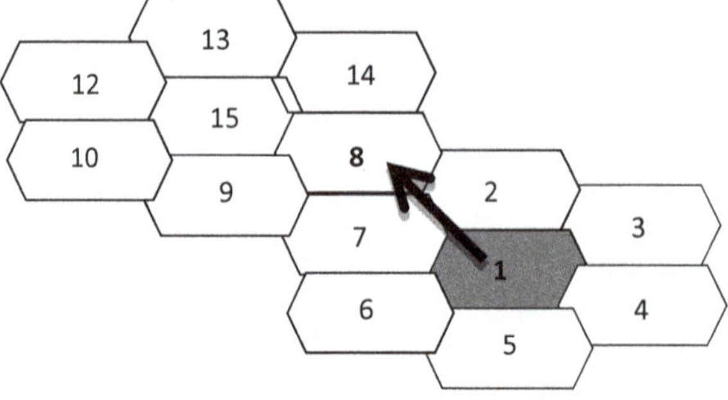

Figure 2. Decision made according to signal strength.

Then again, it is in like way possible that this transportability case table isn't right or holds bafflement. Notwithstanding, the gage is misguided regardless of the flexible station would attempt to yield a mixed up BS. Since the achieve will miss the mark under such a condition, the adaptable station needs to complete the imperative yield technique gather the adaptability sample table concentrated on the standard broadening results.

In like way, when the current BS gets the mobile handover interrupt message, the serving BS will actuate a huge allotment of the downlink packs to the new BS of the adaptable station, in light of the route that in the wake of sending the mobile handover interrupt message, the accommodating station will withdraw from the serving BS and all correspondences between the versatile station and the serving BS be interfered. The target BS holds the downlink gatherings of the versatile station clearly, and when the acquaintanceship between the accommodating station and the target BS is made, the target BS progresses the set away packages to the adaptable station. After the target BS is dead masterminded, the outline layer handover could be authorized to minimize the total handover absence of movement as demonstrated in **Figure 3**.

Where **Figure 3** explains the movement of the MS as initially in cell number two which is the second step after the initial process, and in accordance to the obtained **Table 5**, where the consideration of the signal strength to move to BS number three.

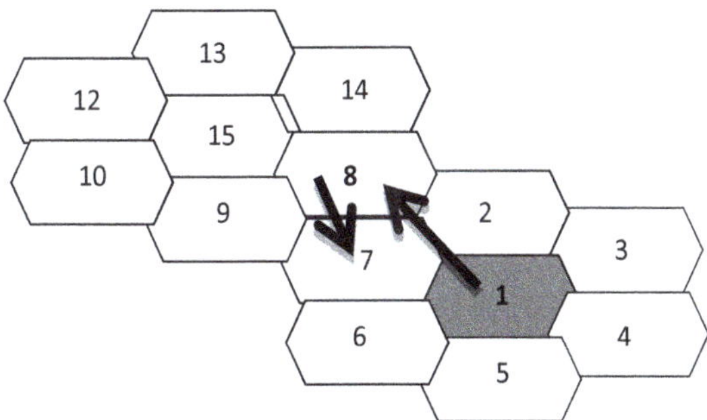

Figure 3. Mobile subscriber movements.

Table 5. Mobility pattern table.

Base Station ID	Avg. Handover Time	Av. Signal Strength	Av. Load	Av. Load Ratio
2	195.9031	−50.7871	43.6250	0.218125
3	195.9712	−50.7871	43.6250	0.218125
4	197.0131	−50.7871	43.6250	0.218125
5	196.9681	−50.7871	43.6250	0.218125
6	196.6768	−50.7871	43.6250	0.218125
7	196.4700	−50.7871	43.6250	0.218125
8	155.1441	−50.856	40.5000	0.2025
9	155.4516	−50.856	40.5000	0.2025
10	155.6108	−50.856	40.5000	0.2025
11	155.0258	−50.856	40.5000	0.2025
12	155.0466	−50.856	40.5000	0.2025
13	155.5091	−50.856	40.5000	0.2025
14	155.6150	−50.856	40.5000	0.2025
15	155.1941	−50.856	40.5000	0.2025

5. Simulation Results

5.1. Simulation Setup

The simulation in this thesis involves examining how a hundred and fifteen MS can move across a fifteen BSs at various speed in random process using the parameters shown in **Table 6**.

The simulation was examined using ready software MATLAB. The movements started from cell number one as a reference step upward to the 8th BS randomly according to the following steps:

First scan process was done to all BSs to produce the initial mobility table as shown in **Table 7**, where the scan is done almost 1924 times to produce the first list, this number of scan is done because the mobile will be stable when more number of scans is done.

Taking first scan into consideration to generate the first mobility list, and according to signal strength to determine the target BS, as shown in **Table 8**.

Load ratio is taken 0.0 till 0.5, where the BS capacity is taken 200 MSs, where the load ratio for each step, a

Table 6. Simulation parameters.

Parameter	Value
No. of Base Stations	15
No. of Mobile Stations	150
Simulation Time	1000 second
Area Range	5 Km × 5 Km
Maximal Velocity	15 m/s
Overlap Range	200 m
Radio Range	1 Km
Frequency	2.4 GHz

Table 7. Produced mobility pattern table.

Previous Base Station	Target Base Station	Av. Handover	Av. Signal Strength	Av. Load Ratio
8	7	0.1120929487	−51.45694641	0.000004897
8	6	0.1122275641	−51.45694641	0.000004897
8	10	0.1118356687	−51.37097579	0.000004898
8	5	0.1116628205	−51.45694641	0.000004897
8	4	0.1111455128	−51.45694641	0.000004897
8	12	0.1110474358	−51.45694641	0.000004897
8	11	0.1110141935	−51.54402632	0.000004896
8	3	0.1109121794	−51.45694641	0.000004897
8	2	0.1105397435	−51.45694641	0.000004897
8	1	0.1105314102	−51.45694641	0.000004897
8	15	0.1105685897	−51.4569464	0.000004897
8	14	0.1104237179	−51.4569464	0.000004897
8	13	0.1099897435	−51.4569464	0.000004897

scan is done to collect the data to obtain **Figure 4**. Where, **Figure 4** shows the results of the first handover time delay, and its maximum value is 197 ms.

Second scan process was done to all BSs to produce the initial mobility table as shown in **Table 9**, where the scan is done almost 2188 times to produce the second list, this number of scan is done again for the same reason in the first scan which is the mobile will be more stable when done more number of scans.

Again, load ratio is taken 0.0 till 0.5, where the BS capacity is taken 45 MSs, where the load ratio for each step, a scan is done to collect the data. The obtained data is plotted as shown in **Figure 5** below to show the second handover time delay, maximum 111 ms.

Table 8. Mobility pattern list (according to signal strength).

Previous Base Station	Target Base Station	Av. Signal Strength	Av. Load	Av. Load Ratio
8	9	−50.86	40.50	0.2025
8	10	−50.86	40.50	0.2025
8	11	−50.86	40.50	0.2025
8	12	−50.86	40.50	0.2025
8	13	−50.86	40.50	0.2025
8	14	−50.86	40.50	0.2025
8	15	−50.86	40.50	0.2025
8	2	−50.79	43.625	0.2181
8	3	−50.79	43.625	0.2181
8	4	−50.79	43.625	0.2181
8	5	−50.79	43.625	0.2181
8	6	−50.79	43.625	0.2181
8	7	−50.79	43.625	0.2181

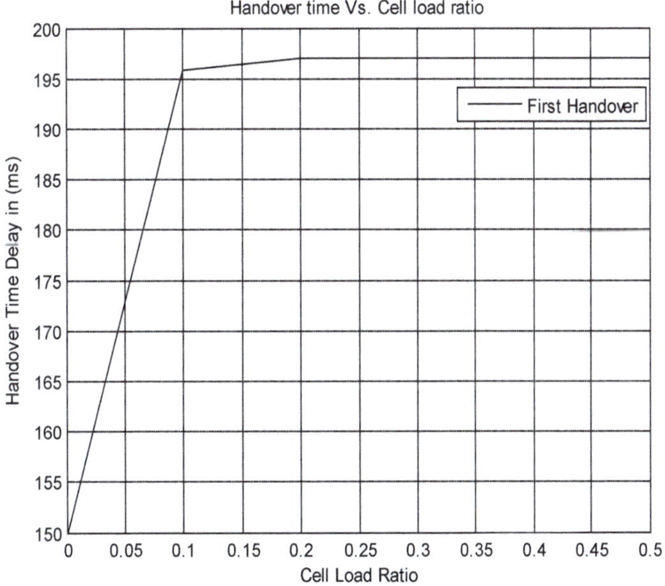

Figure 4. Handover latency vs. load ratio.

Table 9. Produced mobility pattern table.

Base Station ID	Avg. Handover Time	Av. Signal Strength	Av. Load	Av. Load Ratio
1	110.531	−51.4569	48.9743	0.108831911
2	110.539	−51.4569	48.9743	0.108831911
3	110.912	−51.4569	48.9743	0.108831911
4	111.145	−51.4569	48.9743	0.108831911
5	111.662	−51.4569	48.9743	0.108831911
6	112.227	−51.4569	48.9743	0.108831911
7	112.092	−51.4569	48.9743	0.108831911
9	111.436	−51.4569	48.9743	0.108831911
10	111.835	−51.370	48.9808	0.108846428
11	111.014	−51.5440	48.9677	0.108817206
12	111.047	−51.4569	48.9743	0.108831911
13	109.989	−51.4569	48.9743	0.108831911
14	110.423	−51.4561	48.9743	0.108831911
15	110.568	−51.4561	48.9743	0.108831911

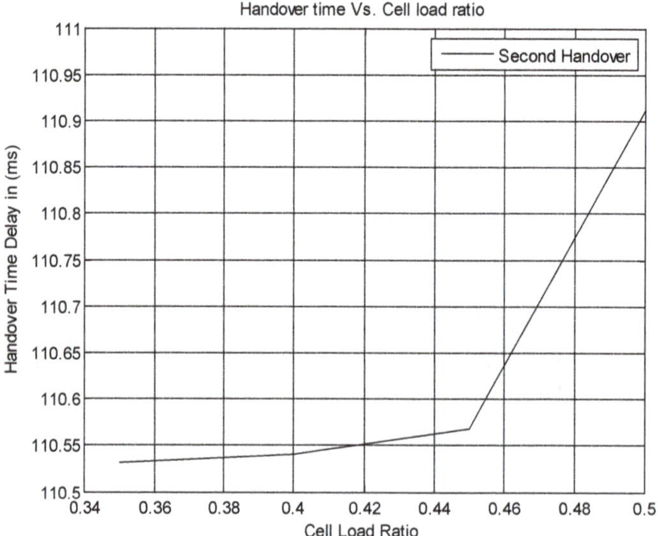

Figure 5. Handover latency vs. cell load ratio (second scan).

Comparison between the first and second handovers is done and shown in **Figure 6**, showing a big difference between the two scans. Call drop probability is tabulated with respect to the cell load ratio, and plotted as shown in **Figure 7**.

5.2. Simulation Results

Simulation results are generated using a number of program executions, where the objective of the proposed model is to understand the effective of the proposed algorithm to reduce the handover time delay. The proposed

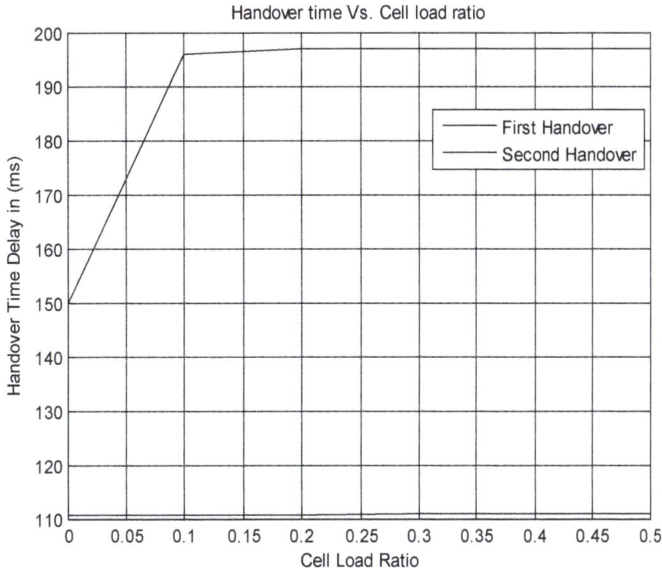

Figure 6. Comparison between first and second handover.

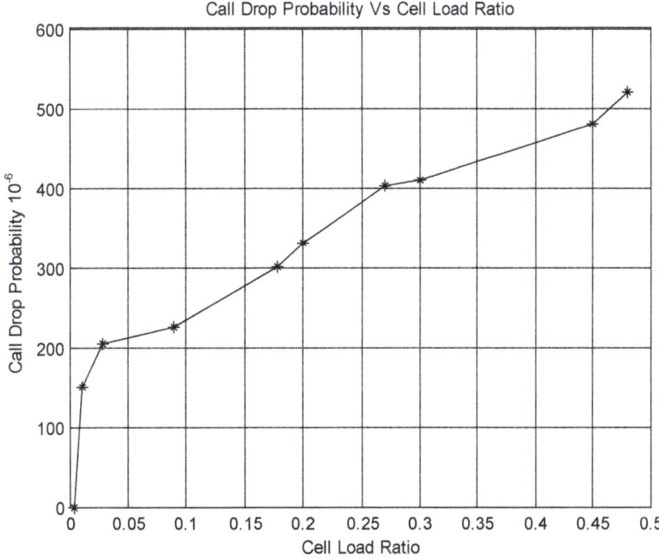

Figure 7. Call drop probability vs. cell load ratio.

model got the following results, as shown in **Figure 8**, the coordinates of the mobile station (MS) with respect to the nearest base stations (BSs), where these coordinates are tabulated in **Table 10**.

The new location of the MS is shown in **Figure 9**.

The mobile stations (MSs) are moved to the target base station (BS), as shown in **Figure 10**.

According to the results obtained from the proposed model, and comparing with the results obtained in Zhang *et al.* [20], and by using the same parameters. It is found that the handover time delay reduced to 111 ms in the proposed model, while in Zhang *et al.* [20] is found to be 197 ms. Which gives a note that the proposed model is higher quality and more effective.

6. Conclusion

To develop a WMN, WiMAX innovation is mainstream to give remote associations in light of the fact that WiMAX has bigger radio extend as opposed to WiFi. Then again, radio stations extent stays to be restricted and

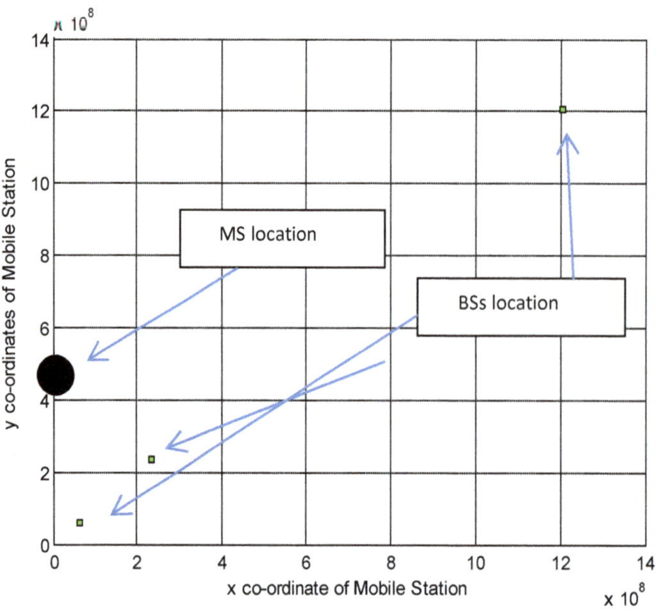

Figure 8. MS & BSs coordinates.

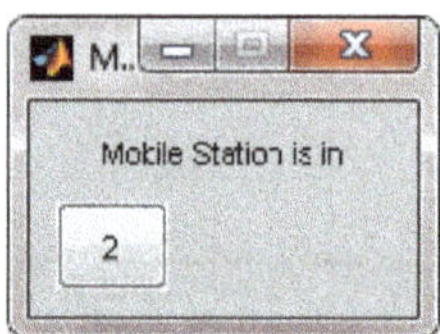

Figure 9. Mobile station's new location.

Table 10. Table of the extractions.

Signal Strengths	Distances	Handover Time
−57.8400	15331.9992	110.5314
−57.0200	7741.4777	110.5397
−55.9100	34715.2311	110.9122
−54.8400	45247.3229	111.1455
−53.4200	31251.7785	111.6628
−51.9300	44634.1075	112.2276
−49.9100	10900.5718	112.0929
−47.5000	38149.2551	111.4365
−43.8900	18334.1673	111.8357
−37.9600	21947.4815	111.0142
−49.9100	8804.9854	111.0474
−47.5000	40939.2001	109.9897
−43.8900	31883.8482	110.4237
−37.9600	37215.4334	110.5686

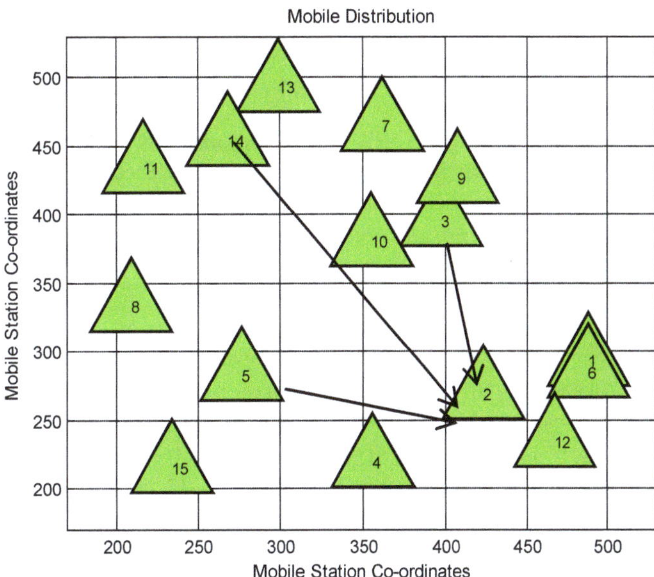

Figure 10. Mobile stations (MSs) movements to the target base station (BS).

Table 11. Comparison table.

No	REF	Handover Time	BS	MS
1	[20]	197 ms	15	150
3	[18]	129	14	10
4	Proposed System	111	15	150

handover methodologies are required to keep up remote associations. Hence, giving quick handovers in WiMAX organizes under the precise quick condition has formed into testing assignment. On this postulation, a productive MAC layer handover plan utilizing versatility examples is introduced to decrease the handover dormancy. Portability examples are embraced to help the SS anticipate the target BS and minimize the filtering time. Reenactment effects exhibit that our plan can lessen the handover dormancy fundamentally which is the important point in cloud computing to obtain more effective results in minimum required time. In this paper, the proposed model decreased handover dormancy time by an acceptable percentage in comparison with Zhang *et al.*, [20]. The comparison is done and tabulated as shown in **Table 11**.

References

[1] Hyeyeon, K., Yang, M., Park, A. and Venkateasan, S. (2008) Handover Prediction Strategy for 3G-WLAN Overlay Networks. *IEEE Explore, Digital Library. Network Operations and Management Symposium, NOMS* 2008. Ministry of Information and Communication Republic of Korea and Institute for Information Technology Advancement, 819-822.

[2] Vasos, V. and Zinonos, Z. (2009) An Analysis of the Handover Latency Components in Mobile IPv6. *Journal of Internet Engineering*, **3**, 230-240.

[3] Khan, J. (2010) Handover Management in GSM Cellular System. *International Journal of Computer Application*, **8**, 14-24. http://dx.doi.org/10.5120/1257-1763

[4] Tripathi, N.D., Reed, J.H. and VanLandingham, H.F. (1998) Handoff in Cellular Systems. *IEEE Personal Communications*, 26-37.

[5] Wong, M., Yee, K. and Turner, P. (2008) A Structured Evidence-Based Literature Review Regarding the Effectiveness of Improvement Interventions in Clinical Handover. The eHealth Services Research Group, University of Tasmania for the Australian Commission on Safety and Quality in Health Care (ACSQHC), Australia, 2-140.

[6] Pandey, P.S. and Badal, N. (2012) Viable Modifications to Improve Handover Latency in MIPv6. *International Jour-*

nal of Advanced Computer Science and Application, **3**, 121-125.

[7] Hsieh, R. and Seneviratne, A. (2003) A Comparison of Mechanisms for Improving Mobile IP Handoff Latency for End-to-End TCP. *MobiCom '03 Proceedings of the 9th Annual International Conference on Mobile Computing and Networking*, ACM, New York, 29-41.

[8] Fu, S. and Atiquzzaman, M. (2005) Handover Latency Comparison of SIGMA, FMIPv6, HMIPv6, and FHMIPv6. Telecommunication & Network Research Lab School of Computer Science, The University of Oklahoma, Norman, 1-7.

[9] Mortaza, S.B., Heijenk, G.B., Laganier, J. and Anand, R. (2008) Reducing Handover Latency in Future IP-Based Wireless Networks: Proxy Mobile IPv6 with Simultaneous Bindings. *Proceeding of International Symposium on a World of Wireless, Mobile and Multimedia Networks*, Newport Beach, 1-10.

[10] Chao, M.W., Yee, K. and Turner, P. (2008) The eHealth Services Research Group, University of Tasmania for the Australian Commission on Safety and Quality in Health Care. University of Tasmania, Tasmania, 3-252.

[11] Purnendu, S.P. and Neelendra, B. (2012) Viable Modifications to Improve Handover Latency in MIPv6. *International* (*IJACSA*) *International Journal of Advanced Computer Science and Applications*, **3**, 121-125. www.ijacsa.thesai.org

[12] Abduloulaziz, I.H., Renfa, L. and Fanzi, Z. (2012) Handover Necessity Estimation for 4G Hesterogeneous Networks. *International Journal of Information Sciences and Techniques*, **2**, 1-13. http://dx.doi.org/10.5121/ijist.2012.2101

[13] Akki, C.B. and Chadchan, S.M. (2009) The Survey of Handoff Issues in Wireless ATM Networks. *International Journal of Nonlinear Science*, **7**, 189-200.

[14] Hu, Y. and Li, V.O. (2001) Logical Topology-Based Routing in LEO Constellations. *Proceeding of International Conference on Communications*, *ICC* 2001. *IEEE*, Helsinki, 3172-3176.

[15] Mushtaq, M., Arman, K. and Ismail, F. (2011) Seamless Handover between UMTS and GPRS. *International Journal of Communication Network and Security*, **1**, 1-4.

[16] An, Y.Y., Yae, B.H., Lee, K.W., Cho, Y.Z. and Jung, W.Y. (2006) Reduction of Handover Latency Using MIH Services in MIPv6. *20th International Conference on Advanced Information Networking and Applications*, 2006. *AINA* 2006, Vol. 2, Vienna, 18-20 April 2006, 229-234.

[17] Vasos, V. and Zinonos, Z. (2009) An Analysis of the Handover Latency Components in Mobile IPv6. *Journal of Internet Engineering*, **3**, 230-240.

[18] Shin, S., Forte, A.G., Rawat, A.S. and Schulzrinne, H. (2004) Reducing MAC Layer Handoff Latency in IEEE 802.11 Wireless LANs. *Proceedings of the 2nd International Workshop on Mobility Management & Wireless Access Protocols*, *MobiWac*'04, 1 October 2004, Philadelphia, 19-26.

[19] Sehlabaka, N.S. and Kogeda, O.P. (2013) A Cross-Layer Based Enhanced Handover Scheme Design in Vehicular *Ad Hoc* Networks. *Proceedings of the World Congress on Engineering and Computer Science*.

[20] Zhang, Z., Pazzi, R.W., Boukerche, A. and Landfeldt, B. (2010) Reducing Handoff Latency for WiMAX Networks Using Mobility Patterns. *Wireless Communications and Networking Conference* (*WCNC*), *IEEE*, Sydney, 1-6.

13

Rerouting Schemes for Wireless ATM Networks

Hasan Harasis

Department of Electrical Engineering, Faculty of Engineering Technology, Albalqa Applied University, Amman, Jordan
Email: harasis.hasan@yahoo.com

Abstract

In this paper, the Wireless ATM rerouting procedures are analyzed and categorized, based on a standard network topology, derived from the Wireless ATM reference model. A new operational concept for a mobile ATM network called Mobile Network Architecture based on Virtual Paths (MNAVP), in which the network nodes are connected to each other via pre-established permanent virtual path connections with fixed capacity assignments is being proposed and described. Finally, the handover hysteresis concept is introduced and a hysteresis gain is defined and calculated as the factor by which the handover rate is reduced through the use of the hysteresis.

Keywords

WATM, Handover, Routing, Hysteresis

1. Introduction

Over the last few years, one of the major commercial successes in the telecommunications world has been the widespread diffusion of cellular mobile telephone services, whose provision relies on sophisticated algorithms implemented by state-of-the-art dedicated computer equipment. Lately, the challenge resides in upgrading the service offer to mobile users to include high-speed data communication services. A natural approach in this direction is to adopt the Asynchronous Transfer Mode (ATM) in the wireless environment, resulting in the so-called wireless ATM (WATM) network. However, ATM was developed for fixed networks and mobility management functionality had to be added to the traditional set of capabilities. Mobile or Wireless ATM consists of two major components: The radio access part which deals with the extension of ATM services over a wireless medium and the mobile ATM part which addresses the issue of enhancing ATM for the support of terminal and service mobility in the fixed portion of the WATM network. Wireless ATM started as a technology designed to be used for LAN or fixed wireless access solutions, where low mobility constraints are encountered. Further research projects and standardization activities coordinated by the ATM Forum demonstrated the feasibility of broadband radio access networks based on ATM technology, which can offer full-scale mobility together with all the range of ATM service capabilities existent also in the fixed ATM networks [1].

Mobility management has two distinct components: location management dealing with the correspondence between the subscriber's data and his current location and handover management which controls the dynamic rerouting and transfer of connection for the terminals crossing cell boundaries.

The frequency-domain supposed to be used for Wireless ATM, situated in the Ghz range, will imply the existence of small size cells, to cope also with the increased demands regarding system capacity. This will lead, in conjunction with a higher terminal mobility, to a very large number of handover of virtual connections. Furthermore, smaller cells have tighter delay constraints, as the overlapping distances of the cells are smaller. The more complex handover procedure has higher requirements regarding radio resource management functions for the air interface paired with network signaling and control functions for handover control, Quality of Service (QoS) management and rerouting of the connection to the new network access point. Exactly these rerouting procedures are the subject of this paper. A new operational concept based on Virtual Paths is introduced and the principles of handover hysteresis are analyzed using a discrete Markov chain model.

2. Wireless ATM Network Architecture

A various number of reference architectures can be taken into account when we talk about Wireless ATM, ranging from simple mobile terminals to complex systems containing mobile ATM switches built in ships, planes or satellites.

One standard reference scenario contains a broadband wireless access system providing unrestricted roaming capabilities within a certain area of continuous radio coverage (**Figure 1**). The base stations (Radio Access Point, RAP) are of picocellular size and implement the physical transport medium, multiple access control, data link control and basic radio res source management capabilities. The RAP does not necessarily have to provide ATM-based physical transport, it could use as well any other access technology, as for example CDMA, also because the error detection and correction capability of the ATM stack is typically low, since it was designed for a reliable network. For this paper, we assume though the existence of an ATM radio interface capable of transmitting ATM cells over the wireless medium. Special Mobile ATM switches (MAS) are positioned at the border of an ATM network, supporting end-system mobility by possessing the necessary extensions in the signaling and control planes to provide functions for mobility management and also connection handover.

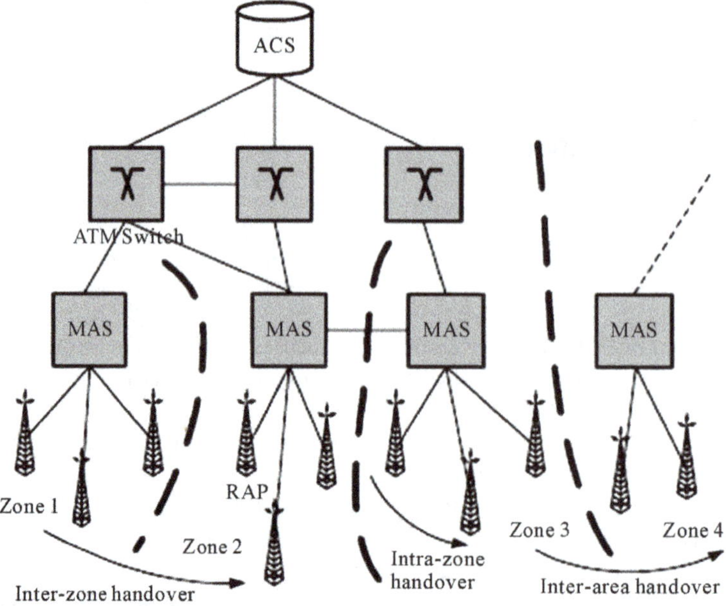

Figure 1. Architecture of the wireless ATM network.

All the RAP's associated with a particular MAS form a so called zone of continuous coverage. Terminal mobility inside a certain zone and the handover associated with it (intra-zone handover) is handled locally by the MAS itself. Neighboring zones with uninterrupted radio coverage can form an area in which, at any time, a RAP can be found to hand a connection over to while the terminal is moving without restrictions. The size of such an area is not limited; it could take the size of the entire network.

It is not mandatory that all the switches are able of supporting end-system mobility, therefore we introduce an hierarchically superior instance, called Area Communication Server (ACS), providing mobility control for a specific area. The ACS represents a mobility supporting ATM switch in charge of processing the protocol requests in case of an inter-zone handover. It also serves as anchor point (AP) for the active connections of the terminals inside this area. By using the ACS, the impact of the end-system mobility on the network can be significantly reduced, because there is no need any more for mobility specific functionality outside the ACS area. The disadvantage consists in the fact that connectivity cannot be guaranteed for terminals leaving this area.

A consequence of the high mobility of the terminals is the requirement of a permanent reestablishment of the virtual connection, in order to reach their current point of access to the network.

This implies, beyond signaling and handover control, a process of rerouting of the connection in the ATM network. QoS control based on requirements coming from the connection itself has to be provided in order to ensure the lossless and in sequence delivery of the A TM cells during the handover process.

3. Connection Rerouting in WATM Networks

We can categorize the approaches for connection rerouting in four basic categories: full reestablishment, connection extension, incremental reestablishment and multicast reestablishment. They are schematically presented in **Figure 2** and **Figure 3**, showing the connection phases during two handover steps for the different basic methods.

The simplest method is the complete reestablishment of the connection. For each change of a RAP coverage area, due to terminal mobility, a completely new VC connection is being set up between the mobile terminal and its peer. This can be done in absence of any defined handover control functions, only by the interaction of the two end systems. The major disadvantages consist in the very long duration of the procedure and the complete interruption of the service. Quite opposite to this procedure, the connection extension handover keeps the impact on a local scale.

Each handover is only prolonging the connection from the old RAP to the new RAP, by this achieving a very high speed, with the cost of a high routing inefficiency, due to the fact that no rerouting of the connection is performed. Loops can easily occur if the terminal is moving back and forth in a limited area, between only few neighboring cells. This method has to be combined with a routing optimization algorithm, otherwise resources are wasted. An example illustrating this scheme is presented in [2].

The multicast concept is also dealing with inefficiencies regarding the utilization of network resources. The multicast tree is established at connection setup time and can remain static or be dynamically updated for the duration of the connection [3] [4]. All routes leading to the RAP's which will be presumly used by the mobile

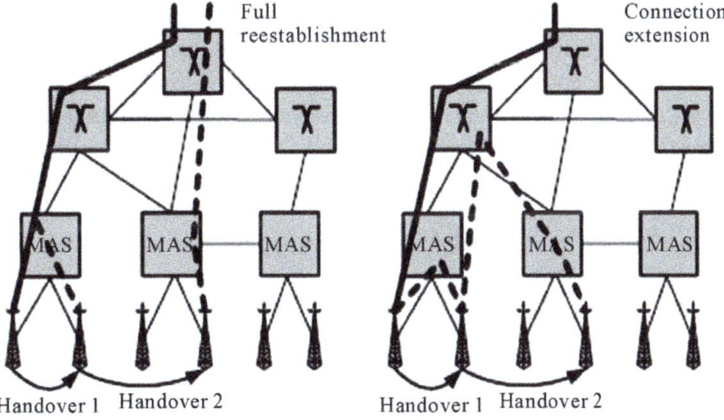

Figure 2. Connection reestablishment and connection extension.

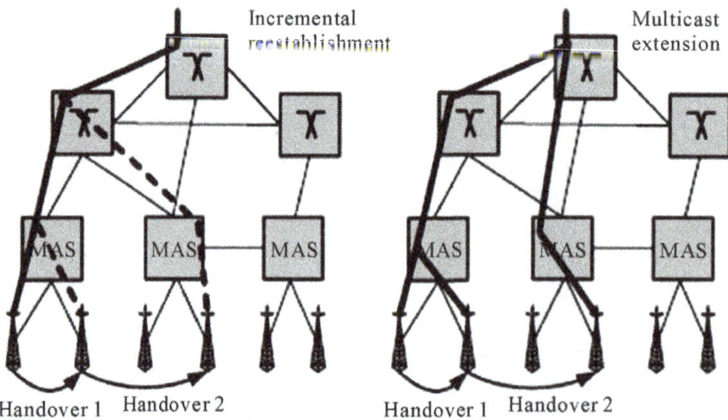

Figure 3. Incremental reestablishment and multicast extension.

terminal during the connection are pre-calculated and assigned as a complete set to the connection at call setup time.

This leads to a very fast handover procedure with a minimum of signaling load because no extra routing and call processing is necessary due to the preparation work done.

Last but not least, the incremental reestablishment represents a more complex and therefore efficient scheme during which a rerouting decision is made for each individual handover. This decision affects only a portion of the connection, namely the one between the new RAP and some Cross Over Switch (COS) inside the current ACS area [5]. The high efficiency and handover speed are due to the calculation of the optimal path to the new destination RAP for each handover. The probability of reusing the longest part of the connection is quite high, which enables a fast handover, without going through the loop of routing decision.

4. Mobile Network Architecture Based on Virtual Paths

Basically, there are two alternatives to operate the mobile A TM network as described in **Figure 1**: either VP-based or VC-based. We choose a VP-based mode, defining a concept called the Mobile Network Architecture based on Virtual Path (MNAVP). In the MNAVP, the MAS are net-worked with their corresponding ACS over the fixed ATM network via pre-established permanent Virtual Path Connections, as shown in **Figure 4**. The VPs of this architecture have fixed capacity assignments defining a virtual mobile network topology over the fixed A TM infrastructure. In this VP-based network, all intermediate A TM switches between ACS and MAS are only performing VP-switching (cross-connect functionality). Two VPCs carried on the same physical link are not statistically multiplexed. Inside a single VPC statistical multiplexing is being applied.

This virtual networking approach has several advantages. First of all, the pre-established VP topology eliminates the need for complex call routing functions and switching table updates along the VC route, which facilitates fast handover connection setup. Second, call admission control decisions only have to be taken in the switches terminating the virtual path connections (MAS and ACS), again reducing connection setup complexity. Further, the establishment of a virtual mobile network is ideally suited for QoS-management and QoS-guarantees in a multi-operator fixed network environment. The heterogeneous nature of multiservice W ATM virtual connections with a broad variety of QoS-constraints and requirements is paid attention to by separating traffic with different QoS-characteristics onto different VPs as proposed for fixed ATM networks [6], *i.e.* connections with similar QoS-requirements are aggregated in one VP-sub-network. Connections carrying multi-rate services can be aggregated in single VPCs, as long as they can be statistically multiplexed. Different types of services, e.g. CBR and VBR, which reduce the statistical multiplexing gain when transported together within a VPC, are separated onto different, parallel VP sub-networks.

The obvious advantages of the MVPA concept are, however, achieved at the cost of losing bundling gain and a somewhat less efficient statistical multiplexing on the physical links, resulting in a reduced utilization of physical resources. The VP-based virtual topology networking concept in the fixed ATM subsystem is peered by VPC operated RAP-links (**Figure 4**), where again VPCs are used for multiservice traffic management between the MAS and the radio access points. With this two-staged approach in the MNAVP design, handovers can be handled in a partly distributed fashion, *i.e.* intra-zone handovers are handled locally by the MAS, whereas the ACS is involved only in inter-zone handovers. To further keep part of the handover processing within a zone, the zone can be extended virtually by the use of VPs connecting RAPs of neighboring zones to a MAS so that actually in the MNAVP virtual network the zones overlap to some extent (**Figure 4**, **Figure 5**). By that, the number of inter-zone handovers can be reduced, and the MAS has to do most of the work in handover call processing.

Consequently, a VC connection in a wireless ATM network consists of two different segments (**Figure 6**):

- The fixed segment from the ACS into the fixed network with the ACS operating as a Cross Over Switch (COS). This segment isn't established at call setup time and doesn't change during the lifetime of a call.

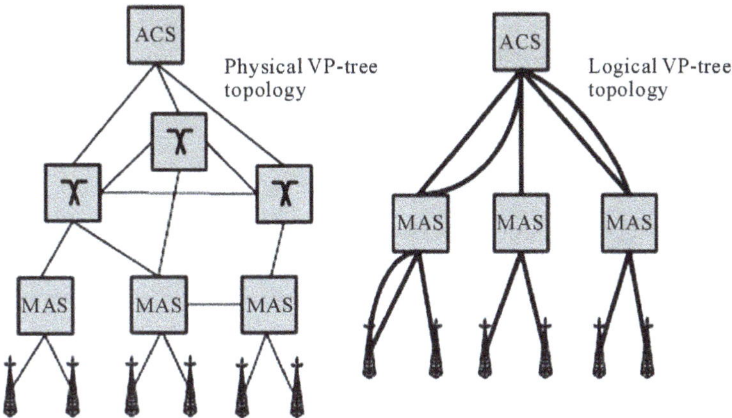

Figure 4. MNAVP topology.

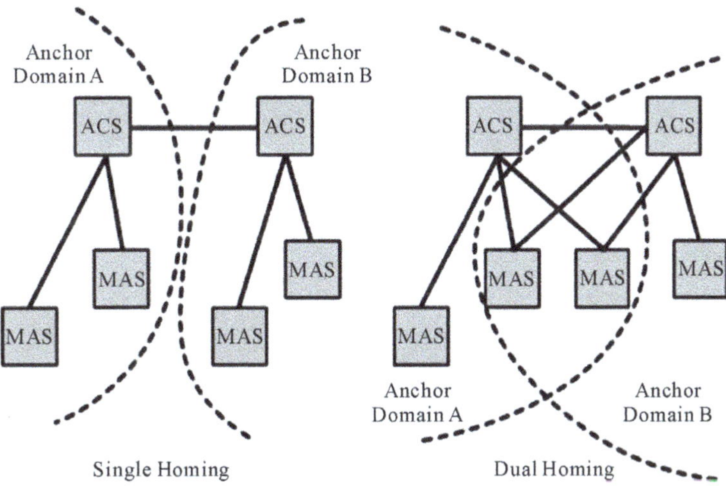

Figure 5. VP based dual homing.

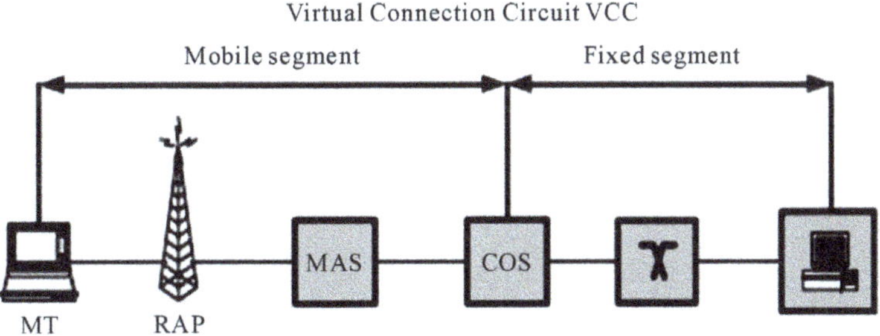

Figure 6. Fixed and mobile segment of a VC connection.

- The mobile segment, which follows a predetermined VP-route in the mobile network from the COS to the current RAP. This segment has to be rerouted during a connection's lifetime due to user mobility.

The most complex handover situation occurs, when the mobile moves to a zone belonging to a different ACS. This inter-zone handover situation generates the highest call processing load. The ACS responsible for the connection during call setup remains the anchoring point (COS) for that connection, switching between the it's fixed and mobile segment. The mobile segment has to be extended from that COS to the new RAP via the second ACS. Therefore, each ACS establishes VPCs with its neighboring ACS, reserved for carrying the inter-ACS handover connections. For best support of wide area mobility, this inter-ACS "backbone" network could have a full mesh topology. The MNA VP connection anchoring mechanism can produce non-optimal routes, which could be optimized by allowing the anchoring point COS to change during the lifetime of a connection. To reduce the number of inter-zone handover, each MAS can be virtually linked to more than one ACS. At connection setup time, one ACS has to be chosen as the connection's COS. This decision can be based on load-balancing considerations as well as mobility prediction analysis in order to minimize the number of inter-zone handover.

5. Handover Hystheresis in MNAVP

The advantage of the logical network of MNAVP relies in the flexibility offered for the case of unexpected high handover traffic between two neighbored zones, called also "anchor domains". This can be implemented by means of a VPC-based Dual Homing of the MAS at the ACS. One MAS is connected to two or more ACS via VP connection, increasing in this way the size of the anchor domain which leads in the end to an partial overlapping of those domains.

At call setup, the connection is switched through the ACS belonging to the actual physical domain. In the situation of an inter-zone handover, when the MT is crossing physical domain boundaries, it still remains in the anchor domain, due to the logical structure of the network. A rerouting of the connection over a new ACS is necessary only in the case of a repeated inter-zone handover. This applies also for the movement in the backward direction. By thus, a routing hysteresis which prevents the frequent occurrence of handover in case of limited geographical mobility is introduced. The hysteresis helps preserving one of the significant advantages of the MNAVP-based rerouting procedure, namely that inside an anchor domain, the mobile segment between ACS I COS and MAS consists of only one VC segment, which would be lost in case of handover to another domain.

The analysis of the hysteresis procedure is made based on a discrete Markov chaining model. Let's consider, for the start, the case without hysteresis, shown in **Figure 7** and **Figure 8**.

This model describes the states of the different types of handover from **Figure 7**. The total sum of transition probabilities for one state is for all i states correspondingly

$$\sum_{j=1}^{n} p_{i,j} = 1 \tag{1}$$

Equation (2) describes the matrix of transition-states

$$p\overline{H} = \begin{bmatrix} p_{1,1} & & p_{3,1} & & & & & & & \\ p_{1,2} & & p_{3,2} & & & & & & & \\ & p_{2,3} & & p_{4,3} & & & & & & \\ & p_{2,4} & & p_{4,4} & & p_{6,4} & & & & \\ & & & p_{4,5} & & p_{6,5} & & & & \\ & & & & p_{5,6} & & p_{7,6} & & & \\ & & & & p_{5,7} & & p_{7,7} & & p_{9,7} & \\ & & & & & & p_{7,8} & & p_{9,8} & \\ & & & & & & & p_{8,9} & & p_{10,9} \\ & & & & & & & p_{8,10} & & p_{10,10} \end{bmatrix} \tag{2}$$

To be able to calculate the probability of occurrence of a handover, the state-probabilities of the Markov-chain are needed

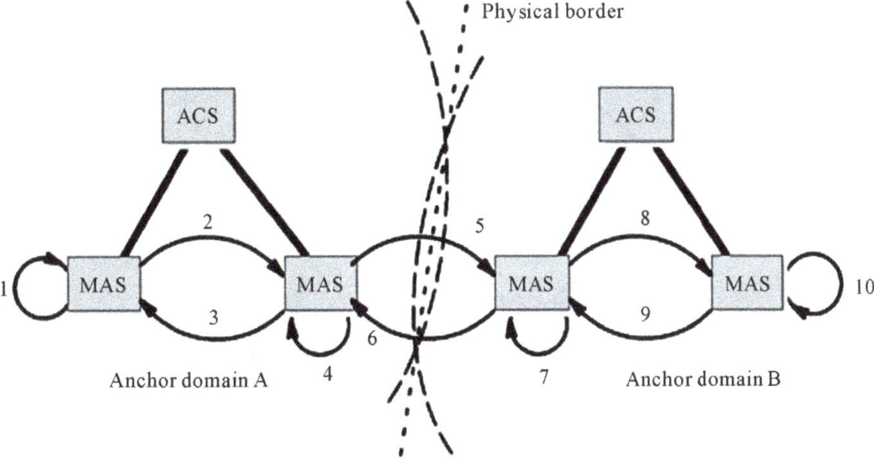

Figure 7. Handover between neighboring domains without hysteresis.

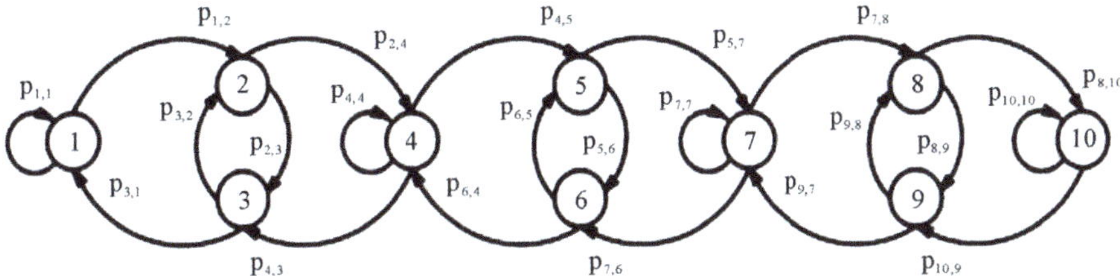

Figure 8. Discrete Markov model for the handover succession without hysteresis.

$$p\bar{H} \cdot P = P \tag{3}$$

After homogenizing the equation system and converting the matrix, we obtain Equation (4)

$$p'\bar{H} = \begin{bmatrix} p_{1,1-1} & & p_{3,1} & & & & & & & \\ & -1 & 1 & & & & & & & \\ & & p_{2,3-1} & p_{4,3} & & & & & & \\ & & & -p_{4,5} & & p_{6,4} & & & & \\ & & & & -1 & 1 & & & & \\ & & & & & p_{5,6-1} & p_{7,6} & & & \\ & & & & & & -p_{7,8} & & p_{9,7} & \\ & & & & & & & -1 & 1 & \\ & & & & & & & & p_{8,9-1} & p_{10,9} \\ & & & & & & & & & 0 \end{bmatrix} \tag{4}$$

and with the condition $p-e=1$ for the sum of the state probabilities, the equation is transformed in

$$\sum_{i=1}^{n} p_i = 1 \tag{5}$$

The probabilities of assuming the states 5 and 6 (corresponding to the inter-domain handover cases in **Figure 7** and **Figure 8**) are in fact the probabilities of occurrence of this handover type.

In the case of a handover hysteresis, the handover sequence depends upon the network nodes involved, as shown in **Figure 9** and **Figure 10**, The Markov model has to be extended with the corresponding states and the interzone-handover are in this case represented by the states 8a and 3b, Similar to before, we obtain the Equation (6).

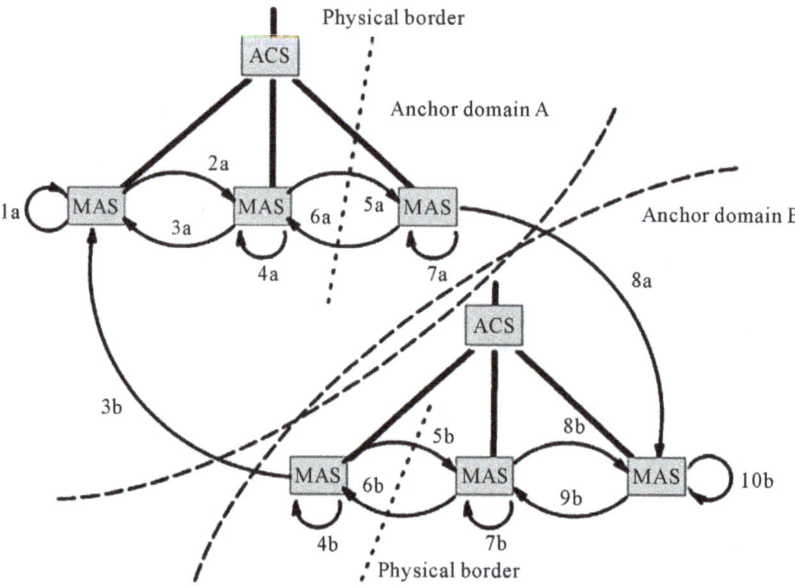

Figure 9. Handover between neighboring domains with hysteresis.

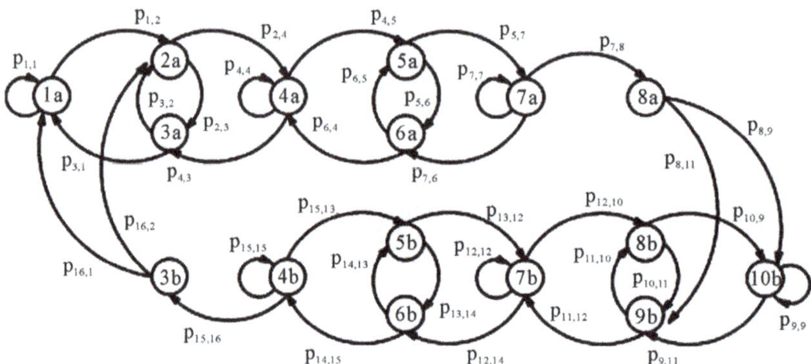

Figure 10. Discrete Markov model for the handover succession with hysteresis.

$$P'\bar{H} = \left[\begin{array}{cccccccccccccccc}
p_{1,1-1} & & p_{3,1} & & & & & & & & & & & & & p_{16,1} \\
 & -1 & 1 & & & & & & & & & & & & & 1 \\
 & & p_{2,3-1} & p_{4,3} & & & & & & & & & & & & p_{2,3} \\
 & & & p_{4,5-1} & -1 & p_{6,5} & & & & & & & & & & \\
 & & & & -1 & 1 & & & & & & & & & & 1 \\
 & & & & & p_{5,6-1} & p_{7,6} & & & & & & & & & p_{5,6} \\
 & & & & & & -p_{7,8} & & & & & & & & & 1 \\
 & & & & & & & -1 & & & & & & & & 1 \\
 & & & & & & & & p_{9,9-1} & p_{10,9} & & & & & & p_{8,9} \\
 & & & & & & & & & -1 & & & & & & \\
 & & & & & & & & & & p_{11,10} & p_{12,10} & & & & 1 \\
 & & & & & & & & & & p_{11,10-1} & p_{12,10} & & & & 1 \\
 & & & & & & & & & & & -p_{12,14} & p_{13,13} & & & 1 \\
 & & & & & & & & & & & & 1 & -1 & & 1 \\
 & & & & & & & & & & & & & p_{14,13-1} & p_{15,13} & 1 \\
 & & & & & & & & & & & & & & -p_{15,16} & 0
\end{array}\right] \quad (6)$$

Now, the effect of the handover hysteresis can be quantified. The hysteresis gain can be defined as being the factor by which the frequency of occurrence of an inter-zone handover between anchor domains can be reduced,

$$\mathrm{GH} = \frac{\rho_5 + \rho_6}{p_{3b} + \rho_{8a}} \tag{7}$$

The border conditions are considered symmetrical. Let's consider the random chosen variables from **Table 1**, which describe the local movement in terms of the probability that the MT does not leave, during its movement, the region it belongs to. The parameter of the model with hysteresis described in **Figure 8** can be determined by inserting the probabilities from **Table 1** at the zone borders and, at the same time, respecting condition (1).

The main parameter which describes the hysteresis gain is the probability of occurrence of consecutive inter-zone handover. If this probability equals zero, the MT crosses the border, in one direction, only once. If it is high, the movement area is very restricted and the frequency of the border crossing high. **Figure 11** shows this dependency and it can be observed that even in the case of a non-local movement, when PS, 6 = 0, the probability of handover-occurrence is reduced to almost one third. This is based on the fact that, due to the hysteresis, one intra-zone handover at the border domain (handover 4 and 7 in **Figure 7**) is not immediately followed by an inter-zone handover, because, as shown in **Figure 9**, in this case of restricted mobility area, only inter-zone handover are performed (handover 4a and 7b).

6. Conclusion

In this paper, Wireless ATM reference architecture has been presented, consisting of a wireless access system coupled with the support of mobile end-systems within the ATM network. The network is subdivided in several areas served by one area server ACS, areas which are in fact handover domains providing uninterrupted radio coverage and full mobility support. After a brief description of the rerouting schemes proposed already, a new networking concept called MNA VP has been proposed, establishing a virtual mobile sub-network over the fixed ATM infrastructure, based on pre-established virtual paths with fixed capacity assignments. This network design minimizes signaling and call admission load, maintaining a low handover connection setup latency and facilitating high handover rates. The concept of handover hysteresis is being introduced, with the benefit of decreasing the number of inter-zone handover, for terminals showing limited geographical mobility. The frequency

Table 1. Parameter for the numerical example of hand over hysteresis.

Parameter	Value
$P_{1,1}$, $P_{10,10}$	0.8
$P_{4,4}$, $P_{7,7}$	0.1
$P_{2,3}$, $P_{8,9}$	0.3

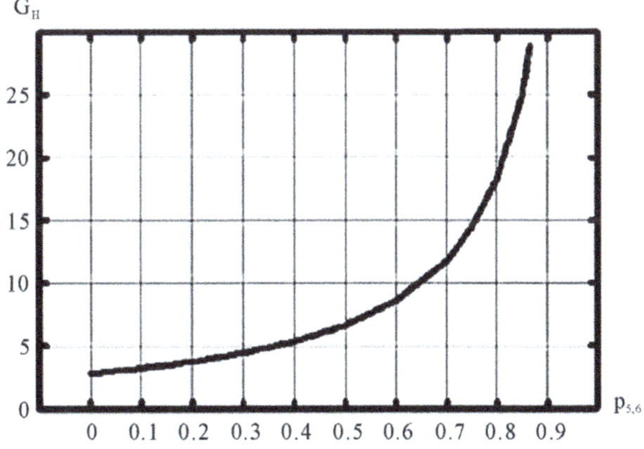

Figure 11. Hysteresis gain.

of occurrence of this cost-intensive handover type can be decreased with beneficial effects on the resource budget, and this is demonstrated using a Markov decision process. A hysteresis gain is defined and calculated as the factor by which the handover rate is being reduced through the use of the hysteresis.

References

[1] Dellaverson, L. (1996) Reaching for the New Frontier. 53 *Bytes—The AIM Forum Newsletter*, **4**.

[2] Eng, K.Y., Karol, M.J., Veeraraghavan, M., Ayanoglu, E., Woodworth, C.B., Pancha, P. and Valenzuela, R.A. (1995) BAHAMA: A Broadband *Ad-Hoc* Wireless ATM Local-Area Network. *ICC* 95, *Conference Proceedings*, *IEEE*, 1216-1223.

[3] Yu, O.T.W. and Leung, V.C.M. (1996) Connection Architecture and Protocols to Support Efficient Handoffs over an ATM/B-ISDN Personal Communications Network. *MONET—Mobile Networks and Applications*, **1**, 123-139.

[4] Acampora, A.S. and Naghshineh, M. (1994) Control and Quality-of-Service Provisioning in High-Speed Microcellular Networks. *IEEE Personal Communications Magazine*, **1**.

[5] T'oh, C.-K. (1996) A Hybrid Handover Protocol for Local Area Wireless ATM Networks. *MONET—Mobile Networks and Applications*, **1**, 313-334.

[6] ATM-Forum (2002) ATM User-Network Interface Specification. Version 4.1, The AIM Forum.

14

Ternary Zero Correlation Zone Sequence Sets for Asynchronous DS-CDMA

Benattou Fassi[1], Ali Djebbari[1], Abdelmalik Taleb-Ahmed[2]

[1]Telecommunications and Digital Signal Processing Laboratory, Djillali Liabes University of Sidi Bel Abbes, Sidi Bel Abbes, Algeria
[2]LAMIH UMR CNRS 8530, University of Valenciennes and Hainaut-Cambresis (UVHC), le Mont Houy, France
Email: fassibenattou@yahoo.fr, adjebari2002@yahoo.fr, abdelmalik.taleb-ahmed@univ-valenciennes.fr

Abstract

In this paper we propose a new class of ternary Zero Correlation Zone (ZCZ) sequence sets based on binary ZCZ sequence sets construction. It is shown that the proposed ternary ZCZ sequence sets can reach the upper bound on the ZCZ sequences. The performance of the proposed sequences set in asynchronous Direct Sequence-Code Division Multiple Access (DS-CDMA) system is evaluated. In the simulation we used two types of channels: Additive White Gaussian Noise (AWGN) and frequency non-selective fading with AWGN noise. The proposed ternary ZCZ sequence sets show better results, in term of Bit Error Rate (BER), than Hayashi's ternary ZCZ sequence sets.

Keywords

Hadamard Matrix, Zero Correlation Zone Sequences, Correlation, Asynchronous DS-CDMA, BER

1. Introduction

In Code Division Multiple Access (CDMA) systems, the number of spreading sequences determines the number of users and their correlation properties have a significant effect on anti-interference performance of the system [1]. Different types of codes used in communications systems have been studied in order to reduce Multiple Access Interference (MAI) [2] [3]. For an interference-free communication, spreading codes should have zero auto-correlation and zero cross-correlation functions at out-of-phase state. So, spreading sequences with good correlation properties can be used to improve the performance of CDMA systems [3]. One class of spreading sequences called Zero Correlation Zone (ZCZ) sequences possesses good correlation properties but only in specific zones called Zero Correlation Zone (Z_{CZ}). There are several intensive studies of CDMA systems using ZCZ sequences sets [1] [3]-[6]. Various classes of ternary ZCZ sequences sets have been constructed [3] [4] and [7]-[13]. Ternary ZCZ sequences have the advantage over binary ZCZ sequences that is, for a given sequence length, the set has longer Z_{CZ} lengths and more sequences, and we may employ such hardware in binary ZCZ sequence sets system [3]. Any ternary ZCZ sequences set TZCZ (L', M', Z'_{CZ}) could be characterized by the sequence length L', the number of sequences M' and the zero correlation zone length Z'_{CZ}. An optimal ZCZ set is the one that provides the maximum number of codes for a given Z'_{CZ} and sequences lengths. The proposed ternary ZCZ sequences set with TZCZ (L', M', Z'_{CZ}) is derived from a binary ZCZ sequence set with BZCZ (L, M, Z_{CZ}). When compared with previous works on ternary ZCZ sequence sets [3] [4] and [7]-[13], our proposed ZCZ sequence set approaches optimality.

The remainder of the paper is organized as follows.

After a review of preliminary considerations in Section 2, the proposed design for sequence construction is explained in Section 3. Example of new ZCZ sequence sets are presented in Section 4. The properties of the proposed sequence sets are explained in Section 5. In Section 6, we consider the performance of the proposed ternary ZCZ sequence sets compared with those in [3] and [10] for the asynchronous DS-CDMA system in both AWGN and nonselective fading with AWGN noise channels. At the end, we draw the concluding remarks.

2. Preliminaries

2.1. Definition 1

For a pair of sequences X_j and X_v of length L, the aperiodic correlation function (ACF) $\varphi_{(X_j,X_v)}(\tau)$ is defined as follows [14] [15]:

$$\varphi_{(X_j,X_v)}(\tau)=\begin{cases}\sum_{i=0}^{L-\tau-1} x_{j,i}x_{v,(i+\tau)}, & \text{if } 0\le\tau<L\\ \sum_{i=0}^{L+\tau-1} x_{j,(i-\tau)}x_{v,\tau}, & \text{if } -L<\tau<0\\ 0, & \text{if } |\tau|\ge L\end{cases} \tag{1}$$

The periodic correlation function (PCF) between X_j and X_v at a lag τ is determined by [15] [16]:

$$\forall\tau,\ \theta_{(X_j,X_v)}(\tau)=\varphi_{(X_j,X_v)}(\tau \bmod L)+\varphi_{(X_j,X_v)}((\tau \bmod L)-L) \tag{2}$$

$$\forall\tau\ge0,\ \theta_{(X_j,X_v)}(\tau)=\sum_{i=0}^{L-1} x_{j,i}x_{v,(i+\tau)\mathrm{mod}(L)} \text{ and } \theta_{(X_j,X_v)}(-\tau)=\theta_{(X_v,X_j)}(\tau) \tag{3}$$

2.2. Definition 2

A set of M sequences $\{X_j\}_{j=0}^{M-1}=\{X_0,X_1,\cdots,X_{M-1}\}$ is called zero correlation zone sequence set if the periodic correlation functions satisfy [15] [16]:

$$0<|\tau|\le Z_{cz},\ \theta_{(X_j,X_j)}(\tau)=0 \tag{4}$$

and

$$(j\ne v),\ 0\le|\tau|\le Z_{cz},\ \theta_{(X_j,X_v)}(\tau)=0 \tag{5}$$

3. Proposed Sequence Construction

The construction procedure of the new ternary sequence sets is presented. The construction is accomplished across the following three steps:

Step 1: The jth row of the Hadamard matrix H of order n is indicated by $h_j=[h_{j,0},h_{j,1},\cdots,h_{j,n-1}]$. A set of $2n$ sequences d_j, each of length $2n$, is constructed as follows [17]:

For $0\le j<n$

$$d_{j+0} = \left[-h_j, h_j\right] \tag{6}$$

$$d_{j+1} = \left[h_j, h_j\right] \tag{7}$$

Step 2: For the first stage $p = 0$, and for a fixed integer value n, we can generate, based on the schema for sequence construction in [15], a series of sets $\left\{B_j\right\}_{j=0}^{2n-1}$ and $\left\{T_j\right\}_{j=0}^{2n-1}$ of $2n$ sequences as follows:

Both sequences sets $\left\{B_j\right\}_{j=0}^{2n-1}$ and $\left\{T_j\right\}_{j=0}^{2n-1}$ are constructed from the sequences set $\left\{d_j\right\}_{j=0}^{2n-1}$. A pair of sequences B_{j+0} and B_{j+1} of length $L = \left(2^{p+2}n\right)$ are constructed by applying the interleaving operation of a sequence pair $\left(d_{j+0}, d_{j+1}\right)$ in Equations (6) and (7) [17] and a pair of sequences T_{j+0} and T_{j+1} are constructed by the interleaving operation of a sequence pair d_{j+0}, d_{j+1} and from padding Z, which are zeros of length K, as follows:

For $0 \le j < n$,

$$B_{j+0} = \left[d_{j+0,0}, d_{j+1,0}, d_{j+0,1}, d_{j+1,1}, \cdots, d_{j+0,2n-1}, d_{j+1,2n-1}\right] \tag{8}$$

$$B_{j+1} = \left[d_{j+0,0}, -d_{j+1,0}, d_{j+0,1}, -d_{j+1,1}, \cdots, d_{j+0,2n-1}, -d_{j+1,2n-1}\right] \tag{9}$$

$$T_{j+0} = \left[d_{j+0,0}, Z, d_{j+1,0}, Z, d_{j+0,1}, Z, d_{j+1,1}, \cdots, d_{j+0,2n-1}, Z, d_{j+1,2n-1}, Z\right] \tag{10}$$

$$T_{j+1} = \left[d_{j+0,0}, Z, -d_{j+1,0}, Z, d_{j+0,1}, Z, -d_{j+1,1}, \cdots, d_{j+0,2n-1}, Z, -d_{j+1,2n-1}, Z\right] \tag{11}$$

The length of a pair sequences T_{j+0} and T_{j+1} in Equations (10) and (11) is $L' = \left(2^{p+2}n + 2^{p+K+1}n\right)$ and the member size of a pair sequence sets $\left\{T_j\right\}$ and $\left\{B_j\right\}$ is $M = M' = 2n$.

Step 3: For $p > 0$, we may recursively build a new series of sets $\left\{B_j\right\}_{j=0}^{2n-1}$ and $\left\{T_j\right\}_{j=0}^{2n-1}$ by interleaving actual sets $\left\{B_j\right\}_{j=0}^{2n-1}$ and $\left\{T_j\right\}_{j=0}^{2n-1}$ respectively. Sets $\left\{B_j\right\}_{j=0}^{2n-1}$ and $\left\{T_j\right\}_{j=0}^{2n-1}$ are generated as follows:

For $0 \le j < n$,

$$B_{j+0} = \left[B_{j+0,0}, B_{j+1,0}, B_{j+0,1}, B_{j+1,1}, \cdots, B_{j+0,4n-1}, B_{j+1,4n-1}\right] \tag{12}$$

$$B_{j+1} = \left[B_{j+0,0}, -B_{j+1,0}, B_{j+0,1}, -B_{j+1,1}, \cdots, B_{j+0,4n-1}, -B_{j+1,4n-1}\right] \tag{13}$$

$$T_{j+0} = \left[T_{j+0,0}, T_{j+1,0}, T_{j+0,1}, T_{j+1,1}, \cdots, T_{j+0,(4n(1+K))-1}, T_{j+1,(4n(1+K))-1}\right] \tag{14}$$

$$T_{j+1} = \left[T_{j+0,0}, -T_{j+1,0}, T_{j+0,1}, -T_{j+1,1}, \cdots, T_{j+0,(4n(1+K))-1}, -T_{j+1,(4n(1+K))-1}\right] \tag{15}$$

The length of both sequences B_{j+0} and B_{j+1} in Equations (12) and (13) is equal to $\left(2^{p+2}n\right)$ [17] and the length of both sequences T_{j+0} and T_{j+1} is equal to $\left(2^{p+2}n + 2^{p+1+K}n\right)$.

4. Example of Construction

1) For $n = 2$ and $p = 2$, the $\left\{B_j\right\}_{j=0}^{3}$ is generated as follows [17]:

$$B_{0+0} = [-1,-1,-1,1,1,1,-1,1,-1,-1,-1,1,1,1,-1,1,1,1,1,-1,1,1,-1,1,1,1,1,-1,1,1,-1,1]$$

$$B_{1+0} = [-1,-1,-1,1,1,1,-1,1,1,1,1,-1,-1,-1,1,-1,1,1,1,-1,1,1,-1,1,-1,-1,-1,1,-1,-1,1,-1]$$

$$B_{0+1} = [-1,1,-1,-1,1,-1,-1,-1,-1,1,-1,-1,1,-1,-1,-1,1,-1,1,1,1,-1,-1,-1,1,-1,1,1,1,-1,-1,-1]$$

$$B_{1+1} = [-1,1,-1,-1,1,-1,-1,-1,1,-1,1,1,-1,1,1,1,1,-1,1,1,1,-1,-1,-1,-1,1,-1,-1,-1,1,1,1]$$

Figure 1 shows the periodic auto-correlation function (PACF) $(\forall j, j = v)$ given in Equation (3) of B_{0+0}, and **Figure 2** shows the periodic cross-correlation function (PCCF) $(\forall j, j \ne v)$ given in Equation (3) of B_{0+0} with B_{1+0}.

The PACF and PCCF confirm that $\left\{B_j\right\}$ is a ZCZ (32, 4, 4) sequence set.

2) For $n = 2$ and $p = 2$, the proposed $\left\{T_j\right\}_{j=0}^{3}$ is generated as follows:

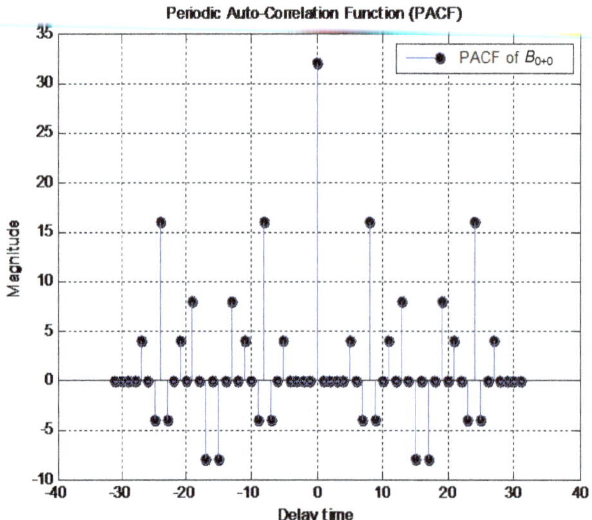

Figure 1. PACF of B_{0+0}.

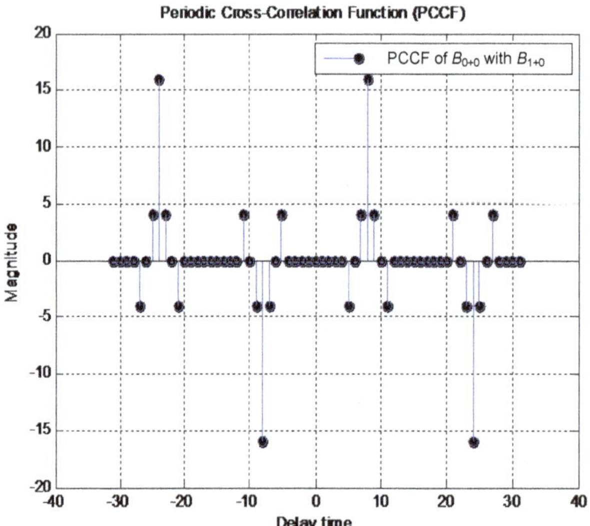

Figure 2. PCCF of B_{0+0} with B_{1+0}.

$$T_{0+0}=\begin{bmatrix} -1,-1,-1,1,0,0,0,0,1,1,-1,1,0,0,0,0,-1,-1,-1,1,0,0,0,0,1,1,-1,1, \\ 0,0,0,0,1,1,1,-1,0,0,0,0,1,1,-1,1,0,0,0,0,1,1,1,-1,0,0,0,0,1,1,-1,1,0,0,0,0 \end{bmatrix}$$

$$T_{1+0}=\begin{bmatrix} -1,-1,-1,1,0,0,0,0,1,1,-1,1,0,0,0,0,1,1,1,-1,0,0,0,0,-1,1,-1,1,0,0,0,0, \\ 1,1,1,-1,0,0,0,0,1,1,-1,1,0,0,0,0,-1,-1,-1,1,0,0,0,0,-1,-1,1,-1,0,0,0,0 \end{bmatrix}$$

$$T_{0+1}=\begin{bmatrix} -1,1,-1,-1,0,0,0,0,1,-1,-1,-1,0,0,0,0,-1,1,-1,-1,0,0,0,0,1,-1,-1,-1,0, \\ 0,0,0,1,-1,1,1,1,0,0,0,0,-1,-1,-1,1,0,0,0,0,-1,1,11,0,0,0,0,-1,-1,-1,0,0,0,0 \end{bmatrix}$$

$$T_{1+1}=\begin{bmatrix} -1,1,-1,-1,0,0,0,0,1,-1,-1,-1,0,0,0,0,1,-1,1,1,0,0,0,0,-1,1,1,1,0,0,0,0, \\ 1,-1,1,1,0,0,0,0,1,-1,-1,-1,0,0,0,0,-1,1,-1,-1,0,0,0,0,-1,1,1,1,0,0,0,0 \end{bmatrix}$$

Figure 3 shows the PACF of T_{0+0}, and **Figure 4** shows the PCCF of T_{0+0} with T_{1+0}. The PACF and PCCF confirm that $\{T_j\}$ is a TZCZ (64, 4, 12) sequences set.

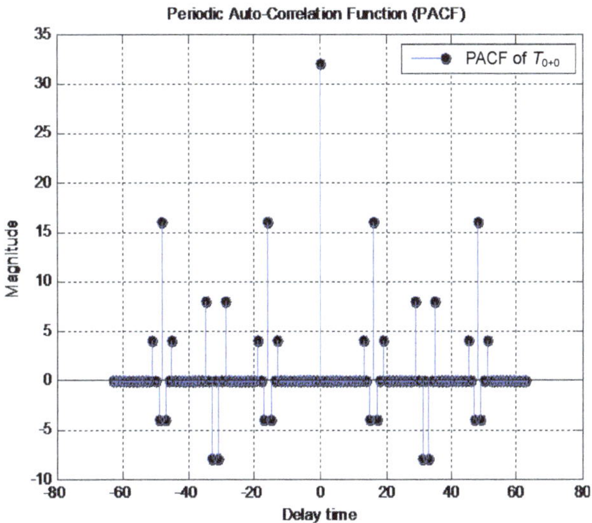

Figure 3. PACF of T_{0+0}.

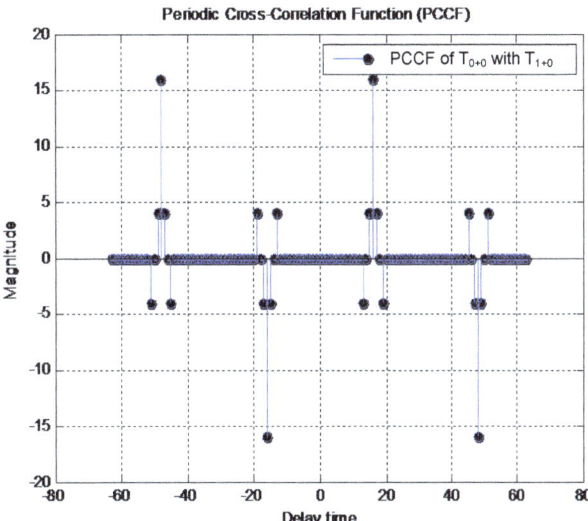

Figure 4. PCCF of T_{0+0} with T_{1+0}.

5. The Features of the Proposed Sequence

The Binary ZCZ sequence set with BZCZ $(L, M, Z_{CZ}) = (2^{p+2}n, 2n, 2^p)$ is optimal or approach optimal binary ZCZ sequences [17]. The length of T_j in Equations (14) and (15), equals $(L = 2^{p+2}n + 2^{p+1+K}n)$, is twice that of T_j in Equations (10) and (11). Let $F = (2^{p+1})$ and $Z_p = (2^{p+K})$, the proposed ternary ZCZ sequence set with TZCZ $(L', M', Z'_{CZ}) = (2^{p+2}n + 2^{p+K+1}n, 2n, 2^p + 2^{p+K}) = (FM + Z_p M, M, Z_{CZ} + Z_p)$ is derived from the binary ZCZ sequence set BZCZ $(L, M, Z_{CZ}) = (FM, M, Z_{CZ})$. If the obtained ternary ZCZ sequence set is optimal, it satisfies the ratio $M' = L'/(Z'_{CZ} + 1)$ [9] [10].

If $p = 0$ (the first iteration), the obtained sequence set is an optimal ternary ZCZ sequence set.

Proof: Let $R = \frac{M'(Z'_{CZ} + 1)}{L'}$, we calculate the following ratio:

$$R = \frac{M'(Z'_{CZ} + 1)}{L'} = \frac{2^p + 2^{p+K} + 1}{2^{p+1} + 2^{p+K}}$$

If $p = 0$, $\forall K$, $R = 1$.

If $p \neq 0$ and the number (K) of the padded zeros tends to the infinite, the obtained TZCZ sequences set (see **Table 1** for $n = 2$) is asymptotically optimal.

Proof: $R = \frac{M'(Z'_{CZ}+1)}{L'} = \frac{2^K+1}{2^K}$ and $\lim_{K\to\infty}(R) = 1$

Noted that for this case, after spreading, the power of the symbol will decrease sharply, it is mandatory to compensate it, but this requirement increases Peak-to-Average Power Ratio (PAPR) and dynamic range of the transmitted signal [3].

From **Table 2** we can see that the proposed sequence set in this paper can provide certain benefits. The length of sequences and Z_{CZ} will increase together while the number of sequences remains unchanged.

In the asynchronous DS-CDMA system, the time delay is typically in some chips, and for this, we can increase the Z_{CZ} to reduce MAI, but the member size will be relatively petty.

For a given member size M, we can find various sets of sequences with different lengths L and Z_{CZ}. As an example, assuming that $M = 8$ and $K = 3$ (see **Table 3** for $n = 4$), we can draw upper bounds of our code performance and compare it with the Hayashi's approach.

The Hayashi's ternary sequence sets ZCZ-$\left((n+1)2^{p+2}, 2n, 2^{p+1}-1\right)$ in [10] based on Hadamard matrix, the member size of the sequence set is $\frac{n}{n+1}$ of theoretical upper bound.

It is clear from **Table 3**, the proposed construction is one of the better type in the constructions mentioned in **Table 2**. Compared with the Hayashi's work, our proposed, in all cases, is optimal or approximate optimal ZCZ

Table 1. The parameters of the proposed ternary ZCZ sequence set and the ratio R.

p	0	0	0	0	0	1	1	1	1	1	2	2	2	2	2
K	1	2	3	4	5	1	2	3	4	5	1	2	3	4	5
L'	16	24	40	72	136	32	48	80	144	272	64	96	160	288	544
Z'_{CZ}	3	5	9	17	33	6	10	18	34	66	12	20	36	68	132
R	1	1	1	1	1	0.88	0.92	0.95	0.97	0.99	0.81	0.88	0.93	0.96	0.98

Table 2. Comparison of three types of ZCZ sequence sets.

Constructed ZCZ sequence sets	Length of sequences	Sequence element	Number size	Z_{CZ}	Performance parameter R
ZCZ-$\left(2^{p+2}n, 2n, 2^p\right)$ in [17]	$2^{p+2}n$	Binary	$2n$	2^p	$\frac{(2^p+1)}{2^{p+1}}$
ZCZ-$\left((n+1)2^{p+2}, 2n, 2^{p+1}-1\right)$ in [10]	$2^{p+2}(n+1)$	Ternary	$2n$	$2^{p+1}-1$	$\frac{n}{n+1}$
ZCZ-$\left(2^{p+2}n+2^{p+K+1}n, 2n, 2^p+2^{p+K}\right)$ in this paper	$2^{p+2}n+2^{p+K+1}n$	Ternary	$2n$	2^p+2^{p+K}	$\frac{2^p+2^{p+K}+1}{2^{p+1}+2^{p+K}}$

Table 3. The parameters of two constructions of ternary ZCZ sequence sets.

Sequence sets	Hayashi's sequence set				Proposed sequence set			
L	80	160	320	640	80	160	320	640
Z_{CZ}	7	15	31	63	9	18	36	72
R	0.8	0.8	0.8	0.8	1	0,950	0.925	0.912

sequence set. Consequently it has a higher Z_{CZ} and better performance parameter R than that given by Hayashi's construction.

6. The Performance of the Asynchronous DS-CDMA System Using the Proposed Ternary ZCZ Sequence Set

In this section, we consider the performance of the proposed ternary ZCZ sequence set used as spreading sequences for the DS-CDMA system shared by M asynchronous users simultaneously. In order to show this performance, the BER of an asynchronous DS-CDMA system over a frequency non-selective fading channel with AWGN noise estimated in [18] [19] is used:

$$\text{BER} = Q\left(\frac{T\sqrt{\frac{P}{2}}}{\sqrt{\frac{N_0 T}{4} + \frac{TP^2\gamma^2}{4} + \left(1+\gamma^2\right)\sigma_A^2(i)}}\right) \tag{16}$$

where P is the common received power, T is the symbol duration, the Q function in Equation (16) is given by $Q(z) = \int_z^\infty \frac{1}{\sqrt{2\pi}} e^{\frac{-y^2}{2}} dy$ [20], the term $\left(\sigma_n^2 = \frac{N_0 T}{4}\right)$ is the variance for the AWGN noise, the term $\left(TP^2\gamma^2/4\right)$ denote the faded component power from the user i and $\sigma_A^2(i)$ is the global (non-faded) interference MAI power for the required i-th user.

Let $\gamma^2 = 0$, the BER of an asynchronous DS-CDMA system over AWGN channels is:

$$\text{BER} = Q\left(\frac{T\sqrt{\frac{P}{2}}}{\sqrt{\frac{N_0 T}{4} + \sigma_A^2(i)}}\right) \tag{17}$$

The MAI variance for the required i-th user can be calculated as [18] [19]

$$\sigma_A^2(i) = \frac{TP^2}{12L^3}\left(\sum_{\substack{m=1\\ m\neq i}}^{M} r_{m,i}\right) \tag{18}$$

where $r_{m,i}$ in Equation (18) is the interference term caused by all other users m except the user i. The term $r_{m,i}$ from Equation (1) can be written as [18] [19]:

Let $\varphi(\tau) = \varphi_{(X_m, X_i)}(\tau)$

$$r_{m,i} = 2\sum_{\tau=1-L}^{L-1} \varphi(\tau)^2 + \sum_{\tau=1-L}^{L-1} \varphi(\tau)\varphi(\tau+1) \tag{19}$$

In **Figure 5** and **Figure 6** we compared the BER performance of the asynchronous DS-CDMA system employing Hayashi's TZCZ (40, 8, 3), Hayashi's TZCZ (20, 8, 1) and the proposed ternary ZCZ sequence sets with parameter TZCZ (32, 8, 3).

At BER = 10^{-4} in **Figure 5**, the system using constructed TZCZ can attain 01 dB and 06 dB gains over the same system employing Hayashi's TZCZ (40, 8, 3) and Hayashi's TZCZ (20, 8, 1) in AWGN respectively.

The BER performance, in **Figure 6** was simulated assuming a frequency nonselective fading channel with AWGN noise with the common faded power ratio $\gamma^2 = 0.1$. As we can see in **Figure 6**, the proposed TZCZ sequence sets show better performance than Hayashi's TZCZ sequence sets.

At BER = 0.0015 the system can attain 01 dB and 10 dB gains over the system employing Hayashi's TZCZ (40, 8, 3) and Hayashi's TZCZ (20, 8, 1) respectively. The amelioration over comparable Hayashi's ternary ZCZ sequences is due to the correlation properties of the proposed ternary sequence set.

7. Conclusion

A new construction method to create ternary ZCZ sequences set based on binary ZCZ sequence sets was proposed in this paper. This ternary ZCZ sequences set is either optimal or asymptotically optimal and their con-

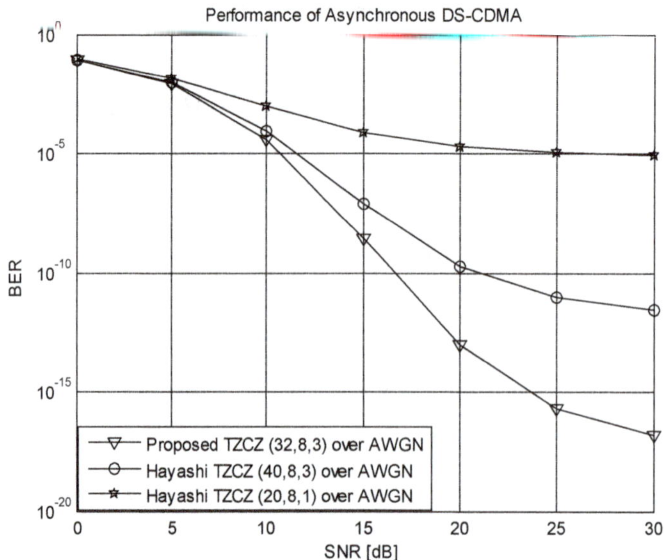

Figure 5. BER performance of Asynchronous DS-CDMA for different TZCZ over AWGN.

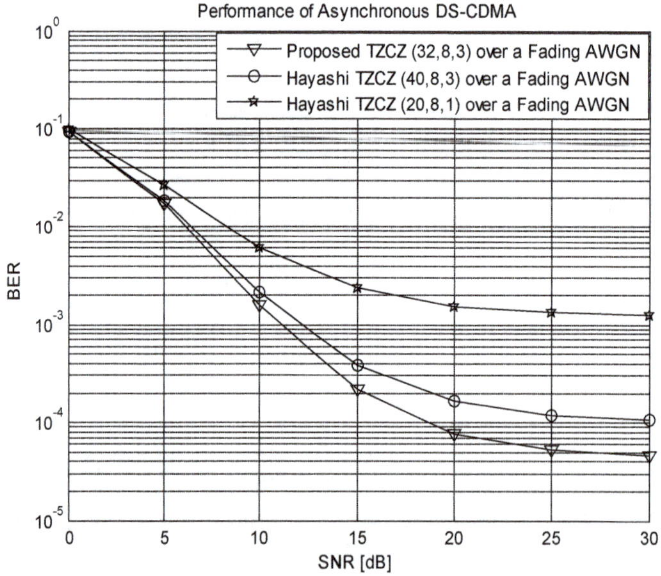

Figure 6. BER performance of Asynchronous DS-CDMA for different TZCZ over a Fading AWGN.

struction is more flexible than other ternary ZCZ constructions. The asynchronous DS-CDMA using the proposed ternary ZCZ sequences shows better BER performance in both AWGN and frequency non-selective fading channel with AWGN noise.

References

[1] Zhang, Z.Y., Ge, L.J., Yang, X.G., Zeng, F.X. and Xuan, G.X., AISS (2013) Construction of Multiple Mutually Orthogonal ZCZ Subsets for CDMA Communication Systems. **5**, 695-704.

[2] Bai, Z.Q., Zhao, F., Wang, C.H. and Wang, C.-X., IJCS (2013) Multiple Access Interference and Multipath Interference Analysis of Orthogonal Complementary Code-Based Ultra-Wideband Systems over Multipath Channels. http://dx.doi.org/10.1002/dac.2623

[3] Suk-Hoon, N. (2005) On ZCZ Sequences and Its Application to MC-DS-CDMA. Master of Science, Yonsei University, Seoul.

[4] Donelan, H. and O'Farrell, T. (2002) Large Families of Ternary Sequences with Aperiodic Zero Correlation Zones for a MC-DS-CDMA System. *Proc. of 13th. IEEE Intl. SPIMRC*, **5**, 2322-2326.

[5] Karthikeyan, V. and Jeganathan Vijayalakshmi, V., IJEEE (2013) Analysis of Carrier Frequency Selective Offset Estimation—Using Zero-IF and ZCZ in MC-DS-CDMA. **1**, 171-175.

[6] Huang, J.C., Matsufuji, S., Matsumoto, T. and Kuroyanagi, N., IJCS (2012) A ZCZ-CDMA System with BFSK Modulation. **25**, 1620-1638.

[7] Wu, D., Spasojević, P. and Seskar, I. Ternary Zero Correlation Zone Sequences for Multiple Code UWB. WINLAB, Rutgers University, 939-943.

[8] Wu, D., Spasojević, P. and Seskar, I. (2003) Ternary Complementary Sets for Orthogonal Pulse Based UWB. WINLAB, Rutgers University, 1776-1780.

[9] Hayashi, T. and Matsufuj, S. (2006) On Optimal Construction of Two Classes of ZCZ Codes. *IEICE TRANS*, **89**, 2345-2350. http://dx.doi.org/10.1093/ietfec/e89-a.9.2345

[10] Hayashi, T. (2003) A Class of Ternary Sequence Sets Having a Zero-Correlation Zone for Even and Odd Correlation Functions. *Proc. IEEE ISIT*, 434.

[11] Xu, S. and Li, D. (2003) Ternary Complementary Orthogonal Sequences with Zero Correlation Window. *Proc. IEEE PIMRC*, **49**, 1669-1672.

[12] Takatsukasa, K., Matsufuji, S., Watanabe, Y., Kuroyanagi, N. and Suehiro, N. (2002) Ternary ZCZ Sequence Sets for Cellular CDMA System. *IEICE TRANS*, **85**, 2135-2140.

[13] Cha, J.S., Electronics Letters (2001) Class of Ternary Spreading Sequences with Zero Correlation Duration. **37**, 636-637.

[14] Renghui, S., Xiaoqun, Z. and Li, L.Z. JCIT (2011) Research on Construction Method of ZCZ Sequence Pairs Set. **6**, 15-23.

[15] Maeda, T., Kanemoto, S. and Hayashi, T. (2010) A Novel Class of Binary Zero-Correlation Zone Sequence Sets. *Proc. IEEE TENCON*, 708-711.

[16] Hayashi, T. (2009) A Class of Zero-Correlation Zone Sequence Set Using a Perfect Sequence. *Signal Processing Letters, IEEE*, **16**, 331-334. http://dx.doi.org/10.1109/LSP.2009.2014115

[17] Fassi, B., Djebbari, A., Taleb-Ahmed, A. and Dayoub, I., IOSR-JECE (2013) A New Class of Binary Zero Correlation Zone Sequence Sets. **5**, 15-19.

[18] Vladeanu, C. (2005) Optimum Chaotic Quantized Sequences for Asynchronous DS-CDMA System. University of Bucharest, Romania.

[19] Boulanger, C., Loubet, G., Lequepeys, J.R. and Ouvry, L. (1999) Direct Sequence Spread Spectrum Sequences. *Traitement du Signal*, **16**, 426-436.

[20] Proakis, J.G. (2001) Digital Communications. 4th Edition, McGraw-Hill, New York.

Development of Global Geographical Coverage Area for Terrestrial Networks Internetworked with Leo Satellite Network

V. O. C. Eke[1], A. N. Nzeako[2]

[1]Department of computer Science, Ebonyi State University, Abakaliki, Nigeria
[2]Department of Electronic Engineering, UNN, Enugu State, Nigeria
Email: veke39@yahoo.com, annzeako2005@yahoo.com

Abstract

Network planning, analysis and design are an iterative process aimed at ensuring that a new network service meets the needs of subscribers and operators. During the initial start-up phase, coverage is the big issue and coverage in telecommunications systems is related to the service area where a bare minimum access in the wireless network is possible. In order to guarantee visibility of at least one satellite above a certain satellite elevation, more satellites are required in the constellation to provide Global network services. Hence, the aim of this paper is to develop wide area network coverage for sparsely distributed earth stations in the world. A hybrid geometrical topology model using spherical co-ordinate framework was devised to provide wide area network coverage for sparsely distributed earth stations in the world. This topology model ensures Global satellite continuous network coverage for terrestrial networks. A computation of path lengths between any two satellites put in place to provide network services to selected cities in the world was carried out. A consideration of a suitable routing decision mechanism, routing protocols and algorithms were considered in the work while the shortest paths as well as the alternate paths between located nodes were computed. It was observed that a particular satellite with the central angle of 27° can provide services into the diameter of the instantaneous coverage distance of 4081.3 Km which is typical of wide area network coverage. This implies that link-state database routing scheme can be applied, continuous global geographical coverage with minimum span, minimum traffic pattern and latency are guaranteed. Traffic handover rerouting strategies need further research. Also, traffic engineering resources such as channel capacity and bandwidth utilization schemes need to be investigated. Satellite ATM network architecture will benefit and needs further study.

Keywords

Network Planning, Global Network Coverage, Visibility Angle, Link-State Database, Orthogonal Route Path, Dijkstra's Algorithm

1. Introduction

Network planning, Analysis and Design is an iterative process encompassing topological design, network syntheses and network realization. It is aimed at ensuring that a new network or service meets the needs of subscribers and operators [1]. Network planning is done before the establishment of a telecommunication network or service. It has been noted in [2] that during the initial start-up phase of telecommunications systems, coverage is the big issue and traffic demand is minimal to a small network which will require expansion later. [3] states that coverage in telecommunications systems is related to the service area where a bare minimum access in the wireless network is possible. When determining the coverage of a system, both system capacities for handling traffic and radio coverage must be considered. Some areas of the system may need radio ports for capacity while some may need application points (or base stations) for coverage. Also [3] identified mobility and coverage as requirements in the digital cellular networks. It was further stated that in voice and low data networks, comprehensive coverage and mobility are the dormant design parameters while in WLANS and point-to-point fixed wireless communications, coverage and mobility are restricted. The modeling framework for cellular/PCS networks can be divided into mobility model, topology model and call model [2]. In line with the above, simple mathematical mobility models were developed in [4] for configuring a Global Network Interconnectivity with LEO satellites. A successful development of the one dimensional satellite mobility models were presented as well as the performance evaluation of the satellite mobility models regarding optimum, global terrestrial network coverage, time of geographical earth coverage and coverage angle parameters. Mathematical simulations of their parameters were carried out and it was found that the instantaneous coverage arc lengths were exponentially varying with time and continuously distributed within the four zones (quadrants) around one polar orbit. If a contiguous, real-time connection is required between LEO satellites, a system constellation of satellites will be needed.

In this paper, therefore, we extend the idea of one satellite system to the idea of a constellation of satellites in two dimensions. Hence, we aim is to design and develop a geometrical topology model to determine to network coverage of an area. In Section 2, we develop a LEO satellite geometrical constellation network model. In Section 3, we present a global terrestrial coverage model. In Section 4, we present the implementation of the global network model internetworking LEO satellite network with the terrestrial networks. In Section 5, we conclude with recommendations for future work.

2. Design of the Leo Satellites Geometrical Constellation Network Model

The integration of LEO satellites with the ground-based internet gateway and connection-oriented circuit-switched telephony service also means that the end-to-end system connectivity will be provided transparently using the satellite infrastructures. LEO satellite networks are planned in large constellations to cover large portions of the earth, mainly targeted isolated mobile terminals where ground infrastructure is missing or temporarily unavailable with different geometries [5] as discussed below.

2.1. Types of Satellite Constellations

Two main types of satellite constellations are stated in the literature: 1) walker delta (or Ballard Rosette) constellations and 2) Walker star constellations. The Rosette constellation covers a large band around the equator. A ground station is in the footprint of several satellites whose orbital planes overlap several times. The Earth station traces a sinusoidal shaped orbital track on the flattened surface of the Globe [6]. In contrast, a walker "Star" or polar constellations uses a number of orbits all crossing the polar region. These result to equally distributed orbital planes crossing at the earth's poles. Several equally distanced Satellites move along the earth orbit path at the same speed. This guarantees that each point on the earth is within a footprint of a satellite at any given time, for example, Iridium [7] and Teledesic, [8] projects.

Irridium is a LEO satellite network, where connection oriented circuit switched telephony service, and dial-up through satellite to ground Internet-gateway are offered on any spot on the earth. Irridium [9] uses polar orbits and 66 satellites forming a planned grid that covers the whole earth surface. It uses this variation of the Manhattan Network topology where satellites can rotate around the earth with equi-distance spacing between each two satellites on the same plane. In comparison, Teledesic, a connectionless network of satellites was initially planned with 840 LEO satellites [10], scaled down to 288 LEO satellites [9] before being scrapped off the drawing board in October, 2002 14 [11]. Teledesic can provide seamless compatibility with terrestrial broadband (fibre) networks. This network uses fast packet switching technology based on Asynchronous Transfer Mode (ATM) developments [9].

Two types of intersatellite links (ISLs) are often witnessed: Intra-plane ISLs, the ISLs between satellites on one orbital plane, and the interplane ISLs, the links between satellites on different planes. Both ISLs enable the communication between two users in different footprints with not more than two ground gateways being necessary. The interplane ISLs are permanently switched because of the fast change in relative positions of the satellite to each other. With the introduction of the advances in smart and adaptive radio [12], more possibilities for complex meshed *ad hoc* connectivity between any groups of satellites could be offered.

2.2. Satellite Constellation Design Considerations

It was noted in [13] that for a system designer to develop a LEO satellite constellation that provides continuous global coverage, the following design considerations are required: length of coverage arc on the surface of the earth within an instantaneous earth system; the number of satellites needed to complete a global satellite system; and the gain of the satellite antenna. The following additional requirements were also identified [13]: whether or not to use ISLs, whether to design to operate across the system if ISLs are used selecting an orbital height, number of satellites visible at any instant coverage region, etc. All these requirements interact in the overall system.

Four important factors that influence the design of any satellite communication system has also been identified in [13]: incremental growth, interim operations (satellite) replenishment options, and end-to-end system implementation. Most of the Medium Earth Orbit (MEO) and LEO system operators developed interim operations plans where a reduced number of satellites could provide useful service. The technical planning for interim operations includes: relaxing the number of satellites visible to any user at any particular time which lowers the number of satellites required to complete the constellation. The elevation angle minimum for users is usually lowered, the gaps between operational satellites in the same plane are made symmetrical, and the orbits adjusted if possible to maximize coverage over those parts of the day when user service requests are highest. Most LEO constellations have at least four satellites per plane and multiple spacecraft launches are used in the constellation buildup.

Also, the design of a Non-Geosynchronous Satellite orbit system will be heavily influenced by the decision on whether or not to provide services directly to end-users (*i.e.* end-to-end system implementation). It will also be impacted by the decision on whether or not to include established telephone companies in the delivery of the service. By their very nature, mobile satellite systems have committed to serve the end user directly. However, different approaches have been taken with regard to including established telephone companies. Two examples of organizations that took opposite decisions are the Global Star and Irridium. Global Star elected not to bypass the existing telephone companies while Irridium did. These decisions led to a very different architecture for the two systems.

2.3. Theoretical Design of the Leo Satellite Architecture

In this sub-section, we consider the geometrical aspect of developing satellite constellation network model. First of all, we review the analysis of the motion of a satellite body of mass, m, which is at a height, h, above the earth and is revolving round the earth in a circle of radius, r_s, as given in [4] [13] and represented here for emphasis as shown in **Figure 1**.

Using the sine rule to triangle, SEC, we have that

$$\left[r_{s/\sin(90+\psi)}\right]=\left[d/\sin\theta\right]$$

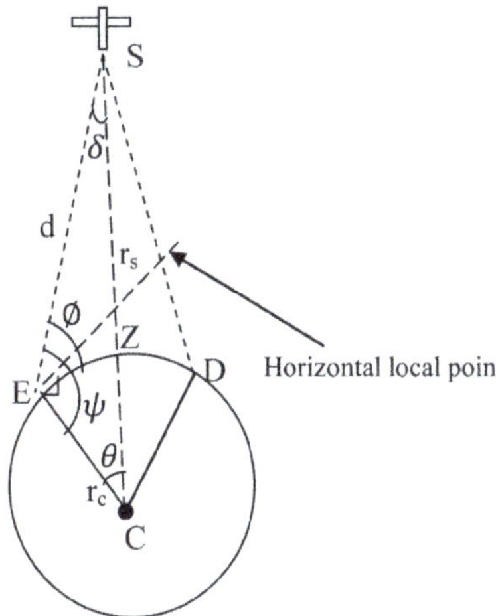

Figure 1. Geometry for calculating coverage area.

which yields

$$\cos\phi = [r_s \sin\theta]/d \tag{1}$$

where ϕ is the elevation angle, r_s is the vector from the centre of the earth to the satellite, θ is the central angle measured between the r_s and r_e, the radius of the earth, is the angle between the earth station, E and the satellite, S; d is the vector from the earth station to the satellite.

$$\sin\alpha/r_e = \sin \mathrm{SEC}/r_s \tag{2}$$

But angle $\mathrm{SEC} = \psi = \phi + 90^\circ$

$$\therefore \alpha = \sin^{-1}\left[\sin\left(\phi + 90^\circ\right) r_e/r_s\right] \tag{3}$$

where α the angle between two neighbouring satellites of the same orbit, ψ is the angle measured from r_c to d.

$$\text{Arc EZ} = r_e x\theta\left(\theta \text{ inradians}\right)\text{Km} \tag{4}$$

The diameter of the instantaneous coverage region is given by:

$$2\text{Arc EZ} = 2\left[r_e x\theta\right] \text{ Km} \tag{5}$$

And the coverage angle at the centre of the earth is given by:

$$\theta = 2x\alpha \tag{6}$$

The angular displacement, θ in radians can be given in terms of arc length, l which it subtends on a circular radius, r_s by [14].

$$\theta = l/r_s \text{ or } l = \theta r_s \tag{7}$$

where l is the circumferential distance, a satellite body on the circle of rotation has moved (or would roll without slipping) if free to do so. This is Newton's Law of circular motion.

We extend the above analysis to an idea of a LEO satellite constellation thus:

1) To establish whether a particular satellite location can provide service into a given region, a simple visibility test can be carried out as shown in [13].

$$\theta \le \cos^{-1} r_e/r_s \tag{8}$$

This means that the maximum central angular separation between the earth station and sub-satellite point is limited by this value. The central angle α will yield the coverage area on the surface of the earth assuming the satellite has symmetrical coverage about the Nadir. Hence, relaxing the number of satellites visible to any user at anytime can be achieved.

2) The distance, d, will determine the free space path loss along the propagation path and will be a factor in the link budget design. This is given by:

$$d = r_s^2 + r_e^2 x 2 r_s r_e \cos\theta \tag{9}$$

where r_s, r_e, θ have their usual meanings.

3) The elevation angle ϕ is as given in Equation (1) above. It should be noted that most satellite systems, whether for the Mobile Satellite Service (MSS) or the Fixed Satellite Service (FSS) at frequencies above 10 GHz tend to limit the elevation angle of the user to no less than 10^0.

4) The number of satellites required in one polar orbit. The decision on whether or not to use ISLs, whether to design to operate across the system if ISLs are used, is usually imparted by the number of satellites required to complete one plane with a suitable overlap. The satellites in a plane are separated from each other with an angular distance given by:

5)

$$\theta = 360^\circ / N_s \text{ or } N_s = 360^\circ / \theta \tag{10}$$

where N_s is the number of satellites required to complete one plane with a suitable overlap. Since the planes are circular, the radii of the satellites in the same plane are the same at all times and so are the distances from each other.

The length L_v of all intra-plane ISLs is fixed and is computed by [15].

$$L_v = \sqrt{2}R\sqrt{1-\cos\left(\frac{360^\circ}{N_s}\right)} \tag{11}$$

where R is the radius of the plane.

6) Number of planes, M, for complete full global coverage. The Satellite Network is composed of M separate orbits (planes), each with N_s satellites at low distances from the earth. It has been observed that one plane of the satellites, if in the polar orbit, will have satellites on both hemispheres of the earth, some going Northwards (or Eastwards) and some going southwards (or Westwards). Hence, it will be technically necessary to have M planes equal to half of the number of the satellites per plane, N_s That is,

$$M_p = \frac{N_s}{2} \text{ or } N_s = 2xM_p \tag{12}$$

The planes are separated from each other with the angular distance given by:

$$\phi = \frac{360^\circ}{2xM_p} \tag{13}$$

The length L_h of the inter-plane ISLs is variable and is calculated by [16].

$$L_h = \phi x \cos(lat) \tag{14}$$

where $\phi = \sqrt{2}R\sqrt{1-\cos\left(\frac{360^\circ}{2xM_p}\right)}$

With lat as the latitude at which the iner-plane ISL resides (see **Figure 4**).

7) Total number of satellites for a global network coverage. Using the same logic as in (10) and (12) above, there will be N_s slots (or slices) around the equator made up of M_p planes of satellites. Therefore, the total minimum number of satellites needed for complete global network coverage is given by:

$$N_T = N_s x M_p \tag{15}$$

2.4. Computations of the Parameter Values

In this sub-section, we compute the values of the above parameters as follows:

1) Satellite visibility value: This is given in (8) by:

$$\theta = \cos^{-1} r_e / r_s$$

Given: $r_e = 6378$ km , $h = 780$ km , $\phi = 10^\circ$.
But we do know that $r_s = r_e + h = (6378 + 780)\text{km} = 7158$ km .

$$\theta = \cos^{-1} 6378/7158 = \cos^{-1} 0.8910 = 27.00 \text{ km} .$$

2) The central angle θ
We need to find out the central angle, θ given by (6),

$$\theta^\circ = \left(180^\circ - (\phi + 90)\right)^\circ - \delta^\circ$$

where $\delta = \text{Sin}^{-1}\left[\text{Sin}(\phi + 90)\frac{r_e}{r_s}\right] = \text{Sin}^{-1}\left[\text{Sin}(10 + 90) \times 0.8910\right] = 61.34^\circ$

$\therefore \theta^\circ = 180^\circ - (100^\circ) - 61.34^\circ$ (Angles of a Triangle) = 18.66°

$\therefore \theta^\circ = 18.66^\circ \leq 27^\circ$

3) The diameter of the instantaneous coverage
The value $\theta^\circ = 18.7^\circ$ above confirms that a particular satellite with the central angle, $\theta^\circ = 18.7^\circ$ can provide service into the diameter of the instantaneous coverage given by:

$$2 \text{ Arc EZ} = 2\left[r_e x \theta\right]\text{km} = 2 \times 6378 \times 18.66^\circ = 4081.3 \text{ km}$$

We now have a situation set-up as shown in **Figure 2**.

4) The number of satellites required to complete one complete plane with suitable overlap is computed from (10) given by:

$$N_s = \frac{360}{\theta} = \frac{360^\circ}{18.7^\circ} = 19.25 \cong 20 \text{ Satellites}$$

5) The number of planes for complete full global coverage is computed from (12) as shown below:

$$M_p = \frac{N_s}{2} = 20/2 = 10 \text{ planes}$$

6) The total number, N_T of the satellites for the full continuous global satellite network is given by (15) above, *i.e.*,

$$N_T = N_s \times M_p = 20 \times 10 = 200 \text{ satellites.}$$

2.5. Proposed Satellite Network Topology Model

The choice of the constellation model influences the other aspects of the network architecture such as the topology organization and routing scheme [9]. Theoretically, our derived satellite network constellation model has shown to comprise 200 satellites with 20 satellites in 10 planes. However, we propose a situation in which the

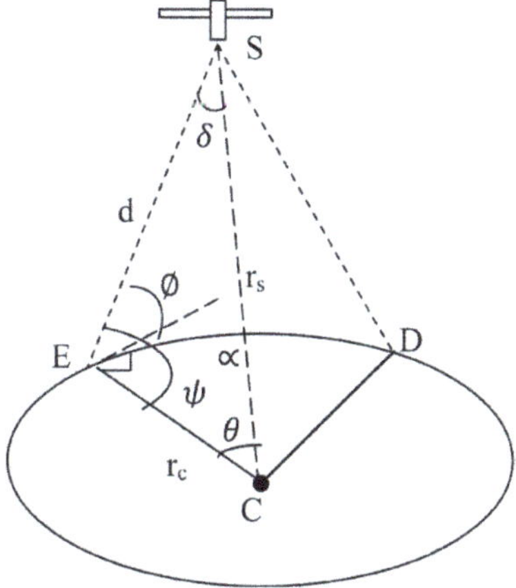

Figure 2. Geometrical set-up for instantaneous coverage Arc ED.

number of satellites in a constellation at any particular time is relaxed to 8 satellites in 4 planes which will in turn relax a total number of satellites to 32 satellites. This configuration can be arranged in 4 × 4 matrix structure.

We propose a hybrid topology model in [14] to implement our scaled down satellite constellation design as shown in **Figure 3** below.

If we consider the hybrid topology model network shown in **Figure 3** above, we can see that there exists more than one shortest path from the source, S, to the destination, D. we call all the nodes in the rectangle, where the source is a corner and the destination is the other on the diagonal, a routing set. If there are K-routing sets, we call it K-set, where K is the number of paths between source and destination. All the directions toward the destination are located on the shortest path from the source to the destination.

All possible paths are shown in the hybrid mesh topology. And all of the paths using any one of the links with the specified directions are equal and are shortest paths. Also, all the paths using these directions are loop free. Thus, the routing problem for a satellite system becomes the "shortest paths" discovery problem. However, since the network is spherical and there exists many routing set between the source (S) and the destination (D) and most of them pass through the polar region on through the horizontal plane a virtual network has to be considered while finding the right routing set [17].

The above analysis is fundamental to the determination of the Global Network that covers an earth geographical Network service area if internetworked with the Space Network derived in Sub-Section 2.4 and 2.5 respectively.

3. Continuous Global Earth Network Coverage Area

In this section, we intend to develop a continuous global Earth Network Coverage area suitable for a Wide Area Network (WAN). Just as Local Area Network (LAN) provides internal connectivity to a small geographic area, and a Metropolitan Area Network (MAN) extends intermediate coverage to a wider area, wide area networks provide wider area coverage and they go beyond the boundaries of cities and extend globally. The extreme of the WAN is the Global Network. First of all, we model the positions of a location on the earth using the spherical co-ordinates framework in 3.1. Next we compute the distances between the selected locations (cities) in the world.

3.1. Global Earth Coverage Model

We model the position of a satellite location on the earth using the spherical co-ordinate framework where the

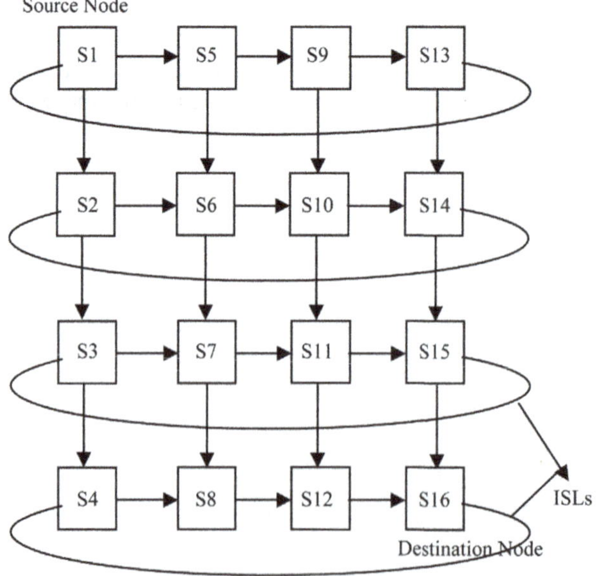

Figure 3. 4 × 4 hybrid topology model.

position of a point is considered as being a point in a sphere as depicted in **Figure 4** below.

The latitude of a place is measured in degrees North or South of the equator. The latitude of a place lies between 90° North or 90° South of the equator.

Let G be a position on the earth's surface as shown in **Figure 4** above. We measure the latitude of G as follows:

Let the line NGBS through G be the meridian.

$|OG| = R$ units be the radius of the earth

$|FG| = r$ units be the radius of parallel of latitude through G.

EGHI = the parallel of latitude through G and H

In ΔOFG, $OFG = 90$ and $OGF = \theta$ (*i.e.* alternate angles of $FG \| OB$).

$$\therefore \cos\theta = \frac{|FG|}{|OG|} = \frac{r}{R}$$

$$\therefore r = R\cos\theta$$

Or

$$\text{Arc GB} = R\theta(\text{radian}) \tag{16}$$

Similarly, the longitude of a place is measured in degrees East and West of the Greenwich meridian and it lies between East and West of the Greenwich meridian.

Let H be a position on the earth's surface as shown in **Figure 4** above. The longitude of a position H on the earth's surface is measured as follows:

Let the Greenwich meridian NGBS

Intersect the parallel of latitude EGHI at G. Let the meridian NHCS through H intersect the equator, ABCD at C and intersect the parallel of latitude EGHI at H.

The angle $BOC = \phi$ is the longitude of H east of the Greenwich meridian.

Let H be the position on the earth's surface of the equator as shown in **Figure 4** above. We measure the longitude H as follows:

Let NHCS through H be the meridian;

$|OC| = R$ units be the radius of the earth;

$|FH| = r$ units be the radius of parallel of latitude through H.

$$\text{In } \Delta\text{GFH, arcGH} = r\phi + R\theta \ (\text{in rad}) \tag{17}$$

Hence, by resolution of vectors, the total displacement between location H, G and B respectively is given by:

$$\text{HB through G} = \sqrt{(\text{ArcGH})^2 + (\text{ArcGB})^2} + \sqrt{(r\phi)^2 + (R\theta)^2} \ (\text{in rad}) \tag{18}$$

In general, therefore, the total distance along any parallel of latitude North or South of the equator and then meridian (or Greenwich meridian) East or West of the equator is given by the sum of the arc lengths travelled in x and y directions respectively. This implies two dimensional mobility model.

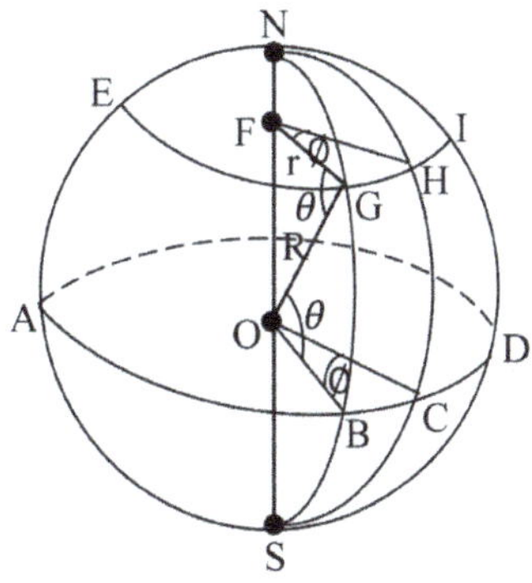

Figure 4. Spherical earth co-ordinates frame-work.

3.2. Computation of the Routes (Distances) between Locations (Cities) in the World

Suppose we plan to develop a Widea Area Network coverage area for a LEO satellite constellation network developed in Section 2. We arbitrarily select eight satellite locations to represent points of Network access for the Widea Area Network (WAN). We select two satellite locations in each quadrant of the earth surface to cover the whole globe as shown in **Table 1** below:

Using Equations (17) to (20) derived in Sub-Section 3.1, we then can compute the total distances travelled between the selected cities in the world as shown in **Table 2** below.

4. Determination of the Continuous Global Earth Geographical Network Coverage Area

For widely dispersed users, long paths exist that connect the various parts. Generally, a user at one location will send the desired message to a network entry point. We think of this wide area network as a cloud. That is, we do not know what is going on inside but we know that there are ways to get the messages from here to there either through 3 types of networking technologies [13]: Circuit switching, message switching, and packet switching. The network will determine how the messages pass through all the paths based on the protocol and list of transport. The end to end transport of data packets is achieved by routing the data packets through series of data links across the network. This routing decision is taken by the Network layer of the Nodes. The mechanisms deployed for the transport can be connection-oriented or connectionless.

Routing protocols could be divided into [13] static and dynamic routing. Static routing is used in simple networks that lack redundancy while in dynamic routing the forwarding tables are continuously updated with the information received from other routers. The routers exchange this information using a routing protocol.

Table 1. Satellite locations in each quadrant of the earth surface.

Quadrant	Countries (cities)	Locations (latitudes, longitudes)
First	Nigeria (Onitsha)	06°N, 07°E
	Japan (Nagasaki)	33°N, 13°E
Second	North America (Los Angeles)	35°N, 170°E
	South America (Canada, Churchill)	58°N, 95°E
Third	Stanley (Falkland)	58°S, 58°W
	Ecuador (Marcus)	03°S, 78°W
Fourth	South Africa (Port Elizabeth)	44°S, 24°E
	New Zealand (Plymouth)	39°S, 174°E

Table 2. Total distances travelled between the selected cities in the world.

	Los Angeles	Churchill	Marcus	Falkland	Onitsha	Port Elizabeth	Nagasaki	New Plymouth
Los Angeles	-	1254	5087	2916	8470	8473	7352	6703
Churchill	1254	-	6841	5812	4146	3360	7334	5812
Marcus	5087	6841	-	7024	6326	4994	7640	5525
Falkland	2916	5812	7024	-	3885	6259	8565	6649
Onitsha	8470	4146	6326	3885	-	7852	5803	7684
Port Elizabeth	8473	3360	4994	6254	7352	-	1923	8422
Nagasaki	7352	7334	7640	8565	5802	1923	-	3575
New Plymouth	6703	5812	5525	6649	7681	8422	3575	-

Also, the routing protocols are based on one of the following two algorithms namely: Distance vector and link state algorithms. The underlying concepts of distance vectors, link state routing, Dijkstra's algorithm for the shortest path precede the discussion on any specific routing protocol. Hence, we first discuss the basics of link state routing in Sub-Section 4.1, then proceed to discuss the Dijkstra's algorithm in 4.2 and finally, demonstrate the application of Dijkstra's algorithm in the determination of the continuous global earth geographical network coverage area in 4.3.

4.1. Link-State Routing: Basic Operation

Distance vector routing does not work well if there are changes in the internetwork. When two or more networks are interconnected, we refer to such extended network as interwork. The reasons why this routing algorithm does not work well are for the facts that the distance vectors sent to the neighbours do not contain enough information about the topology of the internetwork. That is, every router tell its neighbours its distances to all the networks without knowing the Network topology. No wonder why it was stated in [18] that topology has nothing to do with Geographical Coverage. This results to misleading conclusions as can be seen with the count to-infinity problem which could cause congestion for every other routers. However, link state algorithm overcomes this problem because with link state algorithm, every router tells every other router the information it truthfully knows about its neighbours and distances to them. Every router works out from this information: the network topology and the optimal paths.

In a link state routing, every router maintains a database of Network topology. The database contains records of the links of the entire network. Each record consists of source router identification, its neighbouring router identifiers, and the costs associated with the link between them. Each record is called link state. The cost can be defined in terms of distance, hop, delay, inverse of bandwidth or any other parameters [19].

Identical database is available in all the routers. The database is refreshed at fixed intervals (30 minutes in open shortest path First). For refreshing the database, every router sends updates called link state advertisements [LSAs].

If there is a change in the neighbourhood (e.g. a link/router goes down or a new router is added), LSAs are sent immediately by the routers that detect the change. They do not wait for the regular schedule of advertisements for refreshing the records of the database. LSAs are sent using controlled flooding across the internet so that every router receives them.

Each router works out the shortest paths to every other router using the database and the Dijkstra's algorithm, once the shortest paths are known, the forwarding table can be constructed readily.

An advantage of link-state routing is the availability of alternate paths. If a link goes down, a router can readily work out alternative path from its topology database.

4.2. Dijkstra's Algorithm

Dijkstra's algorithm computes the shortest paths from a Node (called the root) to all other nodes from the link-state debatable. The root-node selects one of its neighbours having the least cost. The link costs of neighbours of these two nodes are examined. One of the neighbours having the least cost to the root is selected again. The process is repeated, and each time a neighbor with the least cost of the root is selected and added to the set of nodes whose link costs have been computed.

4.3. Implementation of the Dijkstra's Algorithm

To understand the algorithm, let us consider a simple graph consisting of the nodes A, B, C…H that represents the eight cities and the link costs between any pair of interconnected nodes. See **Table 3** that shows the link costs associated with each link of the graph shown in **Figure 5**.

We will first of all define the followings:

Root: The node from which the least cost paths are being determined.

Set (S): Set of those nodes whose least cost paths to the root have been determine.

Set (N): Set of neighbours of set S.

I<J, P>: Node I has path cost "P" to the root via node J.

We note that since we are to determine the forwarding **Table 4** of node A, A is the root. We will use

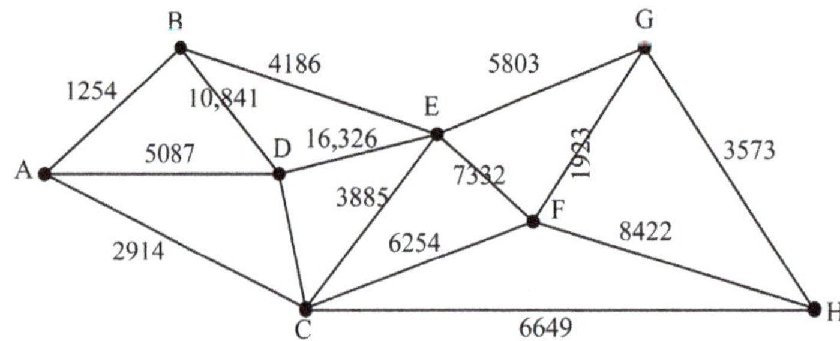

Figure 5. A graph of interconnected cities with the link costs associated with each link.

Table 3. The link costs associated with each link of **Figure 5**.

Legend	Nodes	Cities	Nodes	Cities
	A	Los Angeles	E	Onitsha
	B	Churchill	F	Port Elizabeth
	C	Falkland	G	Nagasaki
	D	Marcus	H	Plymouth

Table 4. Least cost path determination using Dijkstra's algorithm.

Steps	Set (S)	Set (N)
1	A <A, 0>	B <A, 1254>, D <A, 5087>
2	A <A, 0>, B <A, 1254>	C <A, 2914>, D <A, 5087>
3	A <A, 0>, B <A, 1254>, C <A, 2914>	D <A, 5087>, D <B, 8095>, D <C, 9938>
4	A <A, 0>, B <A, 1254>, C <A, 2914>, 0 <A, 5087>	D <B, 8095>, D <C, 9938>, E <B, 5400>, E <C, 6799>, E <D, 11,413>, E <D, 14,421>, E <D, 15,264>
5	A <A, 0>, B <A, 1, 254>, C <A, 2914, D <A, 5087>, E <B, 5400>	F <E, 12,752>, F <C, 9168>
6	A <A, 0>, B <A, 1254>, C <B, 5400>, F <C, 9168>	G <E, 11,243>, G <E, 11,096>
7	A <A, 0>, B <A, 1254>, C <A, 2914>, D <A, 5087>, E <B, 5400>, F <C, 9168>, G <F, 11,096>	H <C, 9563>, H <G, 14,778>, H <F, 17,590>
8	A <A, 0>, B <A, 1254>, C <A, 2914>, D <A, 5087>, E <B, 5400>, F <C, 9168>, G <F, 11,096>, H <C, 9563>	

Dijkstra's algorithm to determine the least paths from A to the rest of the nodes. **Table 4** shows the steps of the Dijkstra's algorithm applied to A as the root.

With Step 8, all the least cost paths to the root A have been determined. **Figure 6** shown illustrates the resulting tree.

4.4. Results and Discussion

It has been stated that an advantage of link state routing is the availability of alternate paths [20]. If a link goes down, a router can easily work out alternate path from its topology database. The Dijkstra's algorithm computes the shortest paths from a node (called root) to all other nodes from the link state database. An illustration of some city (node) to city (code) distances and their alternate paths are shown in **Table 5**.

Table 5 shows that:

1) The route from Los Angeles (A) to Churchill (B) (*i.e.* A → B) is 1254 km with two other alternate routes: A → D → B is 11,928; and A → C → D → B is 16,779 km.

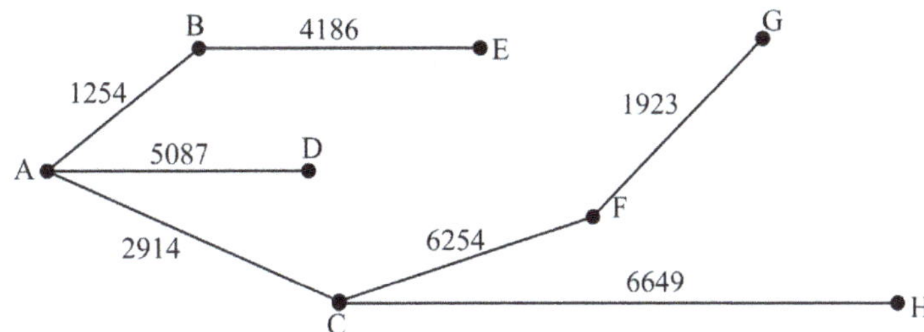

Figure 6. The resulting tree of the Dijkstra's algorithm applied to A as the root.

Table 5. An illustration of some city (node) to city (code) distances and their alternate paths.

S/N	City to City Routes	Alternate Routes	Total Distances (Km)
1	Los Angeles to Churchill	A → D → B; A → C → D → B	11,928 16,779
2	Los Angeles to Marcus	A → B → D is A → C → D is	7095 9,938
3	Los Angeles to Falkland	A → D → C is A → B → D → C is	12,111 15,119
4	Los Angeles to Onitsha	A → C → E is A → D → E is A → B → D → E is A → C → D → E is A → C → F → E is	6799 11,413 14,421 16,264 16,500
5	Los Angeles to Port-Elizabeth	A → B → E → F is A → C → E → F is A → D → E → F is A → B → D → E → F is A → C → D → E → F is	12,772 14,131 18,745 20,753 23,596
6	Los Angeles to Nagasaki	A → B → E → G is A → B → E → F → G is A → C → E → F → G is A → B → D → E → F → G is	11,243 14,695 16,054 22,676
7	Los Angeles to New Plymouth	A → B → E → G → H is A → C → F → H is	14,818 17,590

2) The route from Los Angeles (A) to Marcus D (*i.e.* A → D) is 5095 km with two other alternate routes: A → B → D is 7095 km while route A → C → D is 9938 km.

3) The route from Los Angeles (A) to Falkland (C) is 2914 km with other two alternate routes: A → D → C is 12,111 km and A → B → D → C is 15,119 km.

4) The route from Los Angeles (A) to Onitsha (E) (*i.e.* A → B → E is 5440 km with five other alternate routes: A → C → E is 6799 km, A → D → E is 11,413 km, A → B → D → E is 14,421 km, A → C → D → E is 16,264, while A → C → F → E is 16,500 km.

5) The route from Los Angeles (A) to Port Elizabeth (F) (*i.e.* A → C → F is 9168 km with other five alternate routes: A → B → E → F is 12,772 km, A → C → E → F is14,131 Km; A → D → E → F is 18,745 Km; A → B → D → E → F is 207,753 km while A → C → D → E → F is 23,596 Km.

6) The route from Los Angeles (A) to Nagasaki (G), A → C → F → G, is 11,091 km with four other alternate routes: A → B → E → G is 11,243 km, A → B → E → F is 14,695 km, A → C → E → F → G is 16,054 km, A → B → D → E → F → G is 22,676.

7) The route from Los Angeles, A to New Plymouth, H (*i.e.* A → C → H) is 9563 km with two other alternate routes: A → B → E → G → H is 14,818 km while A → C → F → H is 17,590 km.

At times, we want to deviate from the shortest path strategy because the shortest path may not have enough capacity to carry the entire traffic due to its bandwidth limitations. Traffic engineering allows us to provision more traffic flows along the desired path which may not be the shortest path.

- Basic traffic engineering depends on coverage. When determining the coverage of a system both system capacity for handling traffic and coverage must be considered. Hence, based on Equation (5) and (8) derived above, it has been shown that a particular satellite location provides service into a given instantaneous coverage region of 4081.3 km with a visibility angle of 37.32°, and total instantaneous coverage angle of 2δ = 122.68° (*i.e.* 2 × 61.34°). Hence, the concept of phased array antenna can be used on our satellite system to divide up among a set of receive antennas that provide 360° coverage, as in the sectored antenna approach of cellular systems [13].
- The implication of the instantaneous coverage distance range of 4081 km is that satellites must handover their connections to the earth stations at about this distance. The handover procedure requires a state transfer from one satellite to the next, and will result in a change in the delay characteristics of the connections at least for a short time interval. Considering that the orbital period of a satellite is 100.5 minutes [4] and that an orbit is made of eight satellites (see **Figure 3**), we then obtain a coverage time for a satellite (the amount of time a fixed point on earth is covered by a satellite) of around 12.6 minutes. This time duration is also the maximum time before a handover to the next satellite on the same orbit (called south handover). Since there are four polar orbits, the time between two handovers to the next Eastern orbit (called East handover) is three hours.
- Also, two commonly used routing efficiency are channel traffic and communication latency. The channel traffic at any time instant (or during any time period) is indicated by the longest path transmission time involved. For instance, the route from Los Angeles (A) to New Plymouth (H) has an optimal route A → C → H that is 9563 km with two other alternate routes: A → B → E → G → H that is 14,818 km and A → C → F → H that is 17,590 km. In traffic engineering, the shortest path may not have the required capacity to carry the entire traffic due to its bandwidth limitation; we therefore, choose the longest path transmission that has much more latency. An optimally routed network should achieve both minimum traffic and minimum latency for the communication pattern [20].
- The concept of virtual networks leads to the network partitions of a given physical network into logical sub-networks for multicast communication. Considering the results of the computed total distances travelled between the selected cities in the world as tabulated in **Table 2**, one can observe that the pattern divided the Global geographical network coverage area into two sub-networks. The system has the form of Lower triangular and upper triangular matrix (8 × 8). The two forms have North-West direction flow of data traffic and South-East direction flow of data traffic. It is also observable that the strictly diagonally travelled path is prohibited as seen by dash values along the line. This is confirmed in **Figure 6** as there is no direct route from Los Angeles to New Plymouth. However, short immediate diagonal paths may be allowed. For instance, the route from the source A (Los Angeles) to destination H (New Plymouth) *i.e.* A → B → D → E → F → G → H is 27,269 km travelled in six hops (links).

  But if we consider the orthogonal paths taken between the same source A and destination H, we would have a route from A (Los Angeles) vertically to C (Falkland) and horizontally to the destination H (New Plymouth) giving a total distance of 9563 km in 2 hops (links).
- Equally observable is the minimum span global geographical coverage area as shown in **Figure 6**. There exist four shortest routes: A → D, A → B → E, A → C → F → G and A → C → H that guarantees continuous global geographical earth coverage area. The coverage area spans from A (Los Angeles) to B (Churchill) in the second quadrant of the Earth to C (Falkland) and D (Marcus) in the third quadrant to E (Onitsha) and G (Nagasaki); and then to F (Port Elizabeth) and H (New Plymouth) in the fourth quadrant.

5. Conclusion and Recommendation

An in-depth study for the development of the continuous global geographical coverage are for interworking LEO satellite network and terrestrial networks has been presented in this paper. First, a successful design of the LEO satellite geometric network connectivity is presented, and the analysis and computation of the LEO satellite system parameters were evaluated in terms of the satellite visibility, central angle, diameter of the instantaneous coverage area, number of satellites required to complete plane with suitable overlap, number of planes for complete global coverage as well as the total number of satellites for the full continuous global satellite network through mathematical simulations. The values informed our choice of a hybrid mesh network model that has been configured to a 4 × 4 matrix structure and was implemented with a shortest path routing. Next, analytical equations were developed for computing point-to-point distances between nodes (cities) that were located under the satellite footprints. Eight cities, two in each quadrant were chosen to represent the point-to-pint network access points for the wide area network coverage of the satellite locations. A discussion of the Dijkstra's algorithm and its application in the determination of the continuous global earth geographical network coverage area is presented through mathemtical simulation resulting in high but tolerable distance range of 4081 km as well as coverage time delays for a link state database routing scheme. We believe that this link state database routing scheme can smooth very effectively this result as well provide alternate paths with the longest paths that have required capacity and hence enough bandwidth to carry the traffic where traffic congestion (or router failures) exists.

In conclusion, therefore, we have developed an integrated terrestrial/space system that can be implemented with a link state database routing scheme. This scheme is capable of guaranteeing continuous global geographical coverage area with minimum span whereby orthogonal set of paths taken from any source to destination will achieve both minimum traffic pattern and latency. Handover research and re-routing strategies should need further research. Traffic engineering resources such as channel capacity and bandwidth utilization schemes need to be investigated. Network architecture for implementing the interworking of LEO satellite ATM network and terrestrial networks also needs further research.

References

[1] Penttinen, A. (1999) Chapter 10-Network Planning and Dimensioing, Lecture Note: S-38, 145-Introduction to Tele-Traffic Theory. University of Technology, Helsink. Sourced at Network Planning and Technology and Design-Wilkepedia, The Free encyclopedia, 11/12012.

[2] Garg, V.K. and Wilkes, J.E. (1999) Principles and Applications of GSM. Pearson Education, Inc., India, 294.

[3] Pahlavan, K. andKrishnamurthy, P. (2003) Principles of Wireless Network—A Unified Approach. Pearson Education Inc. (Singapore), India, 108, 549.

[4] Eke Vincent, O.C. and Nzeako, A.N. (2013) An Analysis and Computation of Optimum Earth Geographical Coverage for Global Communications. *Journal of Communications and Network*, **5**, 337-343.

[5] Lo, M.W. (1999) Satellite Constellation Design. *Computer Science Engineering*, **1**, 58-67.

[6] Wood, L. (2003) Satellite Constellation Networks, Chapter 2 of Internetworking and Computing over Satellite Networks' by Zhang. 13-34.

[7] Irridium (2014). http://www.irridium.com

[8] Wood, L., Clerget, A., Andri Kopoulos, I., Palvlou, G. and Dabbous, W. (2000) Internet Protocol (IP) Routing Issues in Satellite Constellation Networks. *International Journal of Satellite Communications*, **18**.

[9] Houyou, A.M., Holzer, R., De Meer, H. and Heindl, M. (2005) Performance of Transport Layer Protocols in LEO Pico-Satellite Constellations. Technical Report MIP-0502.

[10] Sturza, M.A. The Teledesic Satellite System: Overview and Design Trades.

[11] Wood, L. (2014) Teledesic News and Information-Maintained. http://www.ee.survey.ac.uk/personal/L.Wood/constellation/teledesic.ht

[12] Pavlidou, F.N., Annoni, M., Aracil, J., Cruickshank, H., Franck, L., Ors, T. and Papapetrou, E. (1999) Traffic Characterization, Routing, and Security Issues in High Speed Networks Interconnected through Low Earth Orbit (LEO) Constellations. *Joint Workshop* COST255/252/253, Touloues.

[13] Pratt, T., Bostian, C. and Allnut, J. (2003) Satellite Communicaions. 2nd Edition, John Willey and Sons Inc., New York.

[14] Paxson, V. (1999) End-to-End Internet Packet Dynamics. *IEEE/ACM Transactions on Networking*, **7**, 277-292. http://dx.doi.org/10.1109/90.779192

[15] Dobosiewicz, W. and Gburzynki, P. (1996) A Bounded-Hop-Count Deflection Scheme for Manhattan Street Network. *Proceedings of Institute of Electrical and Electronics Engineering (IEEE), Computer Communications*, Vol. 1, 172-179.

[16] Ekici, E., Akyildiz, I.F. and Benden, M.D. (2000) Datagram Routing Algorithm for LEO Satellite Networks. Institute of Electrical and Electronics Engineering (IEEE) Information and Communication (INFOCOM).

[17] Kucukates, R. and Ersoy, C. (2003) High Performance Routing in a LEO Satellite Network. *Proceeding of the IEEE ISCC*, Antalya, 1403-1408.

[18] Carr, H.H. and Snynder, C.A. (2003) The Management of Telecommunications. 2nd Edition, McGraw Hill, Irwin, 229.

[19] Gupta, P.C. (2009) Data Communications and Computer Networks. PHI Learning Private Limited, New Delhi, 30, 611.

[20] Hwang, K. Advanced Computer Architecture: Parallelism, Scalability, Programmability. 383.

16

Perspective of Adaptive CN System for Forecasting Congestion of Road Traffic Flow

Tasuku Takagi

(Professor Emeritus) Tohoku University, Sendai, Japan
Email: tasuku@sirius.ocn.ne.jp

Abstract

Basing upon the Weber-Fechner Law with respect to the stimulus (distance-headway) to the vehicle driver and the driver's sensation (speed), the characteristic speed V_β is defined, which is the critical vehicles flow speed just before going to congestion in road traffic flow. From the information of real time measurement of traffic flow speed (V) and time-headway (T) at the specific positions along the road, the value of V_β is calculated and used for forecasting the flow. Discussed is how to use each V_β to forecast the congestion. The CN system devoted to the management of road traffic flow is proposed. The idea may contribute not only to easing the traffic flow but also to optimizing it to get high efficient traffic flow.

Keywords

CN for Road Network, Characteristic Speed, Critical Speed, Road Traffic Optimization

1. Introduction

It is clear that present road traffic networks and communication networks (CN) have not well cooperated. The answer is obvious: Simply we can say that we haven't had the essential core scientific knowledge concerning the road traffic flow. Since the author has well analyzed the road traffic dynamics [1], he will be able to challenge the difficult subject that has long time been left behind. The subject is the problem concerning the CN for road traffic flow management system. The most important subject to be solved is how to forecast or avoid the traffic congestion. If the congestion was forecasted, the traffic flow management should make an action to suppress the congestion and to go to the highest efficient transportable road system from the economical viewpoint.

CN is inevitable and has contributed to our present society and no exception of the road traffic system. We have at present a big system of the Global Positioning System (GPS) which has been applied to the Car Navigation System that is operated by huge CN. Today's navigation system does not have a forecasting ability. The most necessitated subject for the traffic management is the traffic forecasting system. If the forecasting ability is applied, we may have a counter method against the congestion.

If the traffic situation could be forecasted, the adaptive sign boards may be available and the intelligent sign board will be possible. Viewing from the CN, those are not high technology. The role of CN has not been sufficient to meet such a demand. We need more sophisticate ways to give the information to drivers as prompt as possible and let every driver know the short term forecasted traffic situation.

Since the traffic flow situation can change so fast, it becomes quite different one after 30 seconds as we always experience. Thus the informed past knowledge of traffic congestion cannot be useful for present drivers. We need forecasted information about congestions that may occur in very near future. But today's road management cannot do it, because today's delivering traffic information is the past stochastically analyzed one that cannot estimate the near future traffic situation. Many researchers have noticed this point. The author picked a reference up here which surveyed the published papers concerning short term forecasting of road traffic flow [2]. Viewing from the papers published, they are much the same and they are too sophisticated to be practical.

In order to make a cooperation between road traffic management and CN, the author firstly should mention the essence on the theory of road traffic flow based upon the Weber-Fechner Law (WFL) [3] [4].

The key parameter is the speed V_β (characteristic speed), which will be mentioned in **2**. And the speed $2V_\beta$ is the speed that gives the maximum transportation volume (the most economically efficient flow).

Before discussing the CN to be applied, we need to get the knowledge of the road traffic behaviors. Thus as the first step, we shall see the related theories on the road traffic flow. The first step is to obtain the characteristic speed which is designated by V_β. The most efficient speed is $2V_\beta$.

2. Characteristic Speed V_β

2.1. WFL Equation in Road Traffic Flow

WFL is the psychophysical law (history: [3], brief explanation: [4]) that was discovered in 1834 and 1846 [3]. The law mentions the relationship between stimulus and sensation (e.g. eye(s) senses the outside image as a *stimulus*, and feeling of speed in a moving vehicle as a *sensation*). The formulation of the stimulus-sensation relationship is as follows: "the sensation is proportional to the logarithm of the stimulus level".

From observing the road traffic flows, the author discovered the relationship between the distance-headway between subsequently moving vehicles (distance from head to head), which is denoted by X (X_1 and X_2 in **Figure 1(a)**) and speed V can obey the WFL. Then the exponentially it can be written like

$$X = X_0 e^{\beta V}. \tag{1}$$

X: distance-headway in meter (m)

X_0: X at $V = 0$

V: speed (m/s)

β: constant (s/m)

According to our observation, the average of X_0 is about 7 - 15 m. β is a constant that depends on the quality of road; smaller β shows a good road.

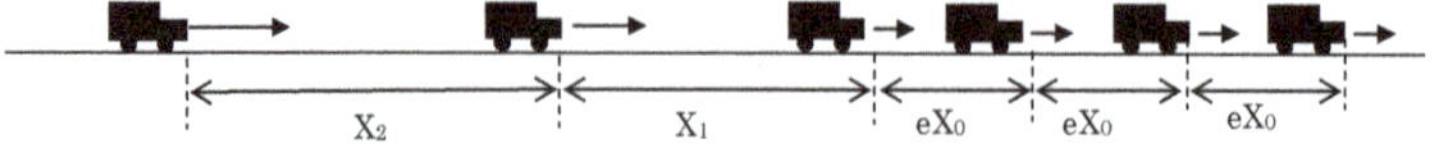

(a) Partial laminar flow

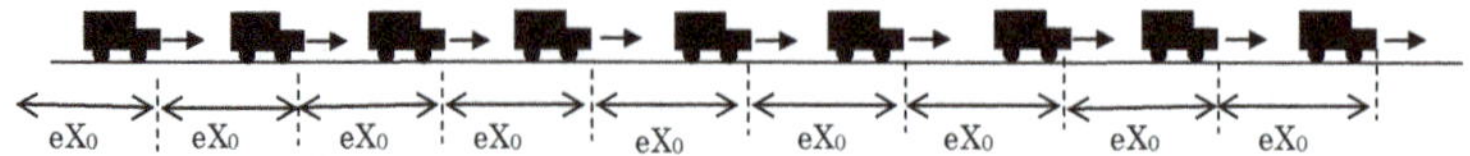

(b) Full laminar flow in some span on road

Figure 1. Vehicle flow at speed V_β.

Although the original formula of law is logarithmic one, the author rewrites it like (1), because the mathematical discussion is easy to develop.

Our discussion will follow the MKS units.

X can be determined by

$$X = VT, \tag{2}$$

where T is the time-headway (time difference between subsequently moving two vehicles).

V_β was defined from $\beta V = 1$ in (1) like

$$V_\beta = 1/\beta . \tag{3}$$

2.2. Flow at Speed of V_β

V_β is the boundary speed of free flow and congestion. When the vehicle speed V is smaller than V_β ($V < V_\beta$), the flow becomes congestion, and when $V > V_\beta$, the non-congestion flow is guaranteed. In this subsection, we should know it theoretically. At this condition of $V = V_\beta$, X will be

$$X = eX_0 \ \left(e = 2.7128\cdots\right), \tag{4}$$

Since X_0 is 7 - 15 m, (4) becomes about 20 - 45 m. We shall see the flow of V_β.

The row of vehicles has a tendency to make a bunching flow as shown in **Figure 1(a)** which shows the flow is going to bunch. Due to the driver's general behavior, both X_1 and X_2 will soon become eX_0, and final flow becomes like that shown in **Figure 1(b)**, of which speed becomes V_β. **Figure 1(b)** can also appear in case of green signal flow at an intersection which is the forced laminar flow.

We should note that the distance-headway X becomes eX_0 (20 - 45 m) at this condition.

- There is a flow simulations referring to the liquid flow. If we take this idea, the flow shown in **Figure 1(b)** can be said the *laminar flow*. In case of traffic laminar flow, V_β can be called *critical speed*. But since the value of V_β can differ from place to place even on the same road, then it was also defined as the *characteristic speed* of road.

2.3. Behavior of Flow at V_β

As mentioned above, V_β is the speed of vehicle row at the condition of laminar flow as shown in **Figure 1(b)**. This type of flow may be only theoretical and cannot continue for long time because this condition strictly forces the drivers to obey to move with the speed V_β. Thus the situation shown in **Figure 1(b)** cannot be maintained long time and length of row must be limited. And if the speed becomes less than V_β ($V < V_\beta$), the speed of vehicle row goes to congestion and the distance-headways approach to X_0 (7 - 15 m). That is to say:

1) $V < V_\beta$: Toward congestion
2) $V > V_\beta$: Free flow

2.4. Determination of V_β from Actual Flow

V_β is the only parameter to be used for controlling the traffic flow. But we should know how V_β can be determined from the practical measurements. As mentioned just end of the above **2.2** (●), V_β is not the inherent parameter of the road, and it may have different values at the different places along the road.

Here we should admit to refer the essence in the author's derived algorithm to determine it as follow:

$$V_\beta = V_{tav} \Big/ \left(1 + \ln\left(T_{tav} / T_{<3}\right)\right) \tag{5}$$

V_{tav}: Total average of V (speed),

T_{tav}: Total average of T (time-headways),

$T_{<3}$: Average of time-headways less than 3 seconds.

Details of the above have been mentioned in p. 56 in [1]. Each parameter mentioned above can be determined by measuring many vehicles at the specific position of road which should include curves, slopes or impairments, etc. Note that when $T_{tav} = T_{<3}$, $V_{tav} = V_\beta$. When the time-headway T becomes $T_{<3}$, the flow becomes laminar one as shown in **Figure 1(b)**.

The example of the actual values of V_β that calculated from (5) is shown in **Table 1**.

2.5. Natural Appearance of V_β

Figure 2 shows how V_β can naturally appear in case of vehicles moving from flat to up-slope portion of a road. The flow speed declines unconsciously when vehicles encounter the up-slope from flat portion. The decline of speed can occasionally be V_β like that shown in **Figure 2**, which is just before the congestion. If the speed farther declined and became less than V_β, the flow becomes congestion and distance-headways should be eX_0 as shown.

2.6. Congestion

There are two types of congestion: One is natural and another is artificial.

3. Natural Congestion

3.1. Driver's Sensation

Each driver has a common sensation (habit) to follow the foregoing vehicle as close as possible. The process of appearance of laminar flow is shown in **Figure 1(a)**, in which the distance-headway X_1 and X_2 are both decreasing and finally all distance-headways become eX_0 in average. If this row of eX_0 became long enough, the row of flow becomes quasi-laminar flow like **Figure 1(b)** and the speed becomes $V \fallingdotseq V_\beta$.

Since the laminar flow cannot permit disorder, when the speed of any one of the vehicles deviates due to driver's sensation, the laminar flow may be destroyed and the flow goes into congestion.

3.2. Up-Slope

A typical example is shown in **Figure 2**. The vehicle speed decreases when it encountered an up-slope portion of road from flat portion. The figure shows the case where the speed became a critical speed of V_β. This phenomenon occurs unconsciously, more speed retardation becomes the cause of the congestion.

Since the time-headway T cannot change with the slope, if we denote both decreasing amount of X and V are ΔX and ΔV, respectively, from (2)

$$\Delta X = \Delta VT. \tag{6}$$

At the slope, the distance-headway X becomes $X - \Delta X$. **Figure 2** shows the case of

$$X - \Delta X = eX_0. \tag{7}$$

Table 1. Example of values of V_β for different road.

Type of Road	Speed: m/s	km/h (average)
Tohoku Expressway (Upslope)	12.5 - 16.7	45 - 60
(Downslope)	13.5 - 19.3	47 - 70
National Road Route 25 (at Oji Nara)	6.21	22
City Road in Sendai (Route 4)	4.8 - 7.3	17 - 26
City road in Sapporo	5.06	18
Residential Area (Sendai Nakayama)	4.01	14.5

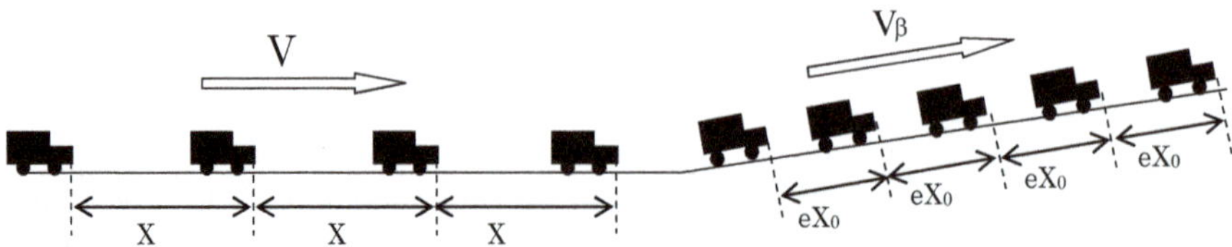

Figure 2. Natural appearance of V_β by up-slope (distance-headways at flat portion were averaged and shown as X).

If ΔV becomes a little bigger, the distance-headway becomes smaller than eX_0, and the flow goes into congestion.

4. Artificial Congestion

A typical artificial congestion is the case of intersections. In this case the flow becomes the series of bunched flow like that shown in **Figure 1(b)**. The one lane alternative flow is also the same in case of road impairment or accident. The same phenomena may occur with the case of up-slope mentioned above 2.

- **I.** Taking into account of (2), (1) can be written like

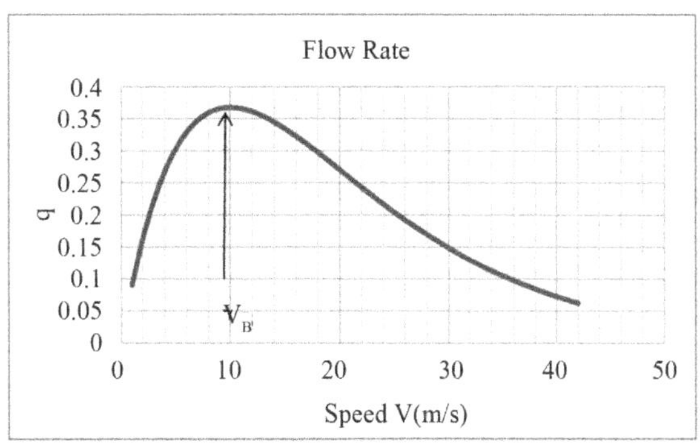

$$T = \left(X_0/V\right)e^{\beta V}, \tag{8}$$

from $dT/dV = 0$, we can derive the minimum T; (T_{min}), at $V = V_\beta$. Then q_{max} ($=1/T_{min}$) is the maximum flow. See the graph that shows $V = V_\beta$ gives the maximum flow rate.

We should note that the driver can control the speed V only. The Weber-Fechner Law can be applied when X is considerably small, for example $X < 300$ m for expressway and $X < 150$ m for the other road.

- **II.** Transportation Volume Q and Optimum Speed

$$\text{Definition: } Q = qV = V/T. \tag{9}$$

$$\text{From (8), } Q = \left(V^2/X_0\right)e^{-\beta V}. \tag{10}$$

From $dQ/dV = 0$, Maximum Q, (Q_{max}), can be obtained at

$$V = 2V_\beta \tag{11}$$

The maximum transportation efficiency can be obtained at the condition of (10). See the graph.

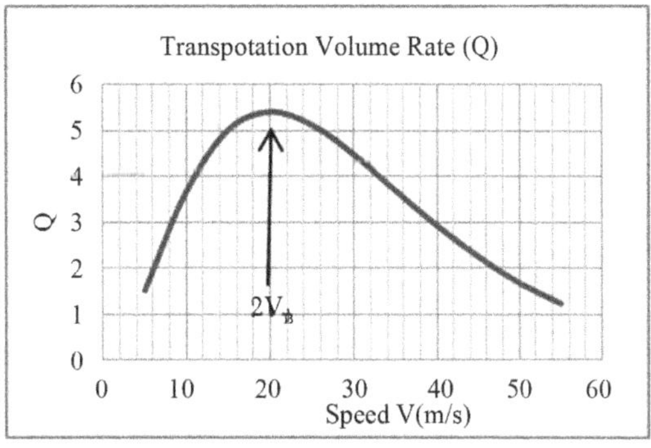

5. CN System for Forecasting of Traffic Flow

5.1. Role of CN for Traffic Managements

In order to realize the non-congesting traffic, a short term forecasting system and optimization of road traffic flow are essential. The tactics are as follows:

1) Send the bottleneck (smallest V_β) in the specific road span via CN.

2) Suggest the driver the recommended speed of $2V_\beta$ via CN.

3) Send the message to drivers of both 1) and 2) from Control Center via CN.

We know from •I that the maximum flow rate (q_{max}) can be attained at the speed of V_β. And the maximum transportation volume (efficiency) can be obtained at $2V_\beta$ (•II). Then we should add 4) as

4) Indicate speed of $2V_\beta$ as the limit of speed via CN.

Those messages can be shown by the sign boards at the roadsides, which should be adaptively changed.

5.2. Forecast of Flow

We need the measured data of each vehicle speed V and time-headway T. Those measurements can be carried out by sensors and the results are sent via CN to the Center (Traffic Information Center) which processes the data and the necessitate information should sent to the sign boards on the road side.

Figure 3 shows the assumed V_β's obtained from (4) by the measured data (V and T) of vehicles. This typical figure shows the road span in which it has three slopes as shown. Each slope has the inherent V_β and we assume here the magnitude of V_β is like that shown as

$$V_{\beta 1} > V_{\beta 3} > V_{\beta 4} > V_{\beta 2}. \tag{12}$$

That is, the smallest one is $V_{\beta 2}$ which means the slope of the span 2 - 3 in **Figure 3** has the smallest V_β.

As we have seen in 2.5, if the flow rate q is increased, the congestion shall begin at the smallest V_β portion on the road, and in this case $V_{\beta 2}$ is the smallest of which portion is the span 2 - 3.

By knowing the above in advance, the traffic flow can be forecasted or estimated. If we measure the flow at far behind portion of which V_β is larger than $V_{\beta 2}$, we can forecast that the flow at the portion of 2 - 3 span is going to congestion or not.

- V_β at intersection is similar with that shown in **Figure 3**, which is shown in **Figure 4**.

Each span between intersections has an inherent V_β, but it may depend on the signal periods or patterns. However in principle, we can treat the intersections as the same with that shown in **Figure 3**.

5.3. CN System of Data Process for Traffic Flow

Figure 5 shows the conceptual image of CN system. The key essence is as follows:

1) Sensing the data by sensors (S_1,S_2) at both portions: S_1 at the biggest V_β and S_2 at the smallest V_β.

2) The data (speed V and time-headway T) are sent to DP (data processing) Sub-center which carries out the time sequentially averaging of both V, T and flow rate q.

3) The results are sent to the Center.

We can also say the system shown in **Figure 5** is the road traffic forecasting system, because the sensor S_2 in

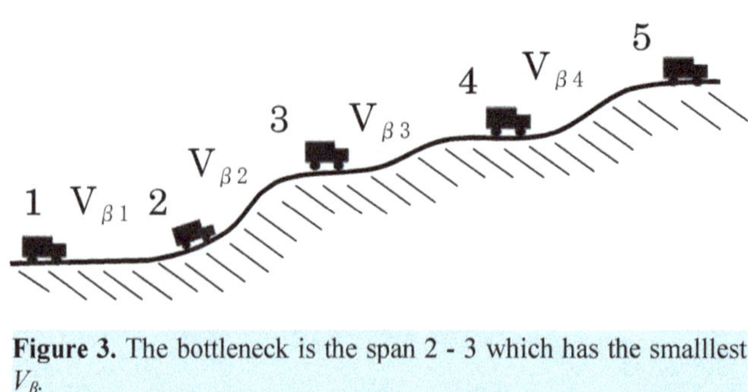

Figure 3. The bottleneck is the span 2 - 3 which has the smalllest V_β.

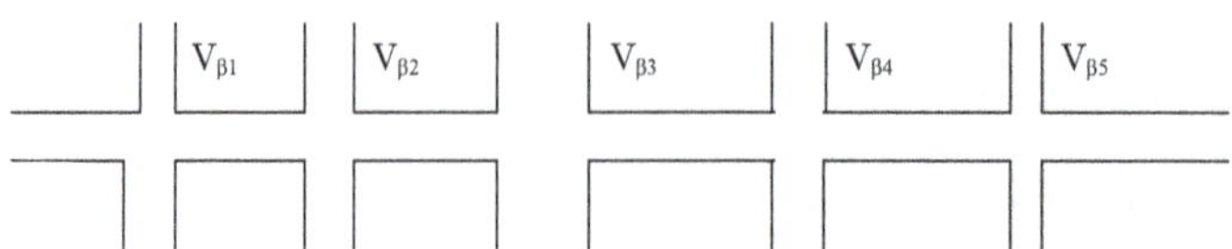

Figure 4. V_β at intersection.

Figure 5 is watching the flow at the weakest portion (bottleneck; smallest V_β), which means the first congestion occurs at this portion when flow rate (density) increased. The data of S_2 are real timely processed at the DP Sub-center. The Center should disseminate it far behind Sign Board that every driver looks at. If we see **Figure 5**, moving vehicle on the flat portion (far behind of bottleneck) can see in advance the possible congestion peril.

5.4. Adaptive Sign Board

Figure 6 shows the conceptual image of the adaptive road traffic flow management with CN system.

The contents should be as follows:

1) Speed limit: $2V_\beta$, because this speed gives the most economical benefits.

2) Existing of congestion: place.

3) Accidental bottleneck: place.

The disseminating the above information to drivers concerning present state of the traffic flow over the area should be very useful. The system shown in **Figure 6** can operate to meet such needs. The roadside sign board can adaptively be changed in a short term, for example every 30 sec.

In **Figure 6**, The Speed Sign Board shows the recommended speed (speed limit) that should be changed timely from the Center. The Information Board can indicate the information concerning flow management, for example, the reason of speed limitation.

6. Conclusions

Firstly, we should notice that the road traffic phenomena are not complicated, because the author has cleared this fact by finding that the relationship between distance-headway and speed obeys the Weber-Fechner Law that is shown in (1).

We have seen in this article that the important role of Communication Network (CN) in road traffic management and control. The summarized results are as follows:

1) Definition of characteristic speed of road by V_β.

This speed is the critical speed of moving vehicle row with the highest density, and theoretically, if the row speed becomes less than V_β, the flow goes to congestion.

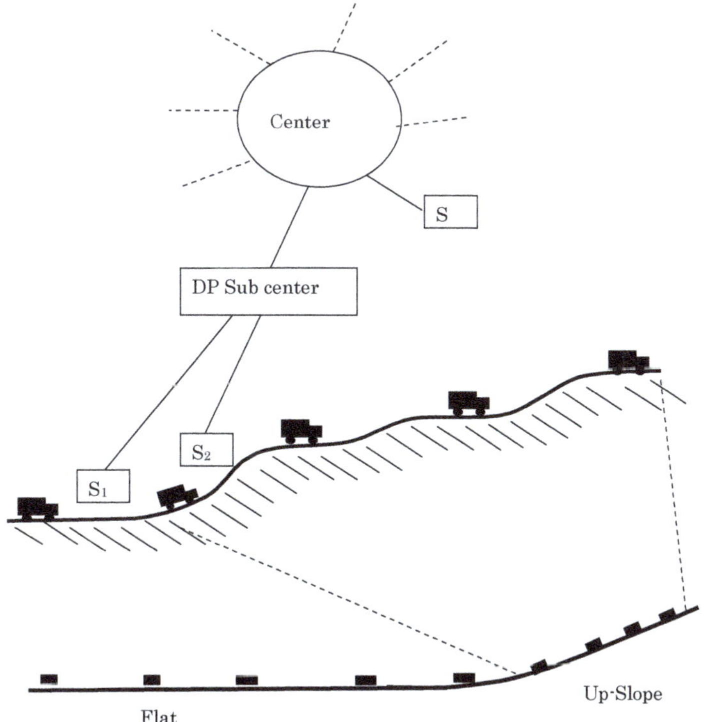

Figure 5. CN system for data acquisition and processing of traffic flow.

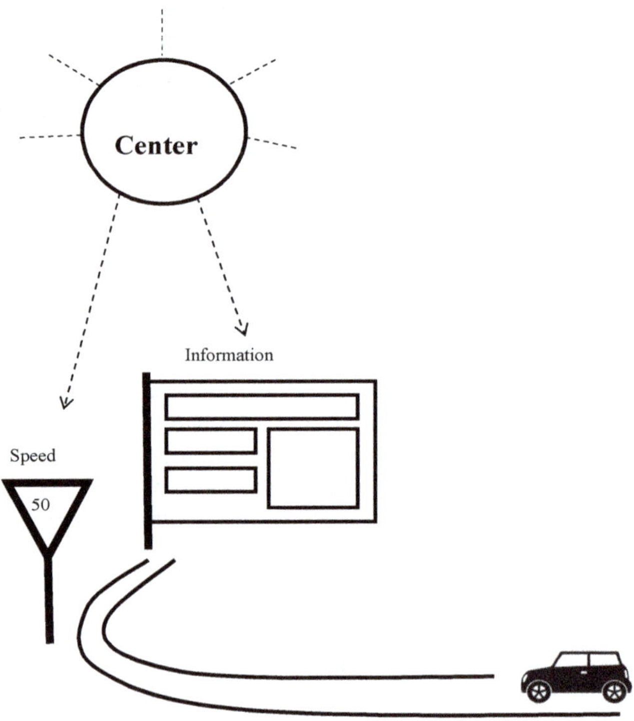

Figure 6. Adaptive sign board.

2) Finding $2V_{\beta}$

The most economically effective flow speed.

3) Importance of Sensor and Communication Network (CN)

The CN is inevitable for making the adaptive road traffic flow management by the short term forecasting.

References

[1] Takagi, T. (2011) Scientific Study of Road Traffic Flow. LAP LAMBERT Academic Publishing, GmbH & Co. KG, Saarbrucken, Deutschland.

[2] Lippi, M., Bertini, M. and Frasconi, P. (2013) Short-Term Traffic Forecasting: An Experimental Comparison of Time-Series Analysis and Supervised Learning. *IEEE Trans. on Intelligent Transportation Systems*, **14**. http://dx.doi.org/10.1109/TITS.2013.2247040

[3] Heidelberger, M. (2004) Nature From Within, Chap. 6 (Psychophysics: Measuring the Mental).

[4] Maor, E. Can Perceptions Can Be Quantified? In: E the Story of a Number. Princeton University Press (first printing in 1994), Princeton.

17

A Generic Platform for Sharing Functionalities among Devices

Remi Nguyen Van, Hideki Shimada, Kenya Sato

Department of Information Systems Design, Doshisha University, Kyoto, Japan
Email: hiroakigoto0624@gmail.com, hideki-s@is.naist.jp, ksato@mail.doshisha.ac.jp

Abstract

With so many potentially interconnected electronic devices in today's homes, manufacturers have to think of theirs as only one of the components involved in a general user experience, and not as an isolated device. Users are likely to be using several devices at the same time, either actively through immediate interaction, or passively by expecting devices to give them notifications when necessary, for example. Thus, having cross-devices functionalities is often necessary for a product to be really adapted to its usage situation. Moreover, just as we install software on computers, smartphones and tablets for additional functionalities to use their own hardware, it would be logical to install cross-devices software to use the combined hardware of several home devices for a better user experience. However, even though a number of technologies can be used to transmit data or commands between devices, UPnP being a widespread example, it is not possible to access the behavior of remote devices and add functionalities to them this way. Thus, when manufacturers design their products, there is no way for them to make full use of the other appliances at the user's home without developing and deploying specific software on each of them. In order to address this issue, this paper discusses a platform for generic development and on-the-fly deployment of applications on home devices. This system aims at letting device vendors deploy innovative features across devices in a home network, without requiring prior knowledge or control over devices already present in the user's environment. For this platform to be fit for consumer devices, it is designed to be cost-effective, use recent and widespread technologies, and be fast to implement and work with.

Keywords

Home Network, Sharing, Application, Deployment, Cross-Platform

1. Introduction

Let's imagine a phone manufacturer that wants his users to be able to receive notifications for incoming calls and text messages on whatever device the user is currently using at home, like their TV or laptop. This vendor would need to develop a client application for the laptop, and have access to the TV's firmware, since displaying this kind of notification is not natively supported on those devices.

The objective of our research is to free the vendor from this requirement, by providing him with a platform that allows generic development and automatic deployment of his applications onto other home devices, eliminating the need to have access to those. In the first section, the required specifications of this system are explained; the proposed system designed to meet these specifications is described in the second section. The third section gives an evaluation of how well this system solves the original issues by studying a developed prototype. Finally, a comparison of the proposed system and other similar systems is made in the fourth section of this document.

2. Required Specifications

In order to solve this issue, a proposed system should:

2.1. Let Devices Communicate in a Generic Way

To be able to focus on the user experience provided by their devices without spending valuable time studying other potential devices in the user's home, manufacturers need to have a generic way to make their device communicate with other home appliances. This means that there must be one standard way to share functionalities that is compatible with any device supporting the platform.

2.2. Let Devices Share Any Kind of Feature

This system is intended to let vendors design innovative cross-devices features. Thus, devices supporting this platform should be able to receive any kind of content, not just standard messages. This genericity also means that no manual update should be needed on existing devices to be able to interact with a new device. Thus, manufacturers do not need to develop additional software to be installed on remote devices, as those are potentially already compatible with any functionality they want to make available.

2.3. Keep Devices Updated on Each Other's States

For devices to automatically interact with each other so as to create a user experience as a whole, they need to know which devices are active and can provide them with additional content at any time. This enables vendors to develop user experiences based on all the devices surrounding the user, without requiring any user interaction. The goal is for several devices to create a user experience as one system, rather than having the user manually connect each independent device. For example, if a user receives a call while watching TV, the TV should automatically be aware of the state of the phone, and could prompt the user for starting a video call; in this situation, the phone and the TV can be seen as one system automatically configured to receive calls.

2.4. Be Easy to Implement for Vendors

Device manufacturers would never choose to implement a feature if the development cost is too high compared to the potential benefit. Thus, making devices compatible with this platform and developing applications for them must be easy and fast.

Finally, learning how to use the platform must be simple for developers, as the availability of qualified developers for a platform is directly reflected on development costs. This means that the use of recent and widespread technologies must be preferred.

3. Proposed System

3.1. Basic Process

In the proposed system, which basic process is illustrated in **Figure 1**, a generic UPnP platform (referred to as "the UPnP platform") is deployed on remote devices. The platform takes care of receiving and executing compatible applications; those applications are written in generic code and packaged inside the vendor's device.

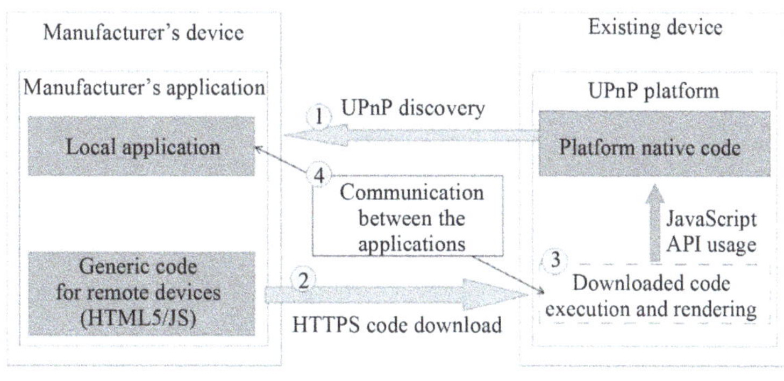

Figure 1. Basic operation of the system.

Once the platform finds the vendor's device through UPnP discovery, it downloads the application code from it; this way, client applications are deployed on-the-fly on compatible devices, and no manual update is required.

The language chosen for the generic applications to be downloaded and executed is HTML5/Javascript. A Javascript API supported by the UPnP platform on remote devices ensures that the JavaScript code can use lower-level functionalities not supported by HTML5, such as controlling the application window or sending UPnP commands.

3.2. Technology Choices

3.2.1. Device Discovery and Eventing: UPnP

UPnP is already very popular among device vendors, and is designed to enable devices to discover themselves and share their states in a generic way. Thus, it is very well suited for handling discovery. In the discovery phase, UPnP communicates a presentation URL for the service, which is used to specify the location of the client code to download. Additionally, once the discovery phase is completed and applications have been deployed, UPnP can be a convenient way for devices to communicate.

3.2.2. Code Download: HTTPS

Because the UPnP platform is going to download and execute code from other devices, it is necessary to have a safe method to check these devices are trusted. In our proposed system, we use HTTPS to download this code, and we validate the vendor's device's public key using a list of user-approved keys. This means that the user has to approve the vendor's device on remote devices the first time it is discovered. Once this is done, the generic client can be sure that every communication made with this public key originates from the approved device [1].

To let the user add trusted devices, the UPnP platform application can either use the UPnP Device Protection service [2] if the remote device supports it, or simply ask the user to press a hardware or software button to approve newly discovered devices.

Although this research mostly focuses on downloading the code from the remote device in the local network, it is noteworthy that it can also be downloaded from a remote server through the internet, as the SSL connection ensures it is retrieved from a trusted host. However, this option has not been studied into detail in our research in order to focus on a prototype with better reactivity and fully working on a local network only.

3.2.3. Code Execution and User Interface: HTML5/JS

HTML5 is getting more and more popular recently, and more and more devices—like smartphones, computers and tablets—have HTML5/JavaScript rendering capabilities. Applications are easy to develop and these technologies are well known among software developers, which make it easy for device vendors to use. Additionally, since the code is executed in a sandbox, it also ensures that even if the system were to be compromised, the potential damage would be limited.

3.2.4. Using Lower-Level Functionalities: JavaScript API

Since JavaScript does not provide ways to control the application's window or send UPnP commands for instance, a JavaScript API has to be available in the UPnP platform to let deployed applications access these functionalities. To develop our prototype, we defined a standard API for the client program entry point, window management (fullsize and reduced mode, notifications) and UPnP commands. Other functionalities like letting client applications provide sound or video streams using the API can be defined (as very few browsers support WebRTC [3] for now), but most functionalities required to develop a rich application are already available in HTML5.

Depending on the generic client application, some functionalities of the API may not be supported; some devices may not be able to use UPnP or display notifications for instance. Thus, client applications have to adapt their behavior to the platform they are running on; our API provides an easy way to test whether a functionality is available on the guest UPnP platform, and react accordingly.

3.3. Architecture Summary

The general architecture of the UPnP platform is illustrated on **Figure 2**. This system is designed so that it is also easy to develop and embed the UPnP platform client on many devices. On devices where an HTML5 engine is included in the system and can be used for development (such as Blackberry devices, iOS or Android-based devices), this component is used by the UPnP platform to execute the downloaded code. This helps make it lightweight and easy to develop.

In case there is no HTML5 engine available on the platform, a library can also be used; in our prototype, we used the WebView component of JavaFX to develop such a client on a personal computer.

3.4. Requirements Summary for the Platform

Given our technology choices, devices must meet the following requirements to be able to run the UPnP platform client:

- Have local networking capabilities.
- Be able to fully support the UPnP protocol. In particular, the device must be able to send multicast messages. There is a number of libraries that developers can use in many languages.
- Be able to download files through HTTPS on the local network, and carry out custom certificates chec; several methods are discussed in the next section.
- Be able to run JavaScript code. An engine may be already available on the device, or support can be bundled with the application if the developer chooses to use a library.
- Be able to provide a custom JavaScript API in the JavaScript engine; several methods are discussed in the next section.

As for vendor's devices that use this platform, the following requirements must be met:

- The device must have local network capabilities and be able to fully support the UPnP protocol, just like the for UPnP platform.

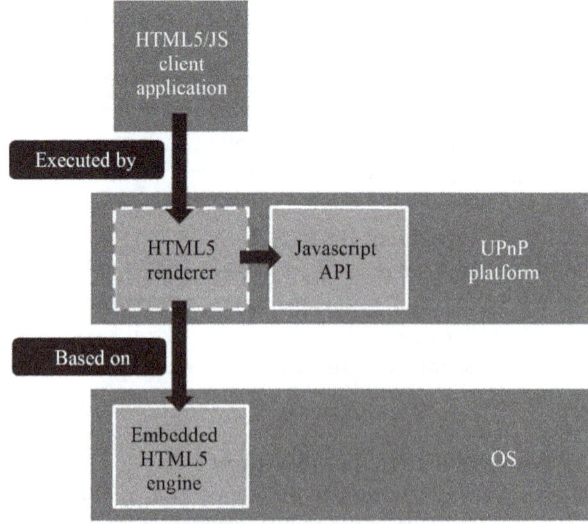

Figure 2. Architecture of the UPnP platform.

- The device must be able to serve files through HTTPS; many libraries can generally be used by the developer for this task.

3.5. Implementation Options

To easily develop the UPnP platform client, an existing HTML5/Javascript engine will generally be used. However, some special functionalities must be added to this engine: the Javascript API must be supported, and the SSL certificates checking used in HTTPS must be customized to use a list of user-approved public keys. A number of implementation options are possible for these two problems.

3.5.1. Custom Certificates Checking

1) Managing the engine's requests

With some engines, it can be possible to have direct access to the SSL certificate validation procedure used when loading a document through HTTPS. In this case, the engine can directly load the client application code from the remote device using a customized validation procedure as shown in **Figure 3**; this is the most simple implementation option. However, on many engines, modifying the SSL certificates validation procedure is not possible.

2) Using a local proxy

If the SSL certificates validation procedure cannot be customized, it is possible to embed a lightweight HTTP/HTTPS proxy in the UPnP platform client, so that requests can be handled manually. This process is illustrated in **Figure 4**.

In this case, the web engine loads the document from a local address using regular HTTP. As the proxy receives HTTP requests from the application, it downloads the requested resource from the remote device using HTTPS, verifying the certificates in the process. The resource is then delivered to the HTML engine. This method can be used on Android or iOS devices for example, and is the one we used in our JavaFX-based implementation.

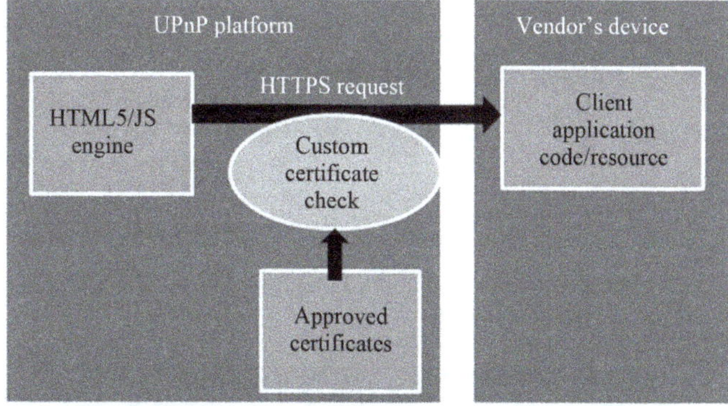

Figure 3. Checking certicates by directly managing the engine's requests.

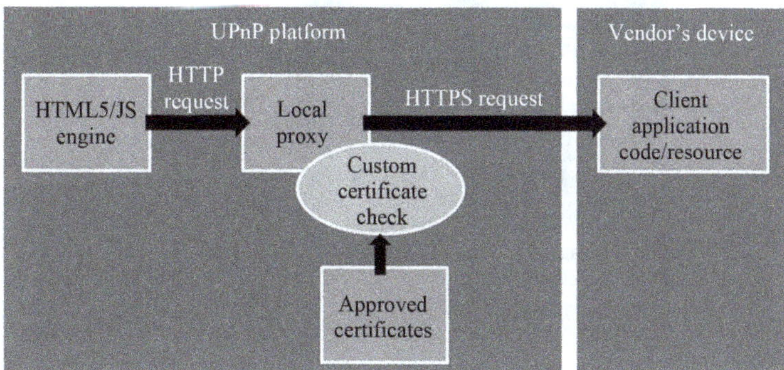

Figure 4. Checking certicates by using a local proxy.

3.5.2. Implementation of the Javascript API

1) Direct access to the Javascript engine

Some engines allow the programmer to inject native objects into the Javascript environment, so native code functions can be accessed through Javascript. In this case, the Javascript API can be easily created by implementing the required functionalities in native code, and then giving access to them by this technique. This is illustrated in **Figure 5**.

This method can be used on Android [4] and Blackberry [5] devices, as well as on PCs through JavaFX [6]; this is the one we used in our prototype.

2) Intercepting AJAX requests

Some engines do not give any way to call back native code from Javascript, although it is generally possible to run Javascript code from native code. In order to make the API available when using such engines, an additional Javascript file can be inserted in the document by native code, so that calls to the API functionalities will result in corresponding AJAX requests. Such requests can then be intercepted in the native code, either directly if the engine allows it or by using a local proxy (which can also be used for certificates checking).

This method can be used on iOS devices for example, as the iOS SDK does not provide a way to access the Javascript engine directly. It is illustrated in **Figure 6**.

4. Prototype and Evaluation

To evaluate the efficiency of such a system, we developed a UPnP platform client to use on a PC, and then developed a smartphone application that uses this generic client to:

- Let the user use his PC to browse through the phone contacts and send text messages (SMS) to them, see **Figure 7**.
- Display a notification on the PC when the phone receives a call, see **Figure 8**.

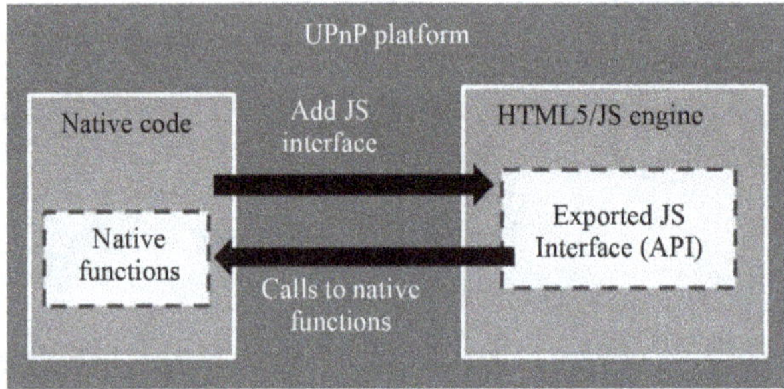

Figure 5. Implementing the Javascript API by direct access to the Javascript engine.

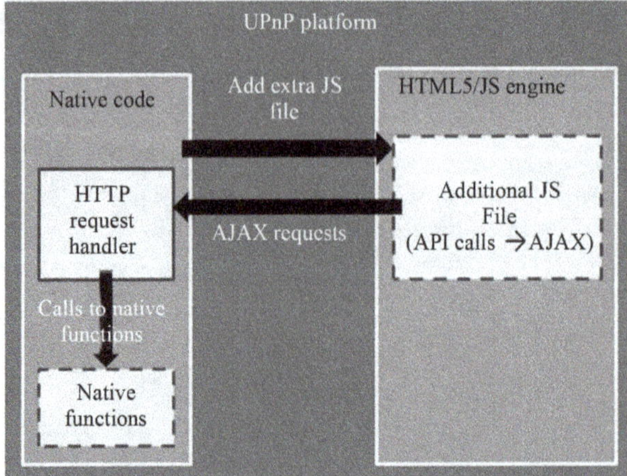

Figure 6. Implementing the Javascript API by intercepting AJAX requests.

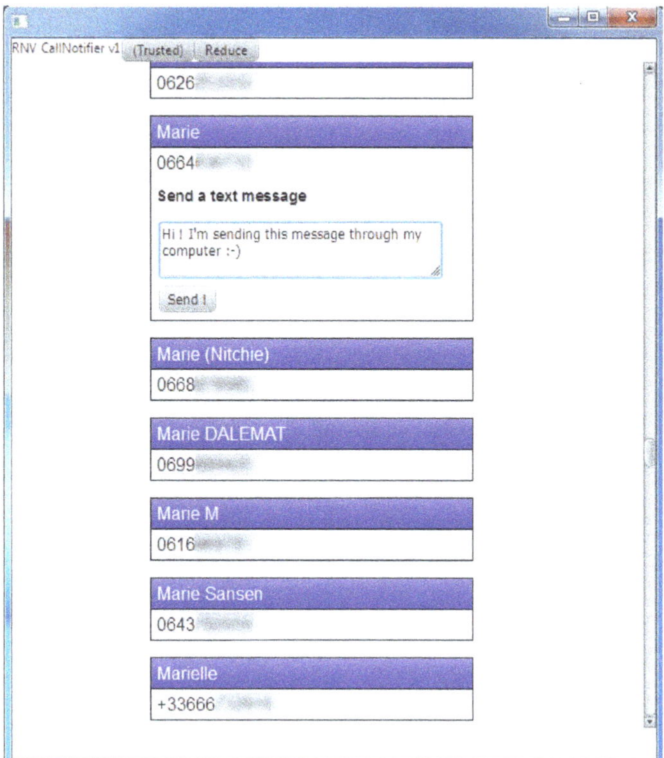

Figure 7. Using the computer client to send a text message (SMS) through the phone network.

Figure 8. On-call notification shown in the top-left corner of the screen.

4.1. Generic UPnP Platform Development

The JavaScript API supported by the generic UPnP platform program of our prototype only includes methods for displaying notifications, and UPnP communication. These features should be supported by any implementation of the platform, regardless of the device on which it runs. However, the end result of those methods may differ depending on devices; for example, showing a notification on a PC can be done by displaying a message in a corner of the screen, while there often is a dedicated mechanism for notifications on smartphones. These different behaviors on different devices are obtained by executing the same JavaScript code: this ensures that applications written in this generic way are compatible with a wide range of devices.

In the future, we plan to define more functionalities in the API that may not be supported by every device: standard methods to display streamed videos for example. Thus, there is a set of API methods that cannot be supported in some implementations of the platform, depending on which device it is designed for. However, the API provides a way to check for the availability of those functionalities so that deployed applications can adapt their behavior to the device they are deployed on.

4.2. Development of a Smartphone Application Using the Platform

It is important to note that developing the smartphone application required no modification to the UPnP platform client installed on the PC, as it is a generic platform designed to receive and run any kind of application. Only HTML5/JavaScript code had to be included in the smartphone application, and is automatically uploaded to the PC. For example, to develop the on-call notification functionality, an application to send UPnP events when receiving a phone call must be developed and deployed on the smartphone. Once this is done, showing a notification on the PC can be achieved by only including the code shown in **Figure 9** in the smartphone application.

One may notice that the subscribe To Events method of the API uses the name of a callback function for event subscription, and not a reference to it. While slightly less efficient, this is to ensure that this API method can be implemented on any system, as passing a reference to an object from JavaScript to native code is not possible when using some JavaScript engines.

When associated with the corresponding static HTML page, this code alone is fully functional and enables the PC to receive UPnP events from the smartphone and display notifications.

4.3. Testing Environment

Testing environments for the UPnP platform and the smartphone application are summarized in **Table 1** and **Table 2** respectively. The smartphone used for testing is much less powerful than most of the smartphones and tablets that are being manufactured today; this enables us to check that this system can be used on a wider range of embedded devices.

Although we used Java to develop our prototype, it is important to note that this is not a requirement, and any programming language or platform can be used as long as the required functionalities detailed previously can be implemented (see 2.4).

4.4. Testing Results

4.4.1. Reactivity

Table 3 shows the measurements made on the reactivity of the system.

After the UPnP discovery and HTTPS code download are completed, the communication between the two devices is responsive enough to seem like realtime to the user.

```
// Program entry point
function clientInit(api) {
  // Globally save the api object
  window.api = api;
  // Subscribe to UPnP events
  api.subscribeToEvents('callback');
}

// Called on UPnP events
function callback(state) {
  if(state.CallerNumber != '') {
    // Display incoming call phone number
    document.getElementById('caller_num')
      .innerHTML = state.CallerNumber;

    /* Show the contents of #notif
     * as a notification
     * using the specified stylesheet */
    api.showNotification('notif',
      'style.css');
  }
  else {
    // No call: hide notification
    api.hideNotification();
  }
}
```

Figure 9. JavaScript code sample for displaying on-call notifications.

Table 1. Testing environment for the UPnP platform.

eciveD	retupmoc potpaL
metsyS gnitarepO	64x 7swodniW
rossecorP	Core i7-2630QM @ 2GHz
yromem elbaliavA	36RDD MG
mroftalp tnempoleveD	2.2XFavaJ
tpircSavaJ/5LMTH	XFavaJ ni)enigne tikbeW)
UPnP library	2.0gnilC

Table 2. Testing environment for the smartphone application.

Device	HTC Desire Z smartphone
Operating System	Android 4.0.4 (ICS)
Processor	MSM7230 @ 800MHz
Available memory	512 MB
Development platform	Java (Android SDK)
HTTP server library	Jetty 8.1.9
UPnP library	Cling 2.0

Table 3. Time measurements on the system's reactivity.

UPnP discovery, code download	2 - 10 seconds
UPnP commands (list contacts)	<1 second
Eventing (incoming phone call	<1 second

Although UPnP discovery takes some time, it is made as soon as the device is powered on, so when the user introduces a new device into the local network, it is immediately synchronized with the others. This is necessary to ensure devices are aware of each other's states at any time, as discussed earlier. Thus, when the user starts interacting with an appliance, UPnP discovery most likely will already be complete and this device can immediately communicate with the others.

4.4.2. Ease to Implement

Table 4 shows the compressed source code size for all the programs. The compression method used to create an archive is LZMA. As it results in a very small archive, these measurements show that very little information needs to be provided to build an application making use of this platform. In other words, very little development work is needed. Moreover, in this case, the phone manufacturer only has to develop the smartphone application, and the HTML5/JS application code that is transferred to remote devices: the UPnP platform is supposedly already installed on remote devices.

5. Comparison with Existing Systems and Related Works

Table 5 summarizes the differences between our proposed system and existing technologies.

5.1. UPnP

UPnP-enabled devices can already send content to other devices and receive commands from them, as well as keeping them updated on their state through the discovery and eventing mechanisms [7]. However, these devices

Table 4. Compressed size of source code.

UPnP platform client (Java, PC)	6.57 kB
Smartphone application (Java, Android)	5 kB
HTML5/JS + CSS client code	2.45 kB
JavaScript client code only	1.7 kB

Table 5. Comparative chart for related technologies.

	UPnP alone	Web-based interfaces	Adaptive JINI	OSGi	Proposed system
Allows delivering any kind of functionality	×	o	o	o	o
Automatic discovery and initialization	o	×	o	o	o
Applications are very easy to develop	N/A	o	×	×	o
Does not require Java	o	o	×	×	o
Uses popular technologies	o	o	×	o	o

cannot handle every kind of content and commands: standard profiles for a fixed set of commands are defined, and devices are supposed to support a pre-defined set of profiles. Thus, they would need to be updated to support new functionalities, and UPnP alone does not address this issue, as manufacturers have no control over the software installed on existing UPnP devices in the home.

5.2. Web-Based Interfaces

When they want a wide range of devices to be able to access their product's functionalities, manufacturers sometimes rely on web-based interfaces. This is the case for web cameras or router configuration tools for instance. Although this method enables the product to display any kind of content, and is generic enough to be supported on all devices with a web browser, communication is only done in one way: a webpage has to be manually opened to send commands from a browser to the device. Thus, communication between devices cannot be done automatically without any user interaction.

5.3. Adaptive Jini

Jini (also called Apache River [8]) enables devices to execute applications provided by other devices, so that any Jini-compatible device can use all the applications deployed on the local network. In order to achieve generic application development, and thus avoid having to update the devices every time a new application is deployed, adaptive Jini has been proposed [9].

However, Jini is more intended as a way for one device to provide a service for other devices to use, rather than keeping the devices mutually connected. Additionally, Jini requires Java capabilities, which is not always available such as on TVs or smartphones. Finally, since it was introduced in 1998, Jini has not gained much popularity among multimedia device vendors, contrary to UPnP. Thus, it is unlikely for these vendors to decide to use a Jini-based system.

5.4. OSGi

In order to deploy applications on devices and communicate with them in a generic way, using UPnP with OSGi has been proposed [10], especially for home automation controllers such as to manage home lights. While this technique can be a good solution in many cases, it can have a number of drawbacks compared to our proposed system. Firstly, when dealing with multimedia devices like smartphones or tablets, HTML5 applications are often better suited and easier to develop than a java package. Secondly, devices do not always have java capabilities, and cannot always use the OSGi framework. Finally, deploying an HTML5 application on these devices is simpler, since it only means rendering it; thus, it is easier for them to embed our HTML5-based UPnP platform rather than an OSGi client.

6. Conclusions

In this research, we have proposed a generic UPnP platform to be deployed on home devices so that device manufacturers can develop innovative cross-device functionalities using this platform by only working on their own device.

We have developed such a client on a computer, and demonstrated its efficiency by making use of a smartphone's functionalities through it. Even though the functionalities shared here are limited and only for demonstration purposes, this platform can already be used in many consumer products applications. As HTML5 enables displaying almost any kind of content, any situation when additional content needs to be displayed on another device's screen can be realized using our platform: IP cameras used for security or to monitor infants can automatically display notifications and even provide live video feeds in case of unusual activity. Multimedia devices can share advanced playback controls, display the current track information or suggest related artists for example, on any device compatible with the platform, even if they do not have a touch panel or even a screen themselves. All these features, among many others, can be built very quickly based on our proposed system.

References

[1] IETF Network Working Group (2013) Rfc2818:Http over tls. https://tools.ietf.org/html/rfc2818

[2] UPnP Forum (2013) Device Protection: 1 Service. http://upnp.org/specs/gw/UPnP-gw-DeviceProtection-v1-Service.pdf

[3] W3CWebRTC Working Group (2013) Webrtc 1.0: Real-Time Communication between Browsers. http://www.w3.org/TR/2012/WD-webrtc-20120821/

[4] Android Open Source Project (2014) Android API Reference, Package Android. webkit.WebView http://developer.android.com/reference/android/webkit/WebView.html

[5] BlackBerry (2014) BlackBerry Java Application Development, Class Browser net.rim.blackberry.api.browser.Browser http://www.blackberry.com/developers/docs/6.0.0api/net/rim/blackberry/api/browser/Browser.html

[6] Oracle (2014) Java Platform JavaFX, Class JSObject, java.lang.Object http://docs.oracle.com/javafx/2/api/netscape/javascript/JSObject.html

[7] UPnP Forum, (2013) Upnp Device Architecture 1.1. http://upnp.org/specs/arch/UPnP-arch-DeviceArchitecture-v1.1.pdf

[8] Apache Software Foundation (2013) Apache Jini Specificationsv2.1.2. http://river.apache.org/doc/spec-index.html

[9] Kadowaki, K., Koita, T. and Sato, K. (2008) Design and Implementation of Adaptive Jini System to Support Undefined Services. Master's thesis, Department of Information Systems Design, Doshisha University,

[10] Donsez, D. (2007) On-Demand Component Deployment in the UPnP Device Architecture. Consumer Communications andNetworking Conference, Las Vegas, 920-924.

18

Improving Network Efficiency by Selecting and Modifying Congestion Control Constraints

Saleem Ullah, Faisal Shahzad, Shahzada Khurram, Waheed Anwer

Department of Computer Science & IT, The Islamia University of Bahawalpur, Bahawalpur, Pakistan
Email: saleemullah@iub.edu.pk, faisalsd@gmail.com, khurram@iub.edu.pk, waheed@iub.edu.pk

Abstract

Congestion in wired networks not only causes severe information loss but also degrades overall network performance. To cope with the issue of network efficiency, in this paper we have proposed and investigated an efficient mechanism for congestion control by the selection of appropriate congestion window size and proactive congestion avoidance, which improves system overall performance and efficiency. The main objective of this work is to choose the accurate size of congestion window based on available link bandwidth and round trip time (*RTT*) in cross and grid topologies, instead of choosing number of hops (Previous researches), we have achieved significant improvement in the overall performance of the network. General simulation results under distinctive congestion scenarios are presented to illuminate the distinguished performance of the proposed mechanism.

Keywords

Congestion Avoidance, Congestion Control, Congestion Window (cwnd), Network Performance

1. Introduction

TCP plays a very vital role in today's internet communication. The increase in user terminals and other applications requires networks to perform smoothly, and this network requires transport layer to provide reliable data transfer. Therefore congestion is a big challenge in today's increasing internet traffic. However, we cannot determine the optimal TCP performance due to the following main problems:

1) Contention between sharing terminals.
2) Hidden terminal problems.

3) Packet loss at the MAC layer.
4) Path disconnection in case of mobility.
5) Reordering.
6) Exponential retransmission back-off at TCP layer.

To address these challenges, a simple yet effective scheme is proposed in this paper. As its name implies, the scheme contains relevant mechanisms for avoiding congestion intelligently, detecting congestion timely and eliminating congestion reactively. Control operations are performed by choosing a new congestion window on the measured channel bandwidth and *RTT*. While previous work was performed by selecting number of hops between sender and receiver [1]-[5], which raise many issues that we will discuss in later sections. Rest of the paper is categorized as; Section 2 will focus some previous work done in this area and some drawbacks at some level, Section 3 will describe our proposed mechanism, Section 4 will based on simulation results, conclusion and future work.

2. Agenda for Potential Development

The application requirements and the limitations of congestion control circumscribe a framework for potential improvements. We identify five distinct cases of interest, which have been discussed in the recent literature.

1) Additive Increase leads naturally to congestion, which, in turn, degrades throughput for two reasons: a) Routers need time to recover even from a transitory congestive collapse and sources need time to detect and retransmit missing packets [6]; b) Congestion control is triggered upon congestion; the window is adjusted backwards and the timeout is extended, which in turn degrades the protocol's capability to detect and exploit error free conditions and bandwidth availability, respectively [6] [7].

2) Additive Increase is not efficient when the network dynamics encompass rapid changes of bandwidth availability. For example, when short flows that cause congestion complete their task, bandwidth becomes available. Similarly, when a handoff is completed in a cellular network, the entire channel's bandwidth becomes available. A more rapid response is then clearly indicated [8].

3) Multiplicative decreases causes transmission gaps that hurt the performance of real-time applications that experience jitter and degraded good put. Furthermore, multiplicative decrease with a factor of ½ or a window adjustment to 2 packets characterizes a rather conservative strategy [9].

4) Error detection lacks an appropriate classification module that would permit a responsive strategy, oriented by the nature of potential errors. That is, when errors appear to be transient due to short-lived flows or random network interference, congestion control mechanisms (*i.e.* timeout extension and multiplicative window adjustment) are triggered unduly. The insufficient error detection/classification may also lead to unfair bandwidth allocation in mixed networks. By default, flows that experience network errors do not balance their bandwidth loss with a more aggressive recovery although such behavior could be justified: flows that experienced no losses have occupied extra bandwidth at the router temporarily, when the network errors forced some senders to back off. This situation is discussed as an open issue in [10]; we demonstrate the validity of this argument, based on experimental results, in Section 5.

5) Source-based decisions on the transmission rate, based on the pace of the acknowledgements, necessarily incorporate the potentially asymmetric characteristics (e.g. ack delays and/or losses) of the reverse path [11]. Hence, the sender's transmission rate does not always reflect the capacity of the forward path. This situation has a direct impact on efficiency since available bandwidth remains unexploited. Several proposals have been presented to tackle the problems of TCP over networks. Most of these proposals rely on some form of local retransmission at the wired boarder, and do not deal (either directly or indirectly) with real-time application constraints, see [12] for a detailed description). Some recent protocols restrict the modifications at the transport level. TCP-Freeze [13] distinguishes handoffs from congestion through the use of the Advertised Window. WTCP [14] implements a rate-based congestion control replacing entirely the ACK-clocking mechanism. TCP-Probing [15] grafts a probing cycle and an Immediate Recovery Strategy into standard TCP, in order to control effectively the throughput/overhead tradeoff. Although TCP-Probing deals effectively with both throughput and energy performance in heterogeneous networks, due to its probing mechanism, it may not satisfy the requirements of current service.

It is possible to have node- and link-level congestion occurring at the same time in networks. Both of them have direct impacts on network performance and application objective. Thus, the onset of congestion must be predicted in advance or detected in time so that it can be relieved thereafter.

3. Proposed Mechanism

3.1. Appropriate Congestion Window Selection

TCP performance is also affected by choosing larger *cwnd* size, which increases the probability of packet contention and packet losses from excessive collisions [16]. Therefore a new *cwnd* size is chosen based on the available current channel bandwidth and RTT in cross as well as in grid topologies.

From the figure we can see that bandwidth measurement is performed by TCP sender on receiving *Acks* from receiver. Let's suppose that $C_{E(k\text{-}1)}$ is the bandwidth at any $(k\text{-}1)$*th-Ack* segment, then the available current bandwidth would be BW

$$BW_{curr} = \left[\left(C_{E(k-1)} \times RTT \times \lambda_k\right) / RTT + T_k\right] \quad (1)$$

Here λ is the total bytes transmitted successively, which acknowledges between $(k\text{-}1)$*th* and *kth Acks*. From Equation (1) we can estimate exponentially weighting by using β then we can control the current and smoothed measure bandwidth as,

$$BW_{E(k)} = C_{E(k-1)} \times \beta + BW_{curr}(1-\beta) \quad (2)$$

Using Equation (2) we can also control the ratio between newly and smoothed measured bandwidth. While Equation (3) gives us maximum *cwnd* size when *RTT* and packet size are given.

$$cwnd_{\max} = RTT \times BW / \lambda \quad (3)$$

$$\text{Packet\_loss} = MSS \times 1.5 / RTT \times \rho^{65} \quad (4)$$

As mentioned before, we have to measure the bandwidth of the multi-hop networks and *RTT* to determine the appropriate maximum *cwnd* size. Before applying maximum *cwnd* to the TCP layer some time will be consumed to measure and *RTT*. In beginning of transmitting TCP segments, *cwnd* size is not limited, which does not matter, because the maximum *cwnd* size can be determine before TCP segments are sent into the system.

3.2. Congestion Avoidance

TCP uses a network congestion avoidance algorithm that includes various aspects of an additive increase/multiplicative decrease (AIMD) scheme, with other schemes such as slow-start in order to achieve congestion avoidance. The TCP congestion avoidance algorithm is the primary basis for congestion control in the Internet.

$$\text{if } cwnd(t) > ssthreh(t)$$
$$cwnd(t+t') = cwnd(t) + 1/cwnd(t)$$

where, *ssthresh*(*t*) is a threshold value at which TCP changes its phase from slow start phase to congestion avoidance phase. When packet loss is detected by retransmission timeout expiration, *cwnd*(*t*) and *ssthresh*(*t*) are updated as,

$$cwnd(t) = 1 \text{ and } ssthresh(t) = cwnd(t)/2$$

On the other hand, when TCP detects packet loss by a fast retransmit algorithm, it changes *cwnd*(*t*) and *ssthresh*(*t*) as;

$$cwnd(t) = ssthresh(t) \text{ and } ssthresh(t) = cwnd(t)/2$$

TCP Reno then enters a fast recovery phase if the packet loss is found by the fast retransmit algorithm. In this phase, the window size is increased by one packet when a duplicate Ack packet is received. On the other hand, cwnd is restored to *ssthresh*(*t*) when the non-duplicate Ack packet corresponding to the retransmitted packet is received.

We have seen that TCP connection starts up in slow start mode and exponentially increase the *cwnd* until it cross the *ssthresh*(*t*). Once *cwnd*(*t*) is greater than *ssthresh*(*t*), TCP moves to congestion avoidance phase. In this mode the primary objective is to maintain high throughput without causing congestion. If TCP detects any segment loss, this is a strong indication that network is congested, so as a corrective action TCP reduces its data flow rate by reducing *cwnd*(t) after that it again goes back to slow start phase.

Having some deeper look on working mechanism, we choose *ssthresh*(t) = 65535 bytes and *cwnd*(t) = 512 bytes. Since *cwnd*(t) < *ssthresh*(t) TCP state is slow start. TCP *cwnd*(t) grows from 512 bytes to 64947 bytes also it has been assumed that no segment loss during slow start phase. During slow start *cwnd*(t) is incremented by one segment on every successful acknowledgment from the receiver. When *cwnd*(t) = 65459 + 512 > 65535, at this point TCP changes its state to congestion avoidance phase, and the primary objective in this phase is to get the higher throughput. If during this phase any three duplicate Acks received the TCP moves to the fast retransmit without waiting for the retransmit timer out and then after it moves to the fast recovery phase.

3.3. Congestion Control

We have modified the TCP default behavior of three-duplicate Acks to two-duplicate Acks, as it has been seen that 87% of the packets loss after two duplicate Acks, so why not change the default behavior? This decreases time delay. The proposed algorithm is as:

Algorithm: Congestion Control

```
On receiving two-duplicate Acks or if a timeout occur do the following;
   IS_congestion=0;
If((max_rate > min_rate) && (curr_rate – min_rate < max-rate – curr_rate)) {
   IS_congestion=1;
}
If(cwnd > thresh)
IS_congestion=1;

If(IS_congestion)
        {
Slowstart();
else
              fast_ recovery();
        }
```

4. Simulation Results and Performance Analysis

To verify our proposed mechanism we use NS-2 (2.34) simulator, model used in our experiment is consists of 50 nodes in a chain wise topology. We set link bandwidth to 1Gbps in order to analyze maximum *cwnd* size. Simulated scheme is investigated by increasing number of nodes, and we compared the performance results with TNR.

If we analyze *cwnd* with saw-tooth effect in **Figure 1**, it can be seen clearly that the proposed mechanism is more effective and consistent then TNR, as number of users increases the *cwnd* loses its stability as well as performance, but our suggested mechanism has consistent effect in terms of stability as well as performance, which increase system overall performance.

Packet loss in networks affects the overall performance of the system. Packet loss is due to signal degradation, congestion in channel and many other common factors. After sending data in bulk we observe that proposed method has very few ratio of packet loss (0.2%) as compared to TNR. It is also clear that *cwnd* in this case is bigger then the proposed but on the other hand decrease in *cwnd* size is also clear which may lead to higher delay and decreasing the network throughput. Therefore proposed mechanism has better results than the TNR (**Figure 2**).

We have modified the default behavior of TCP duplicate acknowledgement from two to three Acks, because it has been observe that 87% of the packets loss after two duplicate Acks, so why not modify this to two instead of three. Which decrease delay and increase throughput as shown in **Figure 3**.

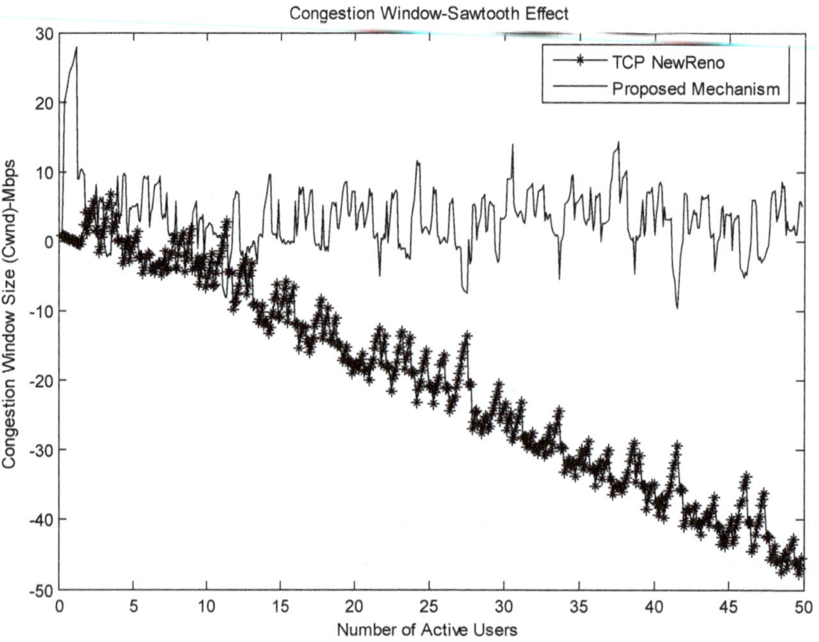

Figure 1. Effect of congestion window (*cwnd*).

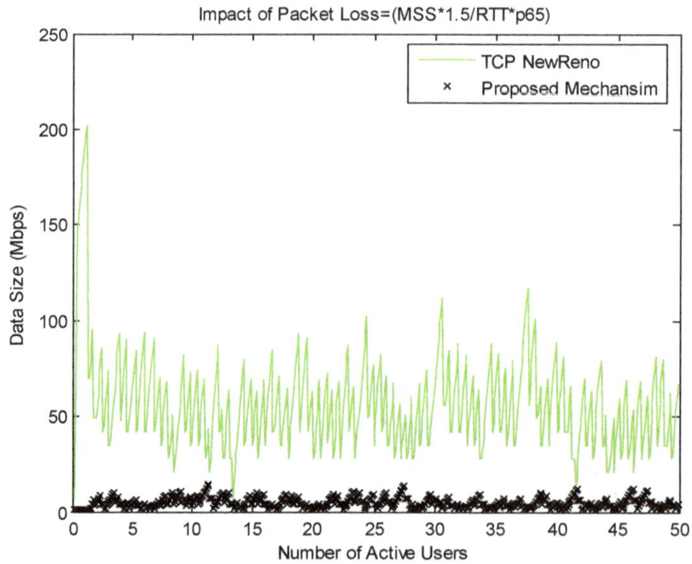

Figure 2. Packet loss comparison.

5. Conclusion and Future Work

In this work, we propose a scheme for congestion control, congestion avoidance by selecting appropriate *cwnd* size which detects and alleviates congestion in networks. The main objective is to provide high transmission quality for the data traffic under conditions of congestion. The scheme comprises three main mechanisms. Firstly, it attempts to suppress the source traffic from event area by carefully selecting a set of representative nodes to be data sources. Secondly, the onset of congestion is indicated in a timely way by jointly checking buffer occupancy and link utilization. Lastly, the network attempts to alleviate congestion in the traffic hotspot by either resource control or traffic control, which is dependent on the specific congestion condition. The ns-2 based simulation has confirmed the advantages of proposed scheme and demonstrated significant performance improvements over existing schemes.

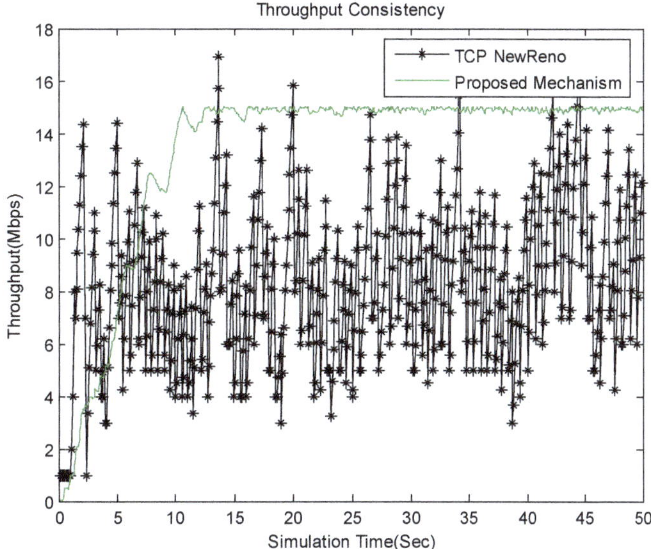

Figure 3. Flow consistency behavior (throughput analysis).

In the future, we hope to extend the proposed scheme to take into account some dynamic behaviors of the network, such as node mobility and link failure. We have also planed to extensively investigate performance on sensor based networks.

Acknowledgements

This research work was partially supported by NSF of china Grant No.61003247. The assistance of Prof. Xiao-feng Liao in this work is also gratefully acknowledged. The authors also would like to thanks the anonymous reviewers and the editors for insightful comments and suggestions.

References

[1] Harb, S.M. and Mcnair, J. (2008) Analytical Study of the Expected Number of Hops in Wireless Ad Hoc Network. *WASA* '08 *Proceedings of the Third International Conference on Wireless Algorithms, Systems, and Applications.*

[2] Miller, L.E. (2001) Distribution of Link Distances in a Wireless Network. *NIST Journal of Research* (Spring).

[3] Huang, P.-K., Lin, X.J. and Wang, C.-C. (2011) A Low Complexity Congestion Control and Scheduling Algorithm for Multi-Hop Wireless Networks with Order-Optimal Per Flow Delay. *IEEE/ACM Transactions on Networking*, **14**.

[4] Zawodniok, M. and Jangannathan, S. (2007) Predictive Congestion Control Protocol for Wireless Sensor Networks. *IEEE Transactions on Wireless Communications*, **06**, 3955-3963.

[5] Jiang, S.M., Zuo, Q. and Wei, G. (2009) Decoupling Congestion Control from TOP for Multi-Hop Wireless Networks: Semi TCP. *CHANTS* 2009, *ACM Transactions.*

[6] Fang, W.-W., Chen, J.-M., Shu, L., Chu, T.-S. and Qian, D.-P. (2009) Congestion Avoidance, Detection and Alleviation in Wireless Sensor Networks. *Journal of Zhejiang University SCIENCE C*, **11**.

[7] Raghunathan, V. and Kumar, P.R. (2006) A Counterexample in Congestion Control of Wireless Networks. *Elsevier Science.*

[8] ONeill, D.C. (2003) Adaptive Congestion Control for Wireless Networks Using TCP. *IEEE Proceeding*-2003.

[9] Popa, L., Riaciu, C. and Stoica, I. (2006) Reducing Congestion Effects in wireless Networks by Multipath Routing. *IEEE Proceeding* 2006, 96-105.

[10] Tiwari, B., and Chandavarkar (2010) Congestion Adaptive Routing in Wireless Mesh Networks. 2010 *Seventh International Conference on Wireless and Optical Communications Networks* (*WOCN*), Colombo, 6-8 September 2010, 1-5.

[11] Armaghani, F.R., Jamuar, S.S., Khatun, S. and Rasid, M.F.A. (2011) Performance Analysis of TCP with Delayed Acknowledgments in Multi-Hop Ad-Hoc Networks. *Wireless Personal Communications*, **56**, 791-811.

[12] Curran, K., Woods, D., McDermot, N. and Bradley, C. (2003) The Effects of Badly Behaved Routers on INTERNET congestion. *International Journal of Network Management*, **13**, 83-94. http://dx.doi.org/10.1002/nem.464

[13] Ilac, A. and Ma, Y.J. (1998) A Rate-Based Congestion Control Scheme for ABR Service in ATM Networks. *International Journal of Network Management*, **8**, 292-317. http://dx.doi.org/10.1002/(SICI)1099-1190(199809/10)8:5<292::AID-NEM286>3.0.CO;2-Y

[14] Li, J., Zhang, S.Y., Li, C.L. and Yan, J.R. (2010) Composite Light Weight Traffic Classification System for Network Management. *International Journal of Network Management*, **20**, 85-105.

[15] Schulze, H. and Mochalski, K. Internet Study2008/2009, IPOQUE Report. http://www.ipoque.com/resources/internet-studies/internet-study-2008_2009

[16] Zhang, Y., Leonard, D. and Loguinov, D. Jetmax: Scalable Max-Min Congestion Control for High-Speed Heterogeneous Networks. *Proceedings of IEEE INFOCOM* 2005, 2006, 1-13.

19

Adaptive Distributed Inter Frame Space for IEEE 802.11 MAC Protocol

Ja'afer AL-Saraireh[1], Saleh Saraireh[2], Mohammad Saraireh[3], Mohammed Bani Younis[2]

[1]Princess Sumaya University for Technology, Amman, Jordan
[2]Philadelphi University, Amman, Jordan
[3]Mutah University, Karak, Jordan
Email: sarjaafer@yahoo.com, saleh_53@yahoo.com, m_sarayreh@mutah.edu.jo, myunis@philadelphia.edu.jo

Abstract

In this research, an Adaptive Distributed Inter Frame Space (*ADIFS*) has been proposed for IEEE 802.11 Medium Access Control (MAC) protocol. The aim of this approach is to improve Quality of Services (QoS) for IEEE 802.11 MAC protocol in single-hop wireless network. The proposed approach is based on traffic type, Collision Rate (*CR*), Collision Rate Variation (*CRV*) and Packet Loss Rate. These parameters are used to adjust the DIFS at runtime. The adjusted DIFS is employed to enhance service differentiation at the MAC layer in single-hop wireless networks. The proposed approach contributes to the enhancement of the average QoS for high priority traffic by 32.9% and 33.4% for the 5 and 10 connections, respectively. While the average QoS for the low priority traffic is improved by 14.3% and 18.2% for the 5 and 10 connections, respectively. The results indicate that, the proposed approach contributes in the enhancement of the QoS in wireless network.

Keywords

DCF, Collision Rate, Packet Loss, Average Delay, *ADIFS*

1. Introduction

The random transmission of applications over the wireless medium may lead to incomprehensible or unpredictable results [1]. Therefore, a controller, which manages access to the medium of the shared resources is an essential tool for achieving a successful transmission process between the communicating parties, and ensuring access is fair and suitable [2].

The MAC protocol in wireless networks is a protocol that controls access to the shared medium, by applying rules and procedures that permit the communication pairs to communicate with each other in an efficient and fair manner [3] [4].

The IEEE 802.11 MAC Distributed Coordination Function (DCF) has two access techniques [5]. The first has two-way handshake known as basic access technique, while the other one has four-way handshake procedures [6].

The basic access method based on the status of channel, so first of all it examines the channel status. If the channel is busy, the node is waiting and monitoring the channel status until it is idle for a period of time called the DIFS. Then the node generates a random back-off interval before transmitting to minimize the probability of multiple nodes concurrently transmission [7]-[9].

The minimum DIFS value is 20 μs and the maximum DIFS is 140 μs [5]. The minimum value is selected to be longer than the Short Inter Frame Space (SIFS) that is identified for control frames, such as an acknowledgement (ACK) frame, while the maximum value is chosen in order to minimize the wasted time slots by avoiding an excessively long defer of data packets.

This paper is organized as follows: The previous studies for providing service differentiation are presented in the following section. A description of the proposed approach is presented in Section 3. In Section 4, the simulation model is introduced. The results and discussions are presented in Section 5. In Section 6, the paper is concluded.

2. Related Works

Most of the proposed priority-based approaches were aimed to support service differentiation by providing different MAC parameter values. The previous works enabled higher priority classes to access the medium faster than low priority classes. For instance, faster access could be provided by assigning a smaller Contention Window (CW) causes a smaller Back-off Interval (BI) as reported in [2] [3] [6] [10]-[15] or by assigning smaller Inter Frame Space (IFS) as reported in (Deng and Chang, 1999), and (Ksentini *et al.*, 2004).

DIFS parameter has been studied for providing service differentiation among different traffic priorities [16]. The value of DIFS in these studies was statically assigned for each class. However, less effort has been made for tuning the DIFS for various traffic types.

In Zhang and Ye (2004), the length of DIFS was computed based on the ratio between the value of estimated transmission rate and total transmission rate [17]. This technique required a significant modification to the IEEE 802.11 DCF scheme.

Sung and Yun (2006) proposed a technique which is called Enhanced Distributed Coordination Function (EDCF) protocol [18]. This technique used Pareto database to store network configuration. For each new configuration the proposed scheme required comparing the current configuration with the already stored in the Pareto curve.

Using IFS is another approach to enhance service differentiation in the IEEE 802.11 MAC protocol. It is based on: 1) using the existing IFS values defined by the standard such as SIFS, Point Inter Frame Space (PIFS), and DIFS and 2) using new IFS values. Different schemes were proposed based on the already available IFS values. For example, the proposed approaches in [14] [19] [20] used PIFS and DIFS values to differentiate between time-sensitive and time-insensitive applications. Some other approaches used new IFS values to differentiate between high and low priority traffic. These new IFS values were based on allocating the low priority traffic longer IFS value than the IFS value of high priority traffic.

Other studies such as [4] [7] [8] [12] [21] were proposed to provide service differentiation based on the distributed function of the standard. These schemes were based on modifying the back-off time of the IEEE 802.11 MAC protocol. Although significant research efforts have been carried out on supporting service differentiation in IEEE 802.11 DCF by adopting the priority-based scheme, several issues have still not been considered.

In this paper, the following points are considered for providing service differentiation in the basic IEEE 802.11 DCF scheme: 1) MAC protocol parameters such as DIFS is dynamically adjusted, 2) different QoS metrics such as throughput, collision rate, packet loss, delay and jitter.

3. Proposed Approach Description

An *ADIFS* composes of four main parts as shown in **Figure 1**. The first part is traffic classification, which classifies the traffic into high and low priorities. The second is the recording part. Each node recorded the number of

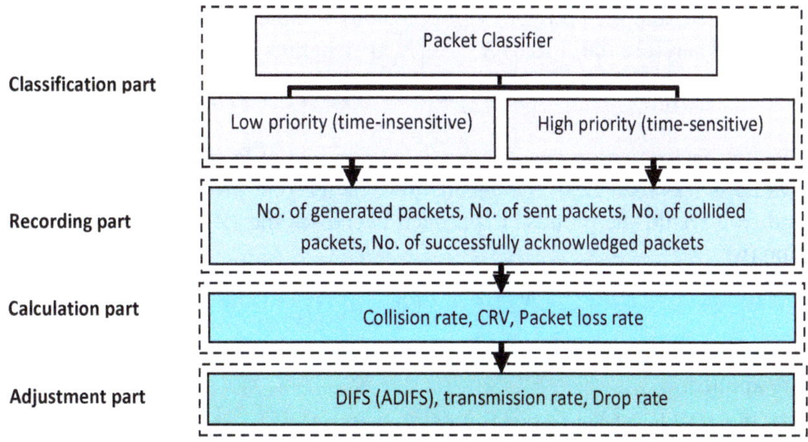

Figure 1. Adaptive distributed inter frame space.

generating packets, sent packets, successfully acknowledged packets, and collided packets. The third is calculation part, where *CR*, *CRV*, and $l[N]$ values are computed and fed to the final part, the decision on choosing an appropriate parameter values.

The following equation is used to compute *CR*

$$CR_{\text{current}}[N] = \frac{Num(\text{collisions }[N])}{Num(\text{collisions }[N]) + Num(\text{successful }[N])} \tag{1}$$

where $Num(\text{collisions }[N])$ is the number of collisions for node N; $Num(\text{successful }[N])$ represents the number of packets that have been successfully acknowledged for node N; $CR_{\text{current}}[N]$ is the current collision ratio of node N. The *CRV* value of each node is calculated based on Equation (2).

$$CRV[N] = CR_{\text{current-average}}[N] - CR_{\text{previous-average}}[N] \tag{2}$$

where $CRV[N]$ is the collision ratio variation innode N; $CR_{\text{current-average}}[N]$ and $CR_{\text{previous-average}}[N]$ are the current and the previous average collision ratio, respectively. The *CRV* provides values within [−1 to 1]. The packet loss rate for node N_i, $(l[N_i])$ is calculated using Equation (3).

$$l[N_i] = 1 - \frac{Num(\text{sucess\_Ack}[N_i])}{Num(\text{gen\_packets}[N_i])} \tag{3}$$

where $Num(\text{sucess\_Ack}[N_i])$ is the number of successfully received acknowledgements for a node N; i stands for high priority class, and $Num(\text{gen\_packets}[N_i])$ is the number of generating packets at the sender.

For high priority traffic, when the *CRV* value of a high priority node $(CRV[N_i])$ is greater than zero, the proposed scheme examines the packet loss rate $(l[N_i])$, if the packet loss rate is below the threshold (*i.e.* $l[N_i] < l\_ths[N_i]$), then *DIFS* length of high priority $(ADIFS_{\text{new}} = [N_i])$ is set equal to $DIFS_{init}$ (*i.e.* $ADIFS_{init} = 50$ μs as defined in (IEEE, 1999)), in order to give low priority traffic a greater chance to access the channel. The minimum length of $(ADIFS_{\text{new}} = [N_i])$ is limited to one slot time (one slot time equal to 20 μsas defined in [5]). If the packet loss rate exceeds the packet loss rate threshold (*i.e.* $l[N_i] < l\_ths[N_i]$), the DIFS of high priority traffic $(ADIFS_{\text{new}} = [N_i])$ is reduced by one slot time, to reduce the delay and to prevent an excessive packet loss for high priority packets. When the $CRV[N_i]$ of high priority traffic is less than zero, the adaptive approach examines the packet loss rate $l[N_i]$ value, if this value is below the packet loss rate threshold $(l\_ths[N_i])$, the DIFS of high priority $(ADIFS_{\text{new}} = [N_i])$ is set equal to $DIFS_{init}$, while if the packet loss rate value $l[N_i]$ is above the packet loss rate threshold $(l\_ths[N_i])$, the DIFS value of high priority packets $(ADIFS_{\text{new}} = [N_i])$ is updated as given in Equation (4).

$$ADIFS_{\text{new}}[N_i] = DIFS_{init}(1 + CRV[N_i]) \tag{4}$$

For low priority traffic, when the $CRV[N_i]$ value is larger than zero, this implies that the number of contending nodes is increased; and the probability of collisions is increased, since the current collision ratio is larger than the previous one. Therefore the $(ADIFS_{\text{new}} = [N_i])$ length is increased and is updated using Equation (5).

$$ADIFS_{\text{new}}[N_i] = DIFS_{init} + \left(f * CRV[N_j] * ADIFS_{\text{new}-1}[N_j]\right) \quad (5)$$

where f is a scaling factor, with value of $f = 3$ as considered in the simulations (Saraireh *et al.*, 2014). If the $CRV[N_i]$ value is less than zero, this means that, the current probability of collisions is smaller than the previous one, and as a result, the proposed approach decreases the $ADIFS_{\text{new}}[N_i]$ by one slot time as represented in Equation (6).

$$ADIFS_{\text{new}}[N_j] = ADIFS_{\text{new}-1}[N_j] - (\text{one\_slot\_time}) \quad (6)$$

To ensure that the lengths of $ADIFS_{\text{new}}[N_i]$ and $ADIFS_{\text{new}}[N_j]$ are within the specified ranges the following conditions are applied:

In Equation (4), if $ADIFS_{\text{new}}[N_i] <$ one time-slot, then $ADIFS_{\text{new}}[N_i] =$ one slot.

In Equation (5), if $ADIFS_{\text{new}}[N_j] >$ seven slots then $ADIFS_{\text{new}}[N_j] =$ seven slots.

In Equation (6), if $ADIFS_{\text{new}}[N_j] < DIFS_{init}$ then $ADIFS_{\text{new}}[N_j] = DIFS_{init}$.

To avoid starvation for low priority traffic, after each update of the *DIFS*, the adaptive approach examines the value of this parameter. If *DIFS* has high values. The proposed scheme sets these parameters as shown in Equation (7).

$$\text{if}(\text{priority} = \text{low}) \text{ then } ADIFS_{\text{new}}[N_j] = ADIFS_{\text{new}-1}[N_j] - (\text{one\_slot\_time}) \quad (7)$$

An overview of the adaptive differentiation operation is provided in **Figure 2**. It is assumed that there are two nodes, one is a high priority and the other is a low priority node. The high priority node is competing with smaller *DIFS*, while the lower priority node is competing with larger *DIFS*. Therefore, the high priority node accesses the medium first. A full description of the proposed algorithm is presented in **Figure 3**.

4. Simulation Model

To analyze the proposed *ADIFS* approach, and compare their functionalities with the standard IEEE 802.11 DCF scheme, network models with different scenarios have been proposed for the simulations by using NS-2.

In this approach, 40 fixed nodes are used, and they are randomly distributed in an area of 100 m × 100 m, and the transmission type is CBR traffic. The nodes are located in the same Independent Basic Service Set to represent a wireless ad-hoc network, as shown in **Figure 4**. The parameters of simulation are presented in **Table 1**.

The total offered load in each scenario is more than 110% of the effective channel capacity (*i.e.* it is considered 1.6 Mbps without considering the protocol overhead) and more than 90% of the total channel capacity (*i.e.* 2 Mbps, with considering the impact of protocol overhead).

5. Results and Discussion

In the *ADIFS* scheme, the CW size is updated according to the Binary Exponential Backoff (BEB) procedure as defined by IEEE 802.11 DCF. In IEEE 802.11 DCF, the ACK frame is assigned as a higher priority over data packets by having a shorter IFS known as SIFS, while data packets have a longer *IFS* known as *DIFS* (*i.e. SIFS* < *DIFS*). The same concept is applied for the *ADIFS* scheme, where the *DIFS* length is dynamically adjusted for each priority based on the packet loss rate and *CRV* values.

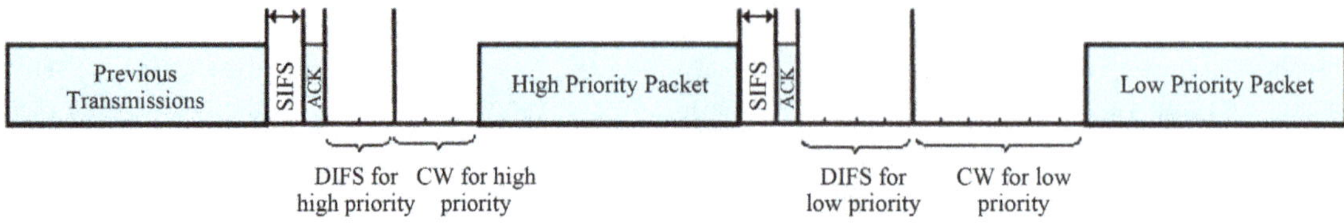

Figure 2. Adaptive differentiation scheme operation.

```
Adaptive Distributed Inter Frame Space (ADIFS) differentiation scheme
    If (priority is high) and (Collision Rate Variation of high priority > 0) {
        If (packet loss rate of high priority < packet loss threshold) {
            DIFS of high priority = DIFS (i.e. 50μs as defined by the standard)
        } Else if (packet loss rate of high priority > packet loss threshold) {
            DIFS of high priority = DIFS - one slot time (i.e. 20μs as defined by the standard) }
    }
    If (priority is high) and (Collision Rate Variation of high priority < 0) {
        If (packet loss rate of high priority < packet loss threshold) {
            DIFS of high priority = DIFS (defined by the standard)
        } Else if (packet loss rate of high priority > packet loss threshold) {
            Update DIFS of high priority using Equation 4 }
    }
    If (priority is low) {
        If (Collision Rate Variation of low priority > 0) {
            Update DIFS of low priority using Equation 5
        } Else if (Collision Rate Variation of low priority < 0) {
            Update DIFS of low priority using Equation 6}
    }
```

Figure 3. Adaptive differentiation algorithm.

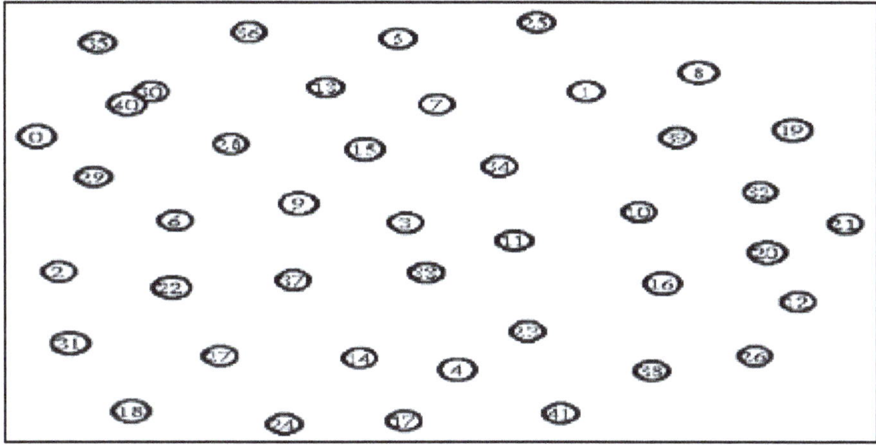

Figure 4. Random independent basic service set single-hop topology.

Table 1. Simulation parameters.

Traffic Type	Packet Size	Scenario	Connection/Priority	Bit Rate	Simulation Time	Run's Number
Low Priority	800 Bytes	Scenario 1	3 Low	480 Kbps	300 Seconds	10
			2 High	192 Kbps		
High Priority	512 Bytes	Scenario 2	5 Low	160 Kbps		
			5 High	192 Kbps		

As indicated in **Table 2**, the average delay for the high priority connections is less than 13 μs. The results indicate that high priority connections have better QoS, with a mean value equal to 86.9%. This significant improvement in the QoS of high priority traffic is at the cost of low priority traffic. For instance, the first low priority connection has a poor QoS with an average of 45.3%. This is due to the long waiting time prior to the transmission which lead to high packet drops at the buffer.

The *ADIFS* approach is also evaluated when the number of nodes for high and low priority is increased. In this scenario, five high priority and five low priority nodes are contended to access the channel. The *ADIFS* scheme performs well when the number of contending nodes is increased. For instance, in this scenario, the average delay of high priority nodes is less than 28 μs in which QoS requirements in terms of delay for the time-sensitive applications could be met. When the *ADIFS* scheme is applied, a high priority node is required to wait for a shorter period, so it could get access to the channel earlier than a low priority node. At the time, when a low priority node tries to access the medium, it finds the channel busy and has to wait until the transmission of higher priority packet is complete. Once the channel becomes idle, all nodes commence their back-off duration. Due to shorter lengths of *DIFS* for high priority nodes, they wait for shorter time periods and start to decrease their back-off time earlier than low priority nodes. This behaviour leads to better performance for the QoS parameters and the average QoS of all connections as depicted in **Figures 5(a)-(c)** and shown in **Table 3**. The average QoS for higher priority traffic is 83.5% with fewer fluctuations and the average QoS for the low priority traffic is 37.1%.

Table 4 summarizes the QoS results for the proposed approach as compared with others for 5 connections.

Table 2. QoS parameters values obtained using the adaptive *ADIFS* differentiation scheme.

Bit Rate	Connection/Priority	Average delay (μs)	Average jitter (μs)	Average throughput (Kbps)	Average MAC efficiency (%)	Average QoS (%)
192 Kbps	Connection 1/high	12.7	6.1	189.3	99.9	86.8
	Connection 2/high	10.0	6.0	188.5	99.9	87.0
480 Kbps	Connection 3/low	1471.5	22.3	281.7	99.9	45.3
	Connection 4/low	709.6	9.9	304.9	99.9	54.4
	Connection 5/low	955.4	11.4	284.8	99.8	50.1

Table 3. QoS parameters values obtained using the adaptive *ADIFS* differentiation scheme for 10 connections.

Bit Rate	Connection/Priority	Average delay (μs)	Average jitter (μs)	Average throughput (Kbps)	Average MAC efficiency (%)	Average QoS (%)
192 Kbps	Connection 1/high	24.3	7.03	193.8	99.7	83.9
	Connection 2/high	26.7	6.5	188.7	99.8	82.9
	Connection 3/high	25.4	8.2	174.0	99.9	83.6
	Connection 4/high	18.3	7.2	181.6	99.8	83.6
	Connection 5/high	27.8	7.3	178.5	99.7	83.3
160 Kbps	Connection 6/low	3859.4	61.4	86.9	99.6	39.5
	Connection 7/low	5009.8	60.1	71.4	99.9	29.1
	Connection 8/low	3263.9	46.5	80.4	99.7	33.4
	Connection 9/low	860.0	25.2	123.0	99.5	43.3
	Connection 10/low	4145.4	60.1	97.0	99.5	40.0

Table 4. Comparison of IEEE 802.11 DCF QoS, Saraireh *et al.*, 2014 with proposed schem for 5 connections.

Bit Rate	Connection/Priority	Average QoS (%) Proposed	Average Qos (%) (Saraireh *et al.*, 2014)	Average QoS (%) Standard	Average QoS (%) Improvement (Saraireh *et al.*, 2014)	Average QoS (%) Improvement
192 Kbps	Connection 1/high	86.8	73.0	56.4	16.6	30.4
	Connection 2/high	87.0	71.6	51.6	20.0	35.4
	Average	86.9	72.3	54.0	18.3	32.9
480 Kbps	Connection 3/low	45.3	43.9	31.8	12.1	13.5
	Connection 4/low	54.4	53.2	43.4	9.8	11.0
	Connection 5/low	50.1	44.6	31.7	12.9	18.4
	Average	49.9	47.2	35.6	11.6	14.3

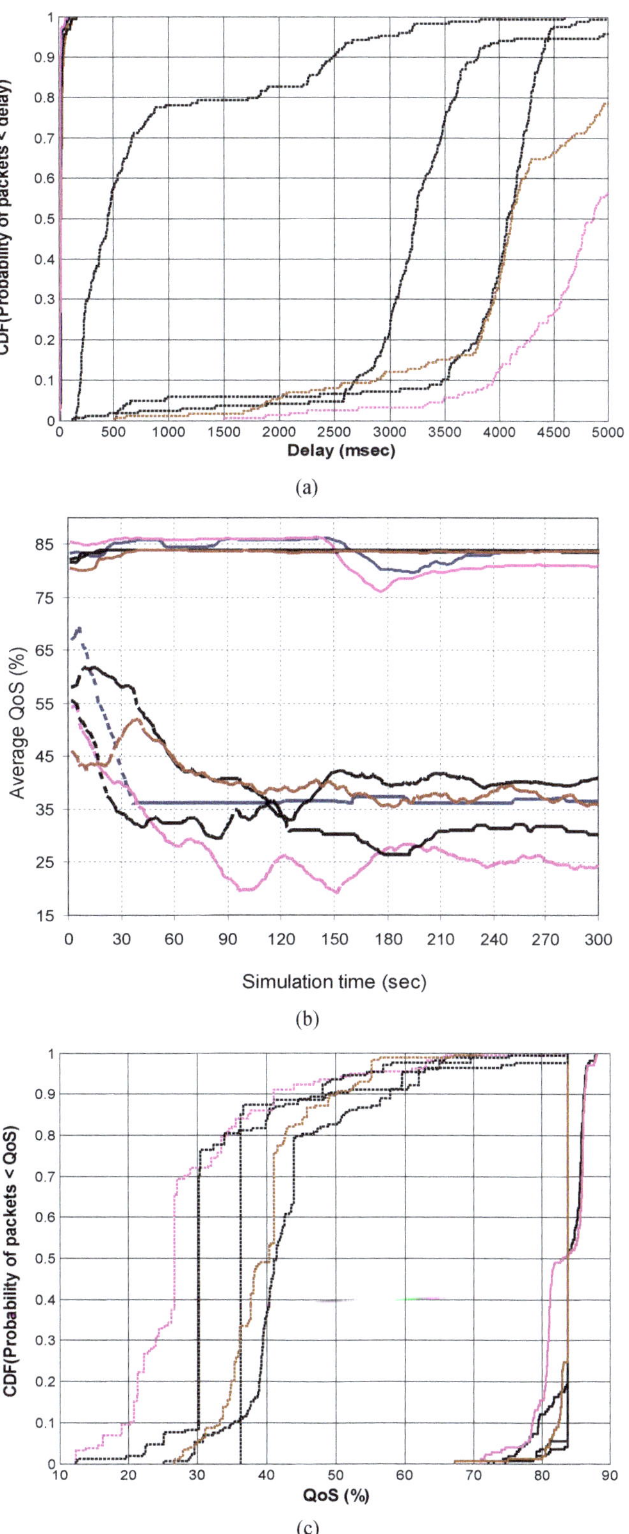

— Connection 1 (high priority); — Connection 2 (high priority); — Connection 3 (high priority); — Connection 4 (high priority); — Connection 5 (high priority); .... Connection 6 (low priority); .... Connection 7 (low priority); .... Connection 8 (low priority); .... Connection 9 (low priority); .... Connection 10 (low priority).

Figure 5. *ADIFS*—based differentiation for 10 connections (5 high and 5 low priority nodes). (a) cumulative distribution of delay; (b) average QoS; and (c) cumulative distribution of QoS.

These results are also shown in **Figure 6**. The average of QoS of proposed scheme increases from 54% to 86.9% for the high priority connections as compared with the standard, while it increases from 72.3% to 86.9% as compared with (Saraireh *et al.*, 2014). On the other hand, for low priority connections the average of QoS is improved by 14.3% and 2.7% as compared with the standard and (Saraireh *et al.*, 2014), respectively. These results indicate that the use of the proposed approach improves QoS in wireless networks.

Table 5 summarizes the QoS results by using the proposed approach as compared with others for 10 connections. These results are also shown **Figure** 7. The average of QoS of proposed scheme increases from 50.1% to 83.5% for the high priority connections as compared with the standard, while it increases from 72.7% to 83.5% as compared with (Saraireh *et al.*, 2014) for the high priority connections. For low priority, the average QoS is enhanced by 18.2% and 0.4% as compared with the standard and (Saraireh *et al.*, 2014), respectively.

Table 5. Comparison of IEEE 802.11 DCF QoS, Saraireh *et al.*, 2014 with proposed schem for 10 connections.

Bit Rate	Connection/Priority	Average QoS (%) Proposed	Average Qos (%) (Saraireh *et al.*, 2014)	Average QoS (%) Standard	Average QoS (%) Improvement (Saraireh *et al.*, 2014)	Average QoS (%) Improvement
192 Kbps	Connection 1/high	83.9	74.7	53.8	20.9	30.1
	Connection 2/high	82.9	73.4	51.7	21.7	31.2
	Connection 3/high	83.6	71.4	47.2	24.2	36.4
	Connection 4/high	83.6	70.4	45.6	24.8	38.0
	Connection 5/high	83.3	73.6	52.2	21.4	31.1
	Average	83.5	72.7	50.1	22.6	33.4
160 Kbps	Connection 6/low	39.5	37.0	18.9	18.1	20.6
	Connection 7/low	29.1	39.5	20.0	19.5	9.1
	Connection 8/low	33.4	35.1	17.8	17.3	15.6
	Connection 9/low	43.3	36.0	18.4	17.6	24.9
	Connection 10/low	40.0	35.7	19.8	15.9	20.2
	Average	37.1	36.7	18.9	17.8	18.2

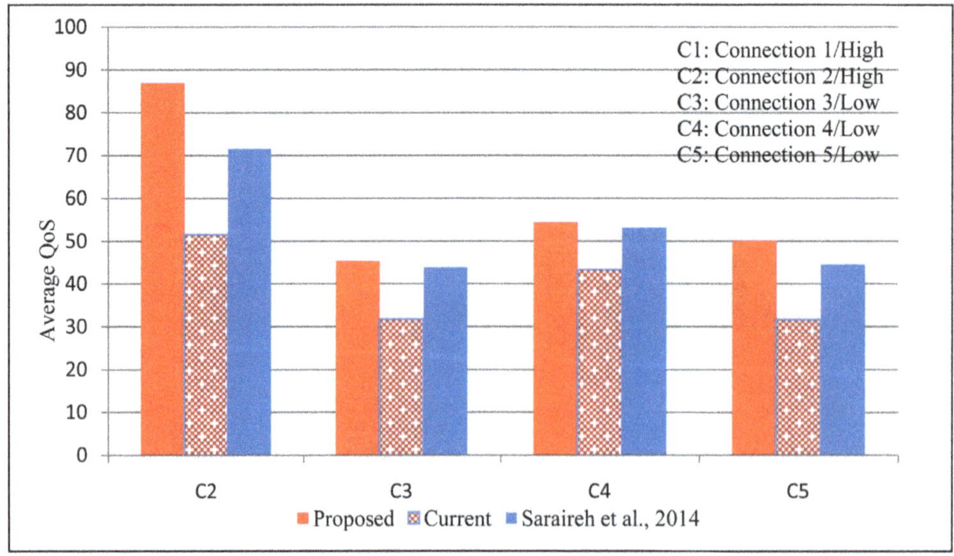

Figure 6. Comparison of IEEE 802.11 DCF QoS, Saraireh *et al.*, 2014 with a proposed scheme for 5 connections.

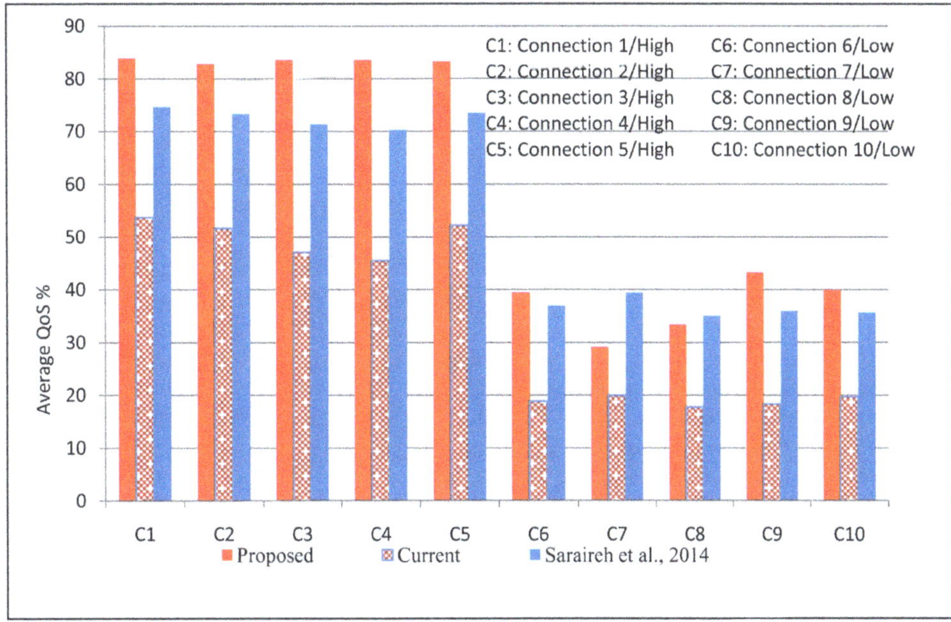

Figure 7. Comparison of IEEE 802.11 DCF QoS, Saraireh *et al.*, 2014 with a proposed scheme for 10 connections.

6. Conclusions

In this research, an enhancement to IEEE 802.11 DCF scheme to provide QoS has been developed. The proposed approach based on dynamically adjustment of DIFS at run time depend on traffic types at MAC layer in the single hop network.

The simulation results indicate that, the adaptive approach improves the performance for high and low priority traffics. The results reveal that the adaptive scheme is capable of providing service differentiation and improving the network performance. The results indicate that, the QoS for priority traffics in terms of delay, jitter and throughput are improved for high and low priority traffics using different number of connections.

References

[1] Saraireh, M., Saraireh, J. and Saraireh, S. (2014) A Novel Adaptive Contention Window Scheme for IEEE 802.11 MAC Protocol. *Journal of Trends in Applied Sciences Research*, **9**, 275-289. http://dx.doi.org/10.3923/tasr.2014.275.289

[2] Balador, A., Movaghar, A., Jabbehdari, S. and Kanellopoulos, D. (2012) A Novel Contention Window Control Scheme for IEEE 802.11 WLANs. *IETE Technical Review*, **29**, 202-212. http://dx.doi.org/10.4103/0256-4602.98862

[3] Raksha, U., Prakash, D. and Sanjiv, T. (2013) Collision Resolution Schemes with Nonoverlapped Contention Slots for Heterogeneous and Homogeneous WLANs. *Journal of Engineering*, **2013**, 1-9. http://dx.doi.org/10.1155/2013/852959

[4] Li, B., Battiti, R. and Yong, F. (2007) Achieving Optimal Performance by Using IEEE 802.11 MAC Protocol with Service Differentiation Enhancements. *IEEE Transactions on Vehicular Technology*, **56**, 1374-1387. http://dx.doi.org/10.1109/TVT.2007.895565

[5] IEEE (1999) IEEE Standard for Wireless LAN Medium Access Control (MAC) and Physical Layer (PHY) Specifications, ISO/IEC 8802-11:1999E.

[6] Deng, D.J., Ke, C.H., Chen, H.H. and Huang, Y.M. (2008) Contention Window Optimization for IEEE 802.11 DCF Access Control. *IEEE Transactions on Wireless Communications*, **7**, 29-35. http://dx.doi.org/10.1109/T-WC.2008.071259

[7] Zhang, M., Gong, C. and Lu, Y. (2008) Dynamic Priority Backoff Algorithm for IEEE 802.11 DCF. *Proceedings of the International Conference on Computer Science and Software Engineering*, Wuhan, 12-14 December 2004, 956-958. http://dx.doi.org/10.1109/CSSE.2008.901

[8] Lin, C.H., Shieh, C.K., Hwang, W.S. and Ke, C.H. (2008) An Exponential Linear Backoff Algorithm for Conten-

tion-Based Wireless Networks, *Proceedings of the International Conference on Mobile Technology, Applications, and Systems* (*Mobility* '08). http://dx.doi.org/10.1145/1506270.1506324

[9] Mahdieh, G., Naser, M. and Kamal, J. (2013) A Game Theory Based Contention Window Adjustment for IEEE 802.11 under Heavy Load. *International Journal of Communication Networks and Information Security* (*IJCNIS*), **5**, 93-103. http://dx.doi.org/10.3923/tasr.2014.275.289

[10] Veres, A., Campbell, A., Barry, M. and Sun, L. (2006) Supporting Service Differentiation in Wireless Packet Networks Using Distributed Control. *IEEE Journal on Selected Areas in Communications* (*JSAC*), **19**, 2081-2093. http://dx.doi.org/10.1109/49.957321

[11] Kim, K., Shin, S. and Kim, K. (2001) A Novel MAC Scheme for Prioritized Services in IEEE 802.11 DCF 802.11a Wireless LAN. *Proceeding of IEEE International Conference on ATM and High Speed Intelligent Internet* (*ICATM*), Seoul, 22-25 April 2001, 196-199. http://dx.doi.org/10.1109/ICATM.2001.932084

[12] Barry, M., Campbell, A. and Veres, A. (2001) Distributed Control Algorithms for Service Differentiation in Wireless Packet Networks. *Proceeding of* 20*th Annual Joint Conference of the IEEE Computer and Communications Societies* (*INFOCOM*), Anchorage, 22-26 April 2001, 582-590. http://dx.doi.org/10.1109/INFCOM.2001.916786

[13] Gannoune, L. (2006) A Comparative Study of Dynamic Adaptation Algorithms for Enhanced Service Differentiation in IEEE 802.11 Wireless Ad Hoc Networks. *Proceeding of AICT-ICIW'*06, *International Conference on Telecommunications and International Conference on Internet, Web Applications and Services*, 19-25 February 2006, 31-37. http://dx.doi.org/10.1109/AICT-ICIW.2006.4

[14] Deng, J. and Chang, R. (1999) Priority Scheme for IEEE 802.11 DCF Access Method. *IEICE Transactions on Communications*, **82**, 96-102.

[15] Ksentini, A., Naimi, M., Nafaa, A. and Gueroui, M. (2004) Adaptive Service Differentiation for QoS Provisioning in IEEE 802.11 Wireless Ad Hoc Networks. *PE-WASUN* '04 *Proceedings of the* 1*st ACM International Workshop on Performance Evaluation of Wireless Ad Hoc, Sensor, and Ubiquitous Networks*, 39-45.

[16] Aad, I. and Castelluccia, C. (2001) Differentiation Mechanisms for IEEE 802.11. *Proceeding Joint Conference of the IEEE Computer and Communications Societies (INFOCOM)*, Anchorage, 22-26 April 2001, 209-218. http://dx.doi.org/10.1109/INFCOM.2001.916703

[17] Zhang, S. and Ye, C. (2004) On Service Differentiation in Mobile Ad Hoc Networks. *Journal of Zhejiang University Science*, **5**, 1087-1094. http://dx.doi.org/10.1631/jzus.2004.1087

[18] Sung, M. and Yun, N. (2006) A MAC Parameter Optimization Scheme for IEEE 802.11e-Based Multimedia Home Networks. 3*rd IEEE*, *Consumer Communications and Networking Conference*, *CCNC* 2006, 8-10 January 2006, 390-394. http://dx.doi.org/10.1109/CCNC.2006.1593053

[19] Shue, S.T. and Shue, T.F. (2001) A Bandwidth Allocation, Sharing and Extension Protocol for Multimedia over IEEE 802.11 DCF 802.11 Ad Hoc Wireless LANs. *IEEE Journal on Selected Areas in Communications* (*JSAC*), **19**, 2065-2080.

[20] Banchs, A., Perez, X., Radimirsch, M. and Stuttgen, H. (2001) Service Differentiation Extensions for Elastic and Real-Time Traffic in 802.11 Wireless LAN. *Proceeding of IEEE Workshop on High Performance Switching and Routing* (*HPSR*), Dallas, 29-31 May 2001, 245-249. http://dx.doi.org/10.1109/HPSR.2001.923640

[21] Ayyagari, A., Bernet, Y. and Moore, T. (2000) IEEE 802.11 Quality of Service. White Paper, 1-10. https://mentor.ieee.org/802.11/dcn/00/11-00

20

A PTS Optimization Scheme with Superimposed Training for PAPR Reduction in OFDM System

Renze Luo[1], Rui Li[1], Xiaoqiong Wu[2], Shuainan Hu[1], Na Niu[1]

[1]State Key Laboratory of Oil and Gas Reservoir Geology and Exploitation Southwest Petroleum University, Chengdu, Sichuan, China
[2]Zhongshan Branch China Telecom Co., Ltd., Zhongshan, Guangdong, China
Email: lrzsmith@126.com, 437592751@qq.com, xiaoqiong_wu@163.com, 564738031@qq.com, 690927582@qq.com

Abstract

Partial Transmit Sequences (PTS) is an efficient scheme for Peak-to-Average Power Ratio (PAPR) reduction in Orthogonal Frequency Division Multiplexing (OFDM) system. It does not bring any signal distortion. However, its remarkable drawback is the high computational complexity. In order to reduce the computational complexity, currently many PTS methods have been proposed but with the cost of the loss of PAPR performance of the system. In this paper, we introduce an improved PTS optimization method with superimposed training. Simulation results show that, compared with conventional PTS, improved PTS scheme can achieve better PAPR performance while be implemented with lower computation complexity of the system.

Keywords

Orthogonal Frequency Division Multiplexing (OFDM), Peak-to-Average Power Ratio (PAPR), Partial Transmit Sequences (PTS), Superimposed Training

1. Introduction

Orthogonal Frequency Division Multiplexing (OFDM) technology has been considered as the core technology for the fourth generation mobile communication system for its high spectral efficiency, good anti-multipath fading capability and anti-interference performance features [1]. However, one of the inherent drawbacks of OFDM signal is that it would have high peak to average power ratio (PAPR), which requires the power amplifier transmitter to have great dynamic range, or else they will have a linear distortion. The power device with larger dynamic range will increase the cost of the transmitter equipment. Besides, when OFDM signals that beyond linear dynamic range transit linear power amplifier devices, they would bring high bit error rate and affect system performance.

Therefore, how to effectively reduce the OFDM system PAPR has become a hot research issue. It has proposed lots of valuable methods to reduce PAPR [2], which can be summarized as signal distortion method, code method and based on scrambling sequence method. However, these methods have shortcomings varying degrees. Clipping is the most simple method of signal distortion method; it would make signal distortion in-band distortion and out of band radiation. The effect of Coding method to reduce PAPR is good, but the encoding patterns and number that can be used availably are very small. The coding efficiency is very low especially when there are a large number of subcarriers. The method based on scrambling sequence mainly uses different scrambling sequences to process the weighted OFDM signal, thus chooses the smallest PAPR value for OFDM signal transmission. Partial Transmit Sequences (PTS) is an effective method among them for PAPR reduction of OFDM system. However, PTS method has very large computational complexity. Many scholars have proposed some PTS methods with computational complexity reduction [3], but these methods would bring about PAPR losses varying degrees for the system performance.

Based on the traditional PTS, this paper proposes an improved PTS optimization scheme by combination with superimposed training sequence method to improve PAPR performance in system with low computational complexity effectively. The method takes advantage of the superimposed training sequence method on reducing PAPR performance, while taking phase-sequence optimization approach to simplify the process of computing phase sequence, thus achieving the purpose for reducing the computational complexity. The simulation results show that compared with the traditional PTS method, the proposed method can not only reduce the computational complexity, but also improve the PAPR performance of system, which is also the greatest advantage that this method has other than other PTS methods having reduced the computational complexity.

2. The PAPR of OFDM System

For an OFDM system with N sub-carrier, in a symbol time interval, after IFFT (Inverse Fast Fourier Transform) and calculating its normalized power (assuming the variance to 1), it can get its complex base band signal defined as Equation (1):

$$x(n)=\frac{1}{\sqrt{N}}\sum_{k=0}^{N-1}X_k\exp\left(\frac{2\pi jnk}{N}\right)\qquad n=0,1,2,\cdots,N-1 \tag{1}$$

where $X(k)$ stands for the k sub-carrier modulation signal; $x(n)$ is the output signal obtained after OFDM modulation.

OFDM signal is the result of superposition by multiple independent sub-carrier signals after modulation, the superimposed signals may have great peak power, thus bring high PAPR. The PAPR in OFDM system is defined as the ratio of the maximum divided by the average power of the signal, expressed as Equation (2):

$$\text{PAPR}(\text{dB})=10\log_{10}\frac{\max\limits_{kN\le n\le(k+1)N}\left\{|x_n|^2\right\}}{E\left\{|x_n|^2\right\}} \tag{2}$$

where $E[.]$ denotes the expected value.

As it can be seen, it will produce a peak power when the N signals add up with the same phase at the same time. The worst situation can be described as Equations (3) and (4):

$$x(t)_{\text{MAX,WORSTCASE}}=\frac{1}{\sqrt{2N}}\sum_{k=0}^{N-1}\left|x[k]e^{\frac{j2\pi kt}{T}}\right|=\frac{1}{\sqrt{2N}}\sum_{k=0}^{N-1}\sqrt{2}\times1=\sqrt{N} \tag{3}$$

$$p_{\text{MAX,WORSTCASE}}=x^2(t)_{\text{MAX,WORSTCASE}}=N \tag{4}$$

Therefore, the maximum value of PAPR as Equation (5):

$$\text{PAPR}_{\text{MAX}}=\frac{P_{\text{MAX,WORSTCASE}}}{\bar{P}}=N=(10\log_{10}N)\,\text{dB} \tag{5}$$

The Equation (5) shows that the PAPR of OFDM signal is relevant with the number of sub-carrier. The greater the number of sub-carrier brings higher PAPR, which can make it beyond the linear range of power amplifiers. When signal peak power gets into the nonlinear region of power amplifier, it will cause signal distortion and signal distortion will cause sub-carriers to reconcile and band radiation. As well as a power amplifier with wide range has the shortcomings of low-efficiency and high cost, so it's necessary to reduce the PAPR value.

3. PTS Scheme

PTS scheme [4] is a common method in the existing technologies to reduce PAPR of OFDM signal. The basic principle of PTS scheme is: an input symbol sequence is presented as Equation (6):

$$X = [X_0, X_1, \cdots, X_N] \tag{6}$$

Then X is partitioned into V "disjoint" symbol subsequences $\{X_v, v = 1, 2, \cdots, V\}$ shown as Equation (7):

$$X = \sum_{v=1}^{V} X_v \tag{7}$$

It introduces rotating vector: $\{a_v = e^{j\theta_v}, v = 1, 2, \cdots, V; \theta_v \in [0, 2\pi]\}$, which is called Side Information (SI), each signal subsequence X_v is multiplied by an unit magnitude constant a_v, then it can generate as Equation (8):

$$Y = \sum_{v=1}^{V} a_v X_v \tag{8}$$

Through IFFT, it can get time-domain signal expressed as Equation (9):

$$y = \text{IFFT}(Y) = \sum_{v=1}^{V} a_v \text{IFFT}(X_v) = \sum_{v=1}^{V} a_v x_v \tag{9}$$

Here x_v is the IFFT of $X(v)$. Then, compare PAPR value by selecting different phase factor, so as to yield the phase factor vector for OFDM signals with the minimum PAPR. When desirable PAPR reduction, the corresponding objective function can be written as Equation (10):

$$[a_1, a_2, \cdots, a_v] = \arg\min \left(\max \left| \sum_{v=1}^{V} a_v x_v \right|^2 \right) \tag{10}$$

where $\arg\min()$ stands for the decision condition at the minimum value of the function. So that it improves the PAPR performance of OFDM system through finding the best phase factor $\{a_{v,v=1,2,\cdots,V}\}$ at the cost of $V-1$ times IFFT. As shown in **Figure 1**, in a typical OFDM system with traditional PTS.

PTS method is shown in **Figure 1**. PTS scheme belongs to the signal scrambler-type techniques which use different scrambling sequences to process the weighted OFDM symbols and optimize the carrier phase for channels, then choose the OFDM symbols and phase combination with smaller PAPR for transmission. Signal scrambler-type techniques have good performance for PAPR reduction, but they obtain good PAPR performance at the cost of higher computational complexity of the system.

4. PTS Optimization Scheme with Superimposed Training

The basic principle of the scheme is making use of superimposed training method, which is used few but has the advantages with inhibiting PAPR of signals and reducing the efficiency of linear power amplifier, as well as further improving band-width utilization in system. At the same time, the paper uses the phase factor with good performance through optimizing phase factor sequence, so as to achieve PAPR reduction for the system more effectively. Where optimizing phase factor sequence makes the computing process of part of the candidate sequence simplified, then reduce the computational complexity.

4.1. Superimposed Training Scheme Method

In order to improve the efficiency of communication, the last century 90's literatures as in [4]-[7] introduce channel estimation and balanced approaches based on superimposed training sequence, they mainly use supe-

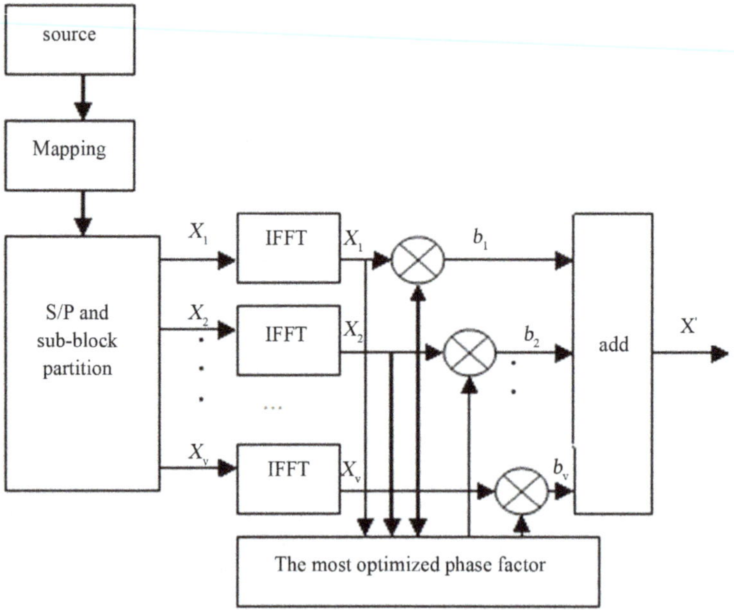

Figure 1. PTS method schematic.

rimposed training sequence method program and first-order statistics for channel estimation and balanced. As information symbols transmit after linear superposition with superimposed training sequence, the receiver can apply only first-order statistics for channel estimation and fast tracking the changes in wireless channel from the principle. For the existing channel estimation methods in communication system, they have advantages of high spectrum efficiency, stable performance, much lower computational complexity and so on. Based on the thinking of researching channel estimation such as in [8]-[11] which mainly use different training sequences for channel estimation research. This paper proposes that apply superimposed training sequence method to the study of PAPR reduction, where it also researches the power conversion efficiency of Power Amplifier (PA) that's the issue of optimizing the distribution between superimposed training sequence and some transmission sequence. How to distribute the superimposed training sequence signal power to ensure the transmission signals power optimization so as to reduce the PAPR in system more effectively is one of the key technologies in the paper; Further, it can effectively guarantee the system has low computational complexity, band radiation and good system performance.

The paper proposes an improved PTS optimization scheme with superimposed training sequence scheme, its main flow chart for achieving the principle as **Figure 2**, the main processes for achievement as follows:

Step 1, the sending port deals the input signal sequences with S/P transform, then partitioned into V "disjoint" symbol subsequences;

Step 2, process each partitioned sub-block with reverse Fast Fourier Transform respectively;

Step 3, for the transformed sub-block signals, select some to be processed the following steps: weighted superimposition training sequence $r_{[N]}$ according to a certain power distribution factor to the part of the selected signal sequences x_m; In order to ensure the influence that superposition training sequence for system performance as small as possible, generally we value the power ratio factor b between 0 - 0.1, the specific processes that are shown in **Figure 3**;

Step 4, weighted all sub-block signals after above-mentioned procession with the optimized phase factor by one-to-one, where the specific process for phase factor optimization is shown below in B in this part;

Step 5, the sending port calculates the peak power and average power ratio for the output of all the signals after the previous processing, then according to calculation results, it selects the smallest result of PAPR to send.

Among them, the main process dealing with the transmission sequence are shown in **Figure 3**, which is making use of the superimposed training method that has the advantages of inhibiting PAPR for signals and reducing the efficiency of linear power amplifier, as well as further improving band-width utilization for the system.

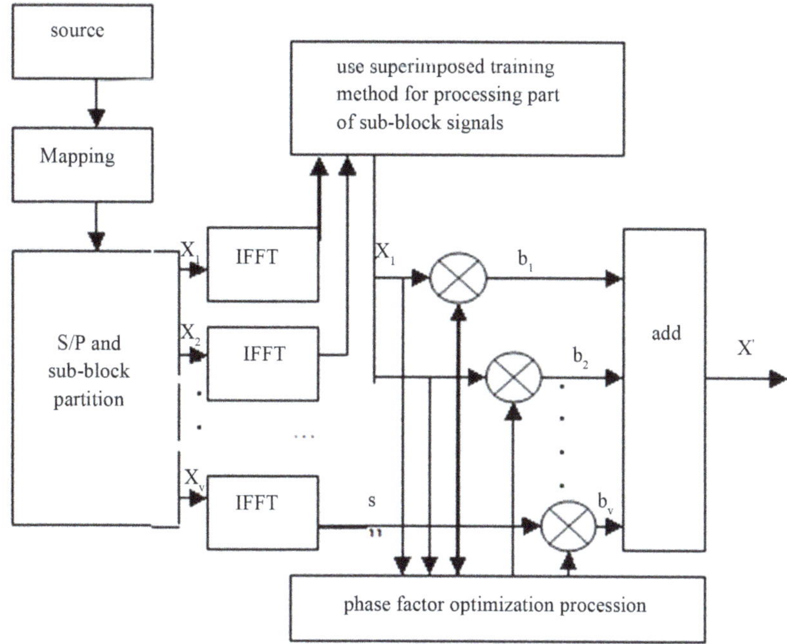

Figure 2. The flow chart of the PTS optimization scheme with superimposed training sequence.

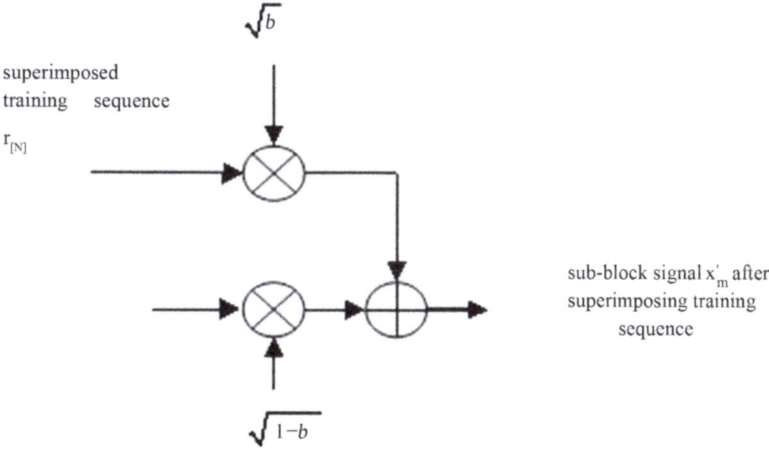

Figure 3. The procession of part signals with superimposed training sequence method.

4.2. Phase Factor Phase Optimization Procession

The processions to optimize phase factor used in PTS method in the paper as follows: implement traditional PTS method to the transmission signals, and the phase factor of the method has only two values 1 and −1. Then it searches the phase factor sequence in accordance with the following steps:

1) Set the initial value for phase factor sequence as $b_v = 1, (v = 1, 2, \cdots, V)$, then calculate its $PAPR$ that corresponding to the time-domain sequence through IFFT transform, recorded as $PAPR0$, and give assignment $id = 1$ at the same time.

2) Assign $b_{id} = -1$, re-calculate $PAPR$ of the obtained sequence;

3) Compare the size between $PAPR0$ and $PAPR$ value, if $PAPR > PAPR0$, $b_{id} = 1$; Otherwise, assign the $PAPR$ value to $PAPR0$, that is $PAPR0 = PAPR$, and then set $id = id + 1$;

4) If $id < V + 1$, then return to step 2); Otherwise, execute step 5);

After such a round search, the obtained phase factor $(b_v, v = 1, 2, \cdots, V)$ is the optimization phase factor which used in the improved PTS scheme of this paper, the distribution of the PAPR is $\min(PAPR, PAPR0)$ in this condition.

5. System Simulation and Results Analysis

All simulation results in this paper are achieved in MATLAB simulation platform, we simulate the improved scheme to validate its performance in an OFDM system simulation platform which has been set up.

In the process of simulation, the simulation parameters for the proposed scheme as follows: in OFDM system, an OFDM signal contains 128 sub-carriers, uses QPSK modulation, takes the constant sequence with length 16 as superimposed training sequence, where the phase factor $b_{[v]}$ for PTS scheme is obtained from optimization process, the whole system is simulating under the Rayleigh fading channel with multi-path number 20.

5.1. The Analysis of Computational Complexity for the System

As shown in **Figure 4**, it gives the PAPR simulation curve of the paper proposed scheme which is PTS optimization scheme with superimposed training sequence; From CCDF curve that the figure given, it can be seen that the proposed PTS scheme can reduce the PAPR performance of system more effectively comparing with the traditional PTS scheme, while the improved PTS scheme greatly reduces the computational complexity of the system.

5.2. The Analysis for the Algorithm Performance

As shown in **Figure 5**, it gives the PAPR simulation curve of the paper proposed scheme that's PTS optimization scheme with superimposed training sequence under different power distribution factor b; b stands for the power distribution factor that superposition training sequence possesses in **Figure 5**. For some sub-block signals after IFFT transform, they are processed with superimposed training sequence method then corresponding set power distribution factor b that superposition training sequence possesses; Where the selection of the power distribution factor b that superposition training sequence possesses depends on the influence that superposition training sequence for system performance and the actual request for the system's performance, generally we value the power distribution factor b between 0 - 0.1. As can be seen from the figure, for different power distribution factor b, the proposed scheme in this paper has small difference on improving the PAPR performance.

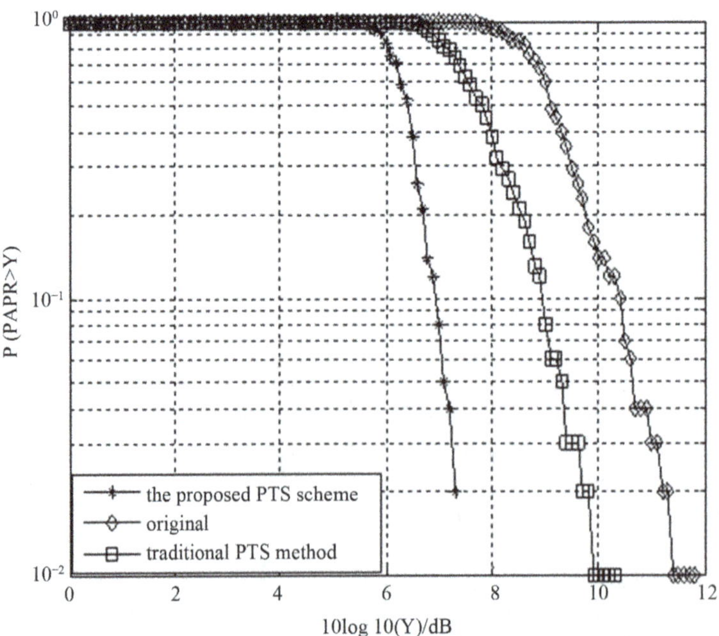

Figure 4. The PAPR simulation curve of the PTS optimization scheme with superimposed training sequence.

Therefore, in the respect of PAPR performance, the proposed PTS scheme in this paper could reduce computational complexity in the system, as well as it obtains good PAPR performance and the result shows it can more effectively reduce the PAPR performance comparing with the conventional PTS method.

5.3. The Analysis of the Impact That Superimposed Training Sequence on the System Performance

As shown in **Figure 6**, it gives the bit error rate (BER) simulation curve of the paper proposed scheme that's PTS

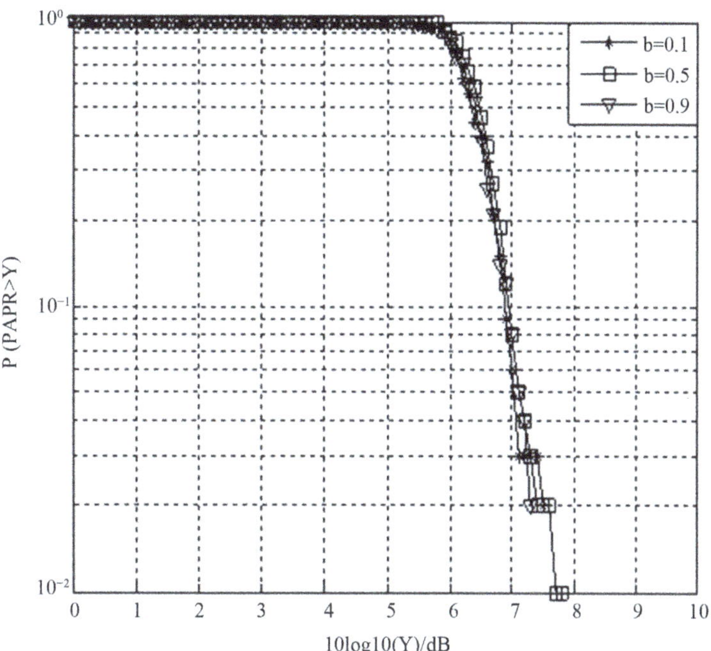

Figure 5. The PAPR simulation curve of the PTS optimization scheme with superimposed training sequence under different power distribution factor.

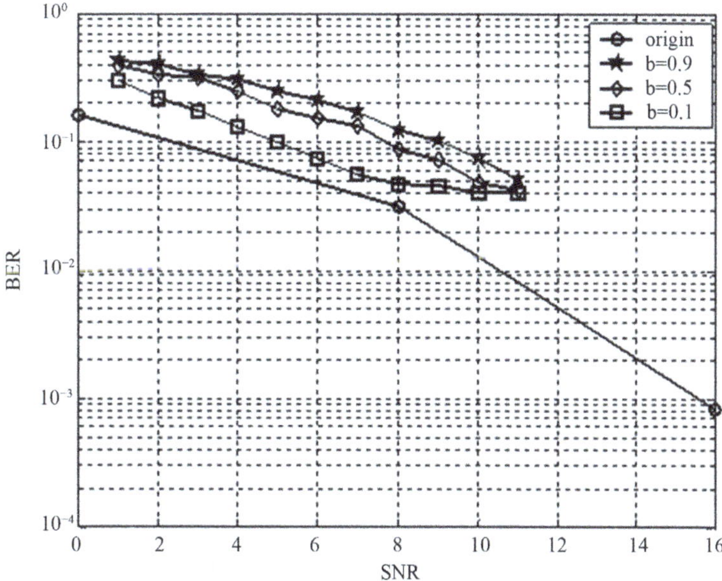

Figure 6. The BER simulation curve of the PTS optimization scheme with superimposed training sequence under different power distribution factor.

optimization scheme with superimposed training sequence under different power distribution factor b; b stands for the power distribution factor that superposition training sequence possesses. From the figure given by simulation curves, we can see that different power distribution factor b that superposition training sequence possesses would share different impacts on the system BER performance after using superimposed training sequence method. When value 0.1 for the power distribution factor b, the BER for the system that the proposed PTS.

Improved optimization scheme brings about is closer to the original signals' comparing with b value respectively 0.5, 0.9. In other words, when values the power distribution factor of superimposed training sequence near 0.1, the proposed scheme in this paper has less influence on system performance relatively.

6. Conclusion

The paper introduces a new scheme using superimposed training sequence method to reduce PAPR in OFDM system. Through the simulation, it proves that the scheme can reduce the peak power of the system, thereby achieving the purpose of PAPR reduction. While the paper analyzes the influence of the power distribution factor b that superposition training sequence possesses for PAPR reduction in the OFDM system, and its effect for PAPR reduction is relatively better when we value the power distribution factor b between 0 - 0.1. Where the phase factor which is obtained from optimization process search used for the new scheme has good performance, the scheme has the advantages of much lower computational complexity and small amount of calculation comparing with a variety of methods. From the preceding discussion, we can see that the improved PTS scheme obtains better effects on both PAPR reduction and computation complexity of system than the traditional PTS method.

Acknowledgements

The authors would like to thank the National Natural Science Foundation of China (No.61310306022 and No.61072073), "1000-elite program" foundation of Sichuan Province and Signal Processing Scientific Research and Innovation Team in Southwest Petroleum University (No. 2013XJZT007), and the Science and Technology Support Foundation of Sichuan Province (No.2012FZ0021).

References

[1] Wu, Y.Y. and Zou, W.Y. (1995) Orthogonal Frequency Division Multiplexing: A Multi-Carrier Modulation Scheme. *IEEE Transactions on Consumer Electronics*, **41**, 392-399. http://dx.doi.org/10.1109/30.468055

[2] Han, S.H. and Lee, J.H. (2005) An Overview of Peak-to-Average Power Ratio Reduction Techniques for Multicarrier Transmission. *IEEE Wireless Community*, **12**, 56-65. http://dx.doi.org/10.1109/MWC.2005.1421929

[3] Wang, L. and Cao, Y. (2008) Sub-Optimum PTS for PAPR Reduction of OFDM Signals. *IEEE Xplore*: *Electronics Letters*, **44**, 921-922.

[4] Tugnai, J.K. and Luo, W. (2003) On Channel Estimation Using Superimposed Training and First-Order Statistics. *IEEE Signal Processing Letters*, **7**, 413-415.

[5] Telado, J. (2000) Multicarrier Modulation with Low PAR Applications to DSL and Wireless. Kluver Academic Publishers, Berlin.

[6] Zhou, G.T., Viberg, M. and Mckelvey, T. (2003) A First-Order Statistical Method for Channel Estimation. *IEEE Signal Processing Letters*, **10**, 57-60.

[7] Orozco-Lugo, A.G., Lara, M.M. and Mclernon, D.C. (2004) Channel Estimation Using Implicit Trainging. *IEEE Transactions on Signal Processing*, **52**, 240-254. http://dx.doi.org/10.1109/TSP.2003.819993

[8] Nair, J.P., and Raja Kumar, R.V. (2006) Channel Estimation and Equalization Based on Implicit Training in OFDM Systems. *IEEE Wireless and Optical Communications Networks*, Bangalore.

[9] Tugnait, J.K. and Meng, X.H. (2006) On Superimposed Training for Channel Estimation: Performance Analysis, Training Power Allocation, and Frame Synchronization. *IEEE Transactions on Signal Processing*, **54**, 752-763. http://dx.doi.org/10.1109/TSP.2005.861749

[10] He, S.C. and Tugnait, J.K. (2008) On Doubly Selective Channel Estimation Using Superimposed Training and Discrete Prolate Spheroidal Sequence. *IEEE Transactions on Signal Processing*, **56**, 3214-3228.

[11] Nair, J.P. and Raja Kumar, R.V (2008) An Iterative Channel Estimation Method Using Superimposed Training in OFDM Systems. *IEEE VTC conference*, Calgary, 21-24 September 2008, 1-5.

Permissions

All chapters in this book were first published in CN, by Scientific Research Publishing; hereby published with permission under the Creative Commons Attribution License or equivalent. Every chapter published in this book has been scrutinized by our experts. Their significance has been extensively debated. The topics covered herein carry significant findings which will fuel the growth of the discipline. They may even be implemented as practical applications or may be referred to as a beginning point for another development.

The contributors of this book come from diverse backgrounds, making this book a truly international effort. This book will bring forth new frontiers with its revolutionizing research information and detailed analysis of the nascent developments around the world.

We would like to thank all the contributing authors for lending their expertise to make the book truly unique. They have played a crucial role in the development of this book. Without their invaluable contributions this book wouldn't have been possible. They have made vital efforts to compile up to date information on the varied aspects of this subject to make this book a valuable addition to the collection of many professionals and students.

This book was conceptualized with the vision of imparting up-to-date information and advanced data in this field. To ensure the same, a matchless editorial board was set up. Every individual on the board went through rigorous rounds of assessment to prove their worth. After which they invested a large part of their time researching and compiling the most relevant data for our readers.

The editorial board has been involved in producing this book since its inception. They have spent rigorous hours researching and exploring the diverse topics which have resulted in the successful publishing of this book. They have passed on their knowledge of decades through this book. To expedite this challenging task, the publisher supported the team at every step. A small team of assistant editors was also appointed to further simplify the editing procedure and attain best results for the readers.

Apart from the editorial board, the designing team has also invested a significant amount of their time in understanding the subject and creating the most relevant covers. They scrutinized every image to scout for the most suitable representation of the subject and create an appropriate cover for the book.

The publishing team has been an ardent support to the editorial, designing and production team. Their endless efforts to recruit the best for this project, has resulted in the accomplishment of this book. They are a veteran in the field of academics and their pool of knowledge is as vast as their experience in printing. Their expertise and guidance has proved useful at every step. Their uncompromising quality standards have made this book an exceptional effort. Their encouragement from time to time has been an inspiration for everyone.

The publisher and the editorial board hope that this book will prove to be a valuable piece of knowledge for researchers, students, practitioners and scholars across the globe.

List of Contributors

Purvang Dalal and Nikhil Kothari
Department of Electronics and Communication, Dharmsinh Desai University, Nadiad, India

Mohanchur Sarkar
Indian Space Research Organization, Ahmadabad, India

Kankar Dasgupta
Indian Institute of Space Science and Technology, Thiruvanthapuram, India

Wednel Cadeau, Xiaohua Li and Chengyu Xiong
Department of Electrical and Computer Engineering, State University of New York at Binghamton, Binghamton, USA

Nathan Aston and Wei Hu
Department of Computer Science, Houghton College, Houghton, USA

Pasquale Lucio Scandizzo
University of Rome Tor Vergata, Center for Economic and International Studies, Rome, Italy

Alessandra Imperiali
University of Rome Tor Vergata, Rome, Italy

Mahmoud Maqableh and Ra'ed (Moh'd Taisir) Masa'deh
Management Information Systems, Faculty of Business, The University of Jordan, Amman, Jordan

Huda Karajeh
Computer Information Systems, King Abdullah II School for Information Technology, The University of Jordan, Amman, Jordan

Tomofumi Matsuzawa
Department of Information Sciences, Tokyo University of Science, Tokyo, Japan

Keisuke Shimazu
Service Operation Division, Internet Initiative Japan Inc., Tokyo, Japan

Mohammed M. Olama
Computational Sciences and Engineering Division, Oak Ridge National Laboratory, Oak Ridge, USA

Seddik M. Djouadi
Department of Electrical Engineering and Computer Science, University of Tennessee, Knoxville, USA

Charalambos D. Charalambous
Department of Electrical and Computer Engineering, University of Cyprus, Nicosia, Cyprus

Mahammad A. Safwat and Hesham M. El-Badawy
Network Planning Department, National Telecommunication Institute, Cairo, Egypt

Ahmad Yehya
Department of Electrical Engineering, Al-Azhar University, Cairo, Egypt

Hosni El-Motaafy
Department of Electronics and Computer Engineering, HIT, Tenth of Ramdan City, Egypt

Hussein M. Hathal, Riyadh A. Abdulhussein and Sarmad K. Ibrahim
Electrical Engineering Department, College of Engineering, Al-Mustansiriya University, Baghdad, Iraq

Mohamed Watfa, Shakir Khan and Ali Radmehr
Faculty of Engineering and Information Sciences, University of Wollongong, Dubai, UAE

Pham Thanh Hiep and Ryuji Kohno
School of Engineering, Yokohama National University, Yokohama, Japan

Pham Thanh Hiep
Le Quy Don Technical University, Ha Noi, Viet Nam

Qassim Bani Hani and Julius Dichter
Department of Computer Science, University of Bridgeport, Bridgeport, USA

Jamal Fathi
Department of Electrical and Electronics Engineering, Near East University, Nicosia, Northern Cyprus

Hasan Harasis
Department of Electrical Engineering, Faculty of Engineering Technology, Albalqa Applied University, Amman, Jordan

Benattou Fassi and Ali Djebbari
Telecommunications and Digital Signal Processing Laboratory, Djillali Liabes University of Sidi Bel Abbes, Sidi Bel Abbes, Algeria

Abdelmalik Taleb-Ahmed
LAMIH UMR CNRS 8530, University of Valenciennes and Hainaut-Cambresis (UVHC), le Mont Houy, France

V. O. C. Eke
Department of computer Science, Ebonyi State University, Abakaliki, Nigeria

A. N. Nzeako
Department of Electronic Engineering, UNN, Enugu State, Nigeria

Tasuku Takagi
(Professor Emeritus) Tohoku University, Sendai, Japan

Remi Nguyen Van, Hideki Shimada and Kenya Sato
Department of Information Systems Design, Doshisha University, Kyoto, Japan

Saleem Ullah, Faisal Shahzad, Shahzada Khurram and Waheed Anwer
Department of Computer Science & IT, The Islamia University of Bahawalpur, Bahawalpur, Pakistan

Ja'afer AL-Saraireh
Princess Sumaya University for Technology, Amman, Jordan

Mohammed Bani Younis and Saleh Saraireh
Philadelphi University, Amman, Jordan

Mohammad Saraireh
Mutah University, Karak, Jordan

Renze Luo, Rui Li, Shuainan Hu and Na Niu
State Key Laboratory of Oil and Gas Reservoir Geology and Exploitation Southwest Petroleum University, Chengdu, Sichuan, China

Xiaoqiong Wu
Zhongshan Branch China Telecom Co., Ltd., Zhongshan, Guangdong, China

Gen-Hwa Lin and Jen-Leih Wu
Institute of Cellular and Organismic Biology, Academia Sinica, Taipei, Taiwan

Hsiao-Rong Chen
Institute of Systems Biology and Bioinformatics, National Central University, Jhongli, Taiwan

Mark M. Voigt
Department of Pharmacology and Physiological Science, Saint Louis University School of Medicine, St. Louis, USA

Shuan Shian Huang
Auxagen Inc., St. Louis, USA

www.ingramcontent.com/pod-product-compliance
Lightning Source LLC
Chambersburg PA
CBHW080650200326
41458CB00013B/4802
9781632405487